Numerical Analysis

Numerical Analysis
Theory and Experiments

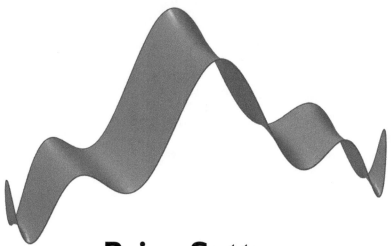

Brian Sutton
Randolph-Macon College
Ashland, Virginia

Society for Industrial and Applied Mathematics
Philadelphia

Publications Director	Kivmars H. Bowling
Executive Editor	Elizabeth Greenspan
Developmental Editor	Gina Rinelli Harris
Managing Editor	Kelly Thomas
Production Editor	David Riegelhaupt
Copy Editor	Gary Davidoff
Production Manager	Donna Witzleben
Production Coordinator	Cally A. Shrader
Compositor	Cheryl Hufnagle
Graphic Designer	Doug Smock

Library of Congress Cataloging-in-Publication Data
Names: Sutton, Brian (Brian David), author.
Title: Numerical analysis : theory and experiments / Brian Sutton
    (Randolph-Macon College, Ashland, Virginia).
Description: Philadelphia : Society for Industrial and Applied Mathematics,
    [2019] | Series: Other titles in applied mathematics ; 161 | Includes
    bibliographical references and index.
Identifiers: LCCN 2019001606 (print) | LCCN 2019005545 (ebook) | ISBN
    9781611975703 (ebook) | ISBN 9781611975697 (print)
Subjects: LCSH: Numerical analysis.
Classification: LCC QA297 (ebook) | LCC QA297 .S877 2019 (print) | DDC
    518–dc23
LC record available at https://lccn.loc.gov/2019001606

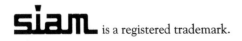

In memory of my grandmother,
Louise Creasey Powers

Contents

Preface

This book is designed to be the primary text for an undergraduate course on numerical analysis. It is appropriate for students who have studied calculus and linear algebra and who have some exposure to differential equations and mathematical proofs. No prior experience with computer programming is expected.

To the student

Numerical analysis is a subject that combines mathematical theory and computer code to solve problems in mathematics, science, and engineering. The focus is on problems that are impossible to solve by hand. For example, even simple and important integrals like $\int e^{-x^2}\,dx$, which comes from the "bell curve" of statistics, may not be expressible as simple formulas. Although exact values may be impossible to obtain, we can often compute accurate approximations to extremely high precision using mathematical ingenuity and a bit of computer code.

If you have never studied numerical analysis, then this book is for you. Prior experience with computer programming or numerical computation is not expected; the MATLAB® computing environment is introduced early, and you will gradually write more sophisticated computer code as your reading progresses. You should be proficient with multivariable calculus and matrices, and some exposure to differential equations is beneficial. Prior experience with mathematical proofs is helpful, but most error bounds follow from a small handful of fundamental theorems.

By working through this text from beginning to end, you will achieve a number of objectives. You will understand why numerical computation is necessary. You will implement and apply algorithms for interpolation, integration, linear systems, zero finding, and differential equations. You will measure error, recognize numerical convergence, and infer rate of convergence from experimental data. You will prove error bounds and explain how the performance of a numerical method is determined by characteristics of the supplied data or the desired solution.

The first part of the book begins to teach you to think like a numerical analyst. You will perform quick computations on the computer, write your first computer code, and evaluate a few methods for accuracy and efficiency. The rest of the book is organized around a few areas of mathematics that benefit from numerical analysis: interpolation, integration, linear systems, zero finding, and differential equations. Within each of these areas, you will develop methods for solving some of the most important problems and evaluate the effectiveness of those methods.

As you begin to compute the impossible, my greatest recommendation is to be active. While reading the text, run the demonstrations on your own computer, conduct additional experiments of your own design, and check the proofs by hand. While working the exercises, if you are not sure whether an approach will succeed, try it on the computer and see. Be guided first and foremost by what works. Theory will follow.

To the instructor

This textbook is based on an undergraduate numerical analysis course that I have taught since 2007. In the years since the first offering, I have made significant changes to better reflect my understanding of how numerical analysts do their work and to include profound developments in the practice of numerical computation. As a result, my students are now solving problems that I considered prohibitively difficult a few years ago, and they have a better understanding of how their methods work. This textbook is intended to bring these developments to a wider audience.

A few distinctive features of the book are described below.

Theory and experimentation. This book is full of numerical experiments in addition to proved theorems. During my own education, I developed an experimental approach to mathematics out of necessity: when encountering an unfamiliar subject, numerical experiments were the quickest and most reliable way to gain insight. From other researchers I learned that experiments can also be quite convincing, even when proofs are available. A kind of graphic that has become very common in research articles is the rate-of-convergence graph, which plots error versus computational resources. For numerical methods that work well, the error steadily falls toward zero as more work is performed, and the rate of convergence is easily inferred. Many examples in this text involve theoretical prediction followed by numerical experimentation, with the rate-of-convergence graph providing a direct comparison.

Local and global methods. A significant motivation for this book is Chebyshev technology, which has seen an explosion of activity in recent years thanks in large part to the Chebfun project initiated by Zachary Battles and Lloyd N. Trefethen [3]. While many older methods work locally—think of Newton's method targeting a single zero or Newton–Cotes quadrature chopping a function into small pieces—a Chebyshev method models a function globally by a single polynomial. This can deliver extremely high accuracy at low cost. I was introduced to this approach during an eye-opening presentation by Trefethen in 2008, and I could never teach numerical analysis the old way again. This book introduces global Chebyshev methods alongside more traditional local methods, in many cases deriving both types of methods from a common core.

Analysis and linear algebra. In this text, I deliberately pursue a marriage between analysis (calculus, differential equations, real and complex analysis, etc.) and linear algebra. Although the table of contents suggests a heavier emphasis on problems from analysis, linear algebra is equally important because it is the foundation of so many solution methods. A common tactic is to project a continuous function onto a finite-dimensional vector space and simultaneously reduce the analysis problem to a linear algebra problem. In particular, definite integration reduces to an inner product; interpolation, differentiation, and antidifferentiation reduce to matrix-vector products; and differential equations reduce to systems of linear (algebraic) equations. The unifying question is, Can we design a matrix to solve the problem for us? Through this pursuit, the student goes beyond the mechanics of matrix computations to recognize and exploit linearity.

In my experience, students enjoy studying numerical analysis. Numerical methods allow them to solve problems that would otherwise be impossible, and the feedback from computer code is immediate and compelling. At the same time, the mathematical theory underlying numerical methods is deep and broad. By completing this course, students will experience the practical implications of theory that may otherwise seem abstract.

Software

The numerical methods in this book are implemented in the MATLAB programming language and are stored in computer files called *M-files*. These files and directions for their installation are available on the web at www.siam.org/books/ot161. Briefly, you will download a package, expand it to produce a folder of M-files, and add the folder to the *search path* in the MATLAB environment. As long as the example code in Chapter 2 executes without error messages, you will know that the software is installed correctly.

Acknowledgments

My collaborator in all things is my wife Megan, who has taught me much about the English language and who has provided essential support and encouragement throughout the writing of this book.

I recognize my MATH 442 students for working through drafts, some rougher than others, of this book.

I thank everyone at SIAM for their improvements to this book and for being so great to work with.

Two conferences, *Chebfun and Beyond* and *New Directions in Numerical Computation*, both located at Oxford University, gave me the opportunity to learn firsthand from many innovators in the area of Chebyshev technology.

The writing of this book was supported by sabbatical leave from Randolph-Macon College and by grants from the Rashkind Family Foundation and the Walter Williams Craigie Teaching Endowment.

List of Notation

Part I

Computation

Chapter 1

Introduction to numerical analysis

To motivate our study, a few problems, specifically chosen to be difficult to solve by hand, are solved numerically in this chapter. The pace is fast as we introduce ideas that will be fully developed later.

1.1 ▪ What is numerical analysis?

This book is about numerical computation. For instance, instead of being content with the statement, "π is the ratio of circumference to diameter," we insist on knowing its numerical value. How do we know that $\pi = 3.14\ldots$?

There are numerous ways to compute π. One approach is via the *Gregory series*

$$\pi = 4\left(1 - \frac{1}{3} + \frac{1}{5} - \frac{1}{7} + \frac{1}{9} - \frac{1}{11} + \cdots\right) = 4\sum_{k=1}^{\infty} \frac{(-1)^{k-1}}{2k-1}.$$

The first few partial sums are the following:
```
>> 4*1
ans =
     4
>> 4*(1-1/3)
ans =
    2.6667
>> 4*(1-1/3+1/5)
ans =
    3.4667
>> 4*(1-1/3+1/5-1/7)
ans =
    2.8952
>> 4*(1-1/3+1/5-1/7+1/9)
ans =
    3.3397
```
These are our first computations in the MATLAB environment. The next chapter will introduce the software in some detail. For now, just follow along by starting the MATLAB application, entering the text that follows >>, and verifying that the system responds as shown here.

3

To obtain a more accurate approximation from the Gregory series, more terms must be included. The 1000th partial sum is computed as follows:

```
>> 4*sum(arrayfun(@(k) (-1)^(k-1)/(2*k-1),1:1000))
ans =
    3.1406
```

The 1001th partial sum is the following:

```
>> 4*sum(arrayfun(@(k) (-1)^(k-1)/(2*k-1),1:1001))
ans =
    3.1426
```

The partial sums alternate between under- and overestimates of π. Considering the previous two computations, the digits 3.14 must be accurate, but further digits would require additional computation. The Gregory series is appealingly simple, but it converges slowly. Later, we will compute π much more efficiently.

This brief attempt at computing π is prototypical. Like many later computations, it involves the following steps:

1. Define the desired quantity.

2. Develop a mathematical procedure for approximating its value.

3. Implement the procedure in computer code.

4. Assess the accuracy of the computed quantity.

At a higher level, the method should be compared and contrasted with competing methods. An ideal numerical method is accurate, efficient, reliable, and easy to use.

Two terms that go hand-in-hand are *numerical method* and *numerical analysis*. A numerical method is an algorithm for solving a mathematical problem involving real numbers. Numerical analysis is the evaluation of such algorithms with regard to their accuracy, efficiency, and robustness.

1.2 ▪ Solving the impossible

Numerical methods are necessary because many mathematical solutions cannot be expressed as simple formulas.

One of the earliest-discovered examples of this phenomenon is the irrationality of $\sqrt{2}$. Because $\sqrt{2}$ is irrational, its decimal representation neither terminates nor repeats. The best we can do is to compute as many digits as desired. In the following example, we compute $\sqrt{2}$ using the *Babylonian method*.

Example 1.1. Compute $\sqrt{2}$ using the iteration

$$x_{n+1} = \frac{1}{2}\left(x_n + \frac{2}{x_n}\right),$$

starting from an initial guess of $x_0 = 1.5$.

Solution. This is an example of an *iterative* method. Starting from an initial guess, the estimate is updated again and again until the desired accuracy is achieved.

```
>> x0 = 1.5
x0 =
    1.5000
>> x1 = 0.5*(x0+2/x0)
```

```
x1 =
    1.4167
>> x2 = 0.5*(x1+2/x1)
x2 =
    1.4142
>> x3 = 0.5*(x2+2/x2)
x3 =
    1.4142
>> x4 = 0.5*(x3+2/x3)
x4 =
    1.4142
```

The decimal expansion 1.4142 is a familiar approximation for $\sqrt{2}$.

Note that if x_n is an underestimate for $\sqrt{2}$, then $2/x_n$ is an overestimate, and vice versa. Hence, their average $(1/2)(x_n + 2/x_n)$ ought to be close to $\sqrt{2}$. More convincingly, if the iteration converges to some limit x, then it must be the case that $x = (1/2)(x + 2/x)$, which is equivalent to $x^2 = 2$. ∎

The irrationality of $\sqrt{2}$ was deeply disturbing to the Pythagoreans of antiquity. Nearly as troubling in the nineteenth century was the insolubility of the quintic equation

$$c_5 x^5 + c_4 x^4 + c_3 x^3 + c_2 x^2 + c_1 x + c_0 = 0.$$

The general solution to this equation cannot be expressed as a formula involving only addition, subtraction, multiplication, division, and roots. In other words, there is nothing analogous to the quadratic formula $(-b \pm \sqrt{b^2 - 4ac})/(2a)$ for the quintic equation, and therefore any solution method must be less direct. Although it may be impossible to find a solution exactly, we may be able to compute as many digits as desired with a numerical method.

A lack of exact formulas causes pain for statistics students as well. The standard "bell curve" is the function

$$f(x) = \frac{1}{\sqrt{2\pi}} e^{-x^2/2},$$

and the integral

$$\int_{-2}^{2} \frac{1}{\sqrt{2\pi}} e^{-x^2/2} \, dx \tag{1.1}$$

underlies various statistical methods that promise 95% confidence. Unfortunately, this integral cannot be found exactly because the antiderivative of $e^{-x^2/2}$ cannot be expressed as a simple formula.

Example 1.2. Approximate the integral (1.1) with a Riemann sum. Use the right-endpoint rule with 100 rectangles.

Solution. The right-endpoint rule for approximating an integral $\int_a^b f(x) \, dx$ is

$$\sum_{i=1}^{n} f(x_i) h,$$

with $h = (b - a)/n$ and $x_i = a + ih$. Using $n = 100$ rectangles, we find the following approximation:

```
>> f = @(x) 1/sqrt(2*pi)*exp(-x^2/2);
>> a = -2;
```

```
>> b = 2;
>> n = 100;
>> h = (b-a)/n;
>> i = 1:n;
>> xs = a+i*h;
>> fs = arrayfun(f,xs);
>> sum(fs*h)
ans =
    0.9545
```

The 95% rule appears to be justified. ∎

Introductory mathematics courses can give the dangerous misimpression that most equations can be solved exactly and most integrals can be computed exactly. The truth is the opposite! In most problems from science and engineering, exact solution formulas are unavailable, and numerical methods are some of our most important tools.

Although born of necessity, numerical analysis is also a beautiful subject. The beauty of numerical methods lies in their generality. Rather than learn separate rules for polynomial, rational, exponential, trigonometric, etc., functions, we develop a single algorithm that works on these functions and many more. Very often, the steps of the algorithm reveal the geometry of the problem, rather than obscuring it beneath pages of symbolic manipulation.

1.3 ▪ Accuracy, precision, and error

The success of a numerical computation is measured largely in terms of the number of digits computed correctly. Here, we introduce some useful terminology.

The *absolute precision* of a decimal number is determined by the place value of its least significant digit. For example, 3.14 is given to the hundredths decimal place, while 3.142 is given to the thousandths decimal place. The *relative precision* of a decimal number is determined by the number of significant digits. For example, 17.42 is stated with four digits of precision.

Let $\hat{x}$ be an approximation for a quantity x. The *error* is

$$x - \hat{x}; \tag{1.2}$$

the *absolute error* is

$$|x - \hat{x}|; \tag{1.3}$$

and the *relative error* is

$$\begin{cases} \dfrac{|x - \hat{x}|}{|x|} & \text{if } x \neq 0, \\ 0 & \text{if } x = 0 \text{ and } \hat{x} = 0, \\ +\infty & \text{if } x = 0 \text{ and } \hat{x} \neq 0. \end{cases} \tag{1.4}$$

To achieve high accuracy is to achieve low error.

It is possible for a number to be high in precision but low in accuracy. For example, 3.143333 is a rather precise figure, having seven digits, but as an approximation of π only the first three digits are accurate.

An advantage of relative error is that it is unitless. On a modern computer, relative error below 10^{-15} is typically considered extremely small, regardless of whether the quantity is measured in millimeters or astronomical units. On the other hand, small relative error is sometimes impractical to achieve. If the true quantity x is nearly equal to zero, then $\hat{x}$ must be extremely close to x in order for the fraction $|x - \hat{x}|/|x|$ to be small. In this case, we may be content with small absolute error.

Example 1.3. Measure absolute and relative error for the approximation $\sqrt{2} \approx 1.4142$.

Solution. The constant $\sqrt{2}$ is known to equal $1.414213562\ldots$. The absolute error for the approximation is

$$|1.414213562\ldots - 1.4142| = 0.13562\ldots \times 10^{-4},$$

and the relative error is

$$\frac{|1.414213562\ldots - 1.4142|}{|1.414213562\ldots|} = \frac{0.000013562\ldots}{1.414213562\ldots} = 0.95900\ldots \times 10^{-5}.$$

In terms of absolute precision, the approximation 1.4142 is stated to four decimal places, and in the language of relative precision, it is stated to five significant digits. Considering the size of the error, the approximation is accurate in its stated digits. ∎

1.4 ▪ Outline of the book

The book is organized into seven parts. The first part focuses on the implementation and analysis of numerical methods with computer code. Each of the remaining parts focuses on a family of mathematical problems and one or more methods for their numerical solution. The seven parts of the book are listed in the following diagram:

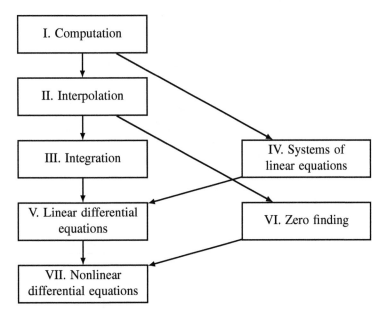

The arrows indicate dependencies. The order I–VII is often best, but one could, for example, treat zero finding before linear differential equations. Also, Chapters 5 and 20 are not prerequisites for any later material.

Notes

The 1992 article "The Definition of Numerical Analysis" by Trefethen is an inspiration for this chapter [72].

Exercises

1. Does your favorite computer algebra system, e.g., Mathematica or Maple, find roots of cubic polynomials exactly? Demonstrate. What about quintic polynomials?

2. Find a rational function that your favorite computer algebra system, e.g., Mathematica or Maple, is unable to integrate exactly.

3. Locate a solution to $x^3 + x^2 + x + 2 = 0$ by trial and error. Achieve two-decimal-place accuracy.

4. Locate a solution to $2x^3 - x^2 + 4x + 1 = 0$ by trial and error. Achieve three-decimal-place accuracy.

5. Approximate $\sqrt{7}$ to three-decimal-place accuracy by hand. Describe your process.

6. Compute $\int_0^2 \frac{1}{\sqrt{2\pi}} e^{-x^2/2} \, dx$ using a Riemann sum with 100 subintervals, using right endpoints for sample points.

7. Compute $\int_0^2 \frac{1}{\sqrt{2\pi}} e^{-x^2/2} \, dx$ using a Riemann sum with 100 subintervals, using left endpoints for sample points.

8. Compute $\int_0^2 \frac{1}{\sqrt{2\pi}} e^{-x^2/2} \, dx$ using a Riemann sum with 100 subintervals, using midpoints for sample points.

9. A *continued fraction* for the golden ratio is

$$\phi = 1 + \cfrac{1}{1 + \cfrac{1}{1 + \cfrac{1}{1 + \cfrac{1}{1 + \cdots}}}}.$$

A *convergent* is obtained by terminating a continued fraction at a finite number of operations. The convergents for the continued fraction above are 1, $1 + 1/1$, $1 + 1/(1 + 1/1)$, ... Compute successive convergents until ϕ is found accurate to five significant digits.

10. A continued fraction for $\sqrt{2}$ is

$$\sqrt{2} = 1 + \cfrac{1}{2 + \cfrac{1}{2 + \cfrac{1}{2 + \cfrac{1}{2 + \cdots}}}}.$$

Use this to compute $\sqrt{2}$ to five significant digits.

Exercises 11–18: A constant and approximation are given. Look up the correct value and then report the absolute precision, relative precision, absolute error, and relative error of the approximation.

11. $\sqrt{2}$; 1.41
12. $\sqrt{2}$; 1.42
13. golden ratio; 1.618
14. golden ratio; 1.616
15. e; 2.718281814723686
16. e; 2.718281828459046
17. Avogadro constant; 6.025×10^{23}
18. Avogadro constant; 6.022×10^{23}

19. An Olympic swimming pool is nominally 50.00 meters long but in reality is allowed to be up to 50.03 meters long. How large of a relative error is allowed?

20. Describe a plausible scenario in which a relative error of 10^{-3} would be acceptable but any larger error would be unacceptable.

21. Describe a plausible scenario in which a relative error of 10^{-4} would be acceptable but any larger error would be unacceptable.

Chapter 2

Programming basics

This chapter introduces the MATLAB programming language and computing environment and demonstrates techniques for implementing numerical methods in computer code.

2.1 ▪ Initializing the MATLAB environment

This book is accompanied by a collection of MATLAB files. Download the package from www.siam.org/books/ot161, start the MATLAB application, and follow the installation directions from the web site. You should see the command prompt:

```
>>
```

To verify that the library is installed correctly, type the following:

```
>> nate
```

The system should respond as follows:

```
Numerical Analysis: Theory and Experiments
```

This confirms that the numerical routines are available.

2.2 ▪ Arithmetic and elementary functions

Initially, the MATLAB environment can be used interactively like a calculator.

The arithmetic operators for addition (+), subtraction (-), multiplication (*), division (/), and exponentiation (^) work as expected.

Example 2.1. Compute $5 + 3 \times 4$ and $(1 + 1/365)^{365}$.

Solution. Below and for the rest of the book, enter the text after >> at the MATLAB prompt and verify that the same result is computed.

```
>> 5+3*4
ans =
    17
>> (1+1/365)^365
ans =
    2.714567482021973
```

(Be sure that you have set the numeric format to "long" as part of the installation process to see all 16 digits on your computer.)   ■

Note that the standard order of operations is obeyed.

There are built-in constants and functions as well, including π, the square root function, exponentials and logarithms, and trigonometric and inverse trigonometric functions.

Example 2.2. Show the built-in value for π and then evaluate $\sin(\pi/4)$.

Solution. The built-in constant for π is called `pi`, and the sine function is called `sin`.

```
>> pi
ans =
   3.141592653589793
>> sin(pi/4)
ans =
   0.707106781186547
```
                                                                                               ■

The display of a number can be controlled using the `sprintf` function, as demonstrated in the following example.

Example 2.3. Compute the weight of an 8.53-kg mass on the surface of Earth, using the gravitational constant $g = 9.81$ m/s^2. Report the answer to the correct number of significant digits.

Solution. The weight equals mg, in which m is the mass.

```
>> m = 8.53;
>> g = 9.81;
>> weight = m*g
weight =
  83.679299999999998
```

Because each given figure is stated to three significant digits, the weight should also be expressed with three significant digits. This can be done with the `sprintf` function.

```
>> sprintf('%4.1f',weight)
ans =
    '83.7'
```

The string `'%4.1f'` is a *format string*. The `%` symbol identifies it as a format string, and the `4` specifies a width of four: ☐☐☐☐. The suffix `.1f` specifies "fixed-point" notation with one-decimal-place precision: ☐☐.☐. The digits are placed into this template: `83.7`.

The alternative to fixed-point notation is exponential notation:

```
>> sprintf('%8.2e',weight)
ans =
    '8.37e+01'
```

The answer should be read as 8.37×10^1. The prefix `%8` identifies a format string and allocates eight characters: ☐☐☐☐☐☐☐☐. The suffix `e` specifies exponential notation and reserves four characters for the exponent: ☐☐☐☐e☐☐☐. The precision specification `.2` reserves two digits to the right of the decimal point: ☐.☐☐e☐☐☐. Finally, the number is printed into this template: `8.37e+01`.   ■

Note that a semicolon (;) at the end of a MATLAB statement prevents the statement's result from being printed to the screen. Try removing the semicolons from the previous example and you will see the values of `m` and `g` echoed by the interpreter.

2.3 ▪ Arrays

An array is an indexed collection of numbers. Column vectors, row vectors, and matrices are particular types of arrays, and the MATLAB language does not draw a distinction between a column vector and a matrix with one column, or between a row vector and a matrix with one row.

Vectors and matrices are surrounded by square brackets. Entries in a row are separated by spaces or commas, and rows are separated by semicolons. The following example illustrates how to construct vectors and matrices.

Example 2.4. Construct the row vector

$$\begin{bmatrix} 8.1 & -3.9 & 4.2 \end{bmatrix},$$

the column vector

$$\begin{bmatrix} 4.4 \\ 9.7 \\ -3.1 \end{bmatrix},$$

and the matrix

$$\begin{bmatrix} 1/\sqrt{2} & -1/\sqrt{2} \\ 1/\sqrt{2} & 1/\sqrt{2} \end{bmatrix}$$

at the command prompt.

Solution.

```
>> [ 8.1 -3.9 4.2 ]
ans =
   8.100000000000000  -3.900000000000000   4.200000000000000
>> [ 4.4; 9.7; -3.1 ]
ans =
   4.400000000000000
   9.699999999999999
  -3.100000000000000
>> [ 1/sqrt(2) -1/sqrt(2); 1/sqrt(2) 1/sqrt(2) ]
ans =
   0.707106781186547  -0.707106781186547
   0.707106781186547   0.707106781186547
```

∎

Because column vectors are common, it is helpful to have a compact notation, and we sometimes write $(x_1, \ldots, x_n)$ in place of

$$\begin{bmatrix} x_1 \\ \vdots \\ x_n \end{bmatrix}.$$

The i, j entry of an array A is accessed with A(i,j). Ranges of entries can be specified with the colon (:) operator: A(i1:i2,j1:j2) refers to a subarray lying in rows $i_1, \ldots, i_2$ and columns $j_1, \ldots, j_2$. There are also shorthand notations for referring to an entire row or column or the last entry of a row or column, as demonstrated in the next example.

Example 2.5. Construct the matrix

$$\begin{bmatrix} 1 & 2 & 3 \\ 4 & 5 & 6 \\ 7 & 8 & 9 \end{bmatrix}$$

and then extract the 1, 3-entry, the second column, the last row, and the submatrix lying in the first two rows and the last two columns.

Solution. The matrix is constructed.

```
>> A = [ 1 2 3; 4 5 6; 7 8 9 ]
A =
      1       2       3
      4       5       6
      7       8       9
```

The various entries and sections can be accessed as follows:

```
>> A(1,3)
ans =
      3
>> A(1:end,2)
ans =
      2
      5
      8
>> A(end,1:end)
ans =
      7       8       9
>> A(1:2,end-1:end)
ans =
      2       3
      5       6
```

Note that end-1 refers to the next-to-last row or column, depending on context. Also, the range 1:end can be abbreviated by a single colon, as below:

```
>> A(:,2)
ans =
      2
      5
      8
>> A(end,:)
ans =
      7       8       9
```

∎

The functions `length` and `size` are useful for finding the dimensions of a vector and matrix, respectively.

Example 2.6. Count the number of entries in

$$\mathbf{x} = \begin{bmatrix} 1 \\ 2 \\ 3 \end{bmatrix}$$

using `length` and the numbers of rows and columns in

$$\mathbf{A} = \begin{bmatrix} 1 & 2 & 3 \\ 4 & 5 & 6 \end{bmatrix}$$

using `size`.

Solution. The vector and matrix are defined.

```
>> x = [ 1; 2; 3 ]
```

```
x =
     1
     2
     3
>> A = [ 1 2 3; 4 5 6 ]
A =
     1       2       3
     4       5       6
```

Function `length` is used to count the number of entries in the vector.

```
>> length(x)
ans =
     3
```

Function `size` finds the numbers of rows and columns in the matrix.

```
>> [m,n] = size(A)
m =
     2
n =
     3
```

∎

Note in the last example how a MATLAB function can return multiple outputs. Each output must be assigned a name in order to inspect it.

An array of zeros may be constructed with routine `zeros`.

Example 2.7. Use `zeros` to construct an array of zeros with the five rows and three columns.

Solution.

```
>> A = zeros(5,3)
A =
     0       0       0
     0       0       0
     0       0       0
     0       0       0
     0       0       0
```

∎

An analogous routine `ones` constructs an array of ones.

A row vector of uniformly spaced numbers can be constructed with the colon (:) operator.

Example 2.8. Construct a row vector with entries $3, 4, 5, 6, 7$. Then construct a row vector with entries $4, 6, 8, 10$.

Solution. The first vector is constructed as follows.

```
>> 3:7
ans =
     3       4       5       6       7
```

For the second vector, the increment of 2 is specified between the minimum and maximum values.

```
>> 4:2:10
ans =
     4       6       8       10
```

∎

The `linspace` function solves the same problem but may be more convenient for numbers that are not integers.

Example 2.9. Construct a column vector with five uniformly spaced numbers spanning the interval $[1, 1.1]$.

Solution.
```
>> linspace(1,1.1,5)'
ans =
    1.000000000000000
    1.025000000000000
    1.050000000000000
    1.075000000000000
    1.100000000000000
```
■

A number of functions for computing with arrays are available, e.g., `sum`, `prod`, and `max`.

Example 2.10. Construct the column vector $(1, 2, 3, 4)$ and then compute the sum of its entries, the product of its entries, and its maximum entry using built-in MATLAB functions.

Solution. The column vector is constructed.
```
>> x = [ 1; 2; 3; 4 ];
```
The requested summaries are computed as follows.
```
>> sum(x)
ans =
    10
>> prod(x)
ans =
    24
>> max(x)
ans =
     4
```
■

2.4 ▪ Matrix arithmetic

When applied to vectors or matrices, the arithmetic operators +, -, and * have their meanings from linear algebra.

In this book, column vectors and matrices are denoted by boldface letters, with vectors typically identified by lowercase letters and matrices by capital letters. A variable identifying a scalar quantity—a single number rather than a vector or matrix—is not in boldface.

Example 2.11. Let

$$\mathbf{x} = \begin{bmatrix} 1 \\ 2 \\ 3 \end{bmatrix}, \quad \mathbf{y} = \begin{bmatrix} 2 \\ 3 \\ 4 \end{bmatrix}, \quad \mathbf{A} = \begin{bmatrix} 1 & 0 & -2 \\ 0 & -1 & 1 \\ -1 & 3 & 0 \end{bmatrix},$$

and compute $2\mathbf{x}$, $\mathbf{x} + \mathbf{y}$, and $\mathbf{A}\mathbf{x}$.

Solution. The vectors and matrices are defined. Note that a matrix input can be broken across lines following a semicolon.

```
>> x = [ 1; 2; 3 ]
x =
      1
      2
      3
>> y = [ 2; 3; 4 ]
y =
      2
      3
      4
>> A = [   1   0  -2  ;
           0  -1   1  ;
          -1   3   0 ]
A =
      1        0       -2
      0       -1        1
     -1        3        0
```

The expressions are computed.

```
>> 2*x
ans =
      2
      4
      6
>> x+y
ans =
      3
      5
      7
>> A*x
ans =
     -5
      1
      5
```

■

An inner product $\langle \mathbf{x}, \mathbf{y} \rangle$ of two vectors, also known as a dot product and denoted $\mathbf{x} \cdot \mathbf{y}$, can be computed as $\mathbf{x}^T \mathbf{y}$, in which $\mathbf{x}^T$ is the transpose of $\mathbf{x}$. In the MATLAB language, the transpose operator is represented by an apostrophe (').

Example 2.12. Define $\mathbf{x}$ and $\mathbf{y}$ as in Example 2.11. Compute the inner product of the vectors. Check the result.

Solution.

```
>> x = [ 1; 2; 3 ];
>> y = [ 2; 3; 4 ];
>> x'*y
ans =
     20
```

The desired result is the sum of products $1 \times 2 + 2 \times 3 + 3 \times 4 = 20$. The computation is correct.   ■

In contrast to the operators *, /, and ^, the "dotted" operators .*, ./, and .^ operate entrywise. Specifically, the i, j-entry of A.*B or A./B or A.^B is simply A(i,j)*B(i,j) or A(i,j)/B(i,j) or A(i,j)^B(i,j), respectively; there is no summing across rows or columns.

Example 2.13. Define **x** and **y** as above, and compute the entrywise product, entrywise quotient, and entrywise power.

Solution.

```
>> x = [ 1; 2; 3 ];
>> y = [ 2; 3; 4 ];
>> x.*y
ans =
     2
     6
    12
>> x./y
ans =
   0.500000000000000
   0.666666666666667
   0.750000000000000
>> x.^y
ans =
     1
     8
    81
```
∎

2.5 ▪ Block matrices

The MATLAB language also supports block matrix notation. A block matrix is built from smaller matrices. For example, from the matrices

$$\mathbf{A} = \begin{bmatrix} 1 & 2 \\ 3 & 4 \end{bmatrix}, \quad \mathbf{B} = \begin{bmatrix} 5 & 6 \\ 7 & 8 \end{bmatrix}, \tag{2.1}$$

$$\mathbf{C} = \begin{bmatrix} 9 & 10 \\ 11 & 12 \end{bmatrix}, \quad \mathbf{D} = \begin{bmatrix} 13 & 14 \\ 15 & 16 \end{bmatrix}, \tag{2.2}$$

we can form the block matrix

$$\begin{bmatrix} \mathbf{A} & \mathbf{B} \\ \mathbf{C} & \mathbf{D} \end{bmatrix} = \begin{bmatrix} 1 & 2 & 5 & 6 \\ 3 & 4 & 7 & 8 \\ 9 & 10 & 13 & 14 \\ 11 & 12 & 15 & 16 \end{bmatrix}.$$

The matrices **A**, **B**, **C**, and **D** are called blocks of the larger matrix. The natural notation works in the MATLAB language:

```
>> A = [  1  2;  3  4 ];
>> B = [  5  6;  7  8 ];
>> C = [  9 10; 11 12 ];
>> D = [ 13 14; 15 16 ];
>> [ A B; C D ]
ans =
     1     2     5     6
     3     4     7     8
     9    10    13    14
    11    12    15    16
```

2.6 ▪ Variables

The notion of a variable in the MATLAB language is different from the traditional meaning in mathematics. In the new conception, a variable is a name bound to a value, which can change over time.

Example 2.14. Run the following code:

```
a = 5
a = a+1
```

Describe what happens.

Solution.

```
>> a = 5
a =
     5
>> a = a+1
a =
     6
```

In the second line, the old value of 5 for a is used in the right-hand side, and then the value of the expression 5+1 becomes the new value of a as specified by the left-hand side. This is different from the traditional interpretation of $a = a + 1$ in mathematics, which is that there is no solution—no value of a which makes the two sides of the equation equal. ∎

The equals symbol = is known as the *assignment operator* because it assigns (or reassigns) the value associated with a variable name. This is different from the *equality operator* (==) which tests whether two quantities are equal. The equality operator is demonstrated in the next section.

2.7 ▪ Conditionals

An *if* statement allows code to be run only when a condition is met. In addition to the equality operator == mentioned above, the inequality operators <, >, <=, and >= are also available for making comparisons.

Example 2.15. Run the following code and comment:

```
if 1==1
  disp('one equals one');
end
if 1==2
  disp('one equals two');
end
```

Solution. The condition 1==1 is true, so the first phrase is displayed.

```
>> if 1==1
     disp('one equals one');
   end
one equals one
```

The condition 1==2 is false, so the second phrase is not displayed.

```
>> if 1==2
     disp('one equals two');
   end
```

∎

An *if-else* block allows one of two different paths to be taken, depending on whether a condition is true or false.

Example 2.16. Run the following code and comment.

```
if pi<3
  disp('pi is less than 3');
else
  disp('pi is greater than or equal to 3');
end
```

Solution. Because the condition pi<3 is false, the code between else and end is executed.

```
>> if pi<3
     disp('pi is less than 3');
   else
     disp('pi is greater than or equal to 3');
   end
pi is greater than or equal to 3                                    ∎
```

Provided with this book is an "inline-if" function called iif. The expression iif(b,@()x,@()y) evaluates to x if b is true or to y if b is false. (The @() notation is described in the next section.)

Example 2.17. Run the following code and comment.

```
iif(sqrt(2)>1.4,@()1,@()0)
iif(sqrt(2)>1.5,@()1,@()0)
```

Solution. Because the condition sqrt(2)>1.4 is true, the first iif expression evaluates to 1.

```
>> iif(sqrt(2)>1.4,@()1,@()0)
ans =
     1
```

Because the condition sqrt(2)>1.5 is false, the second iif expression evaluates to 0.

```
>> iif(sqrt(2)>1.5,@()1,@()0)
ans =
     0                                                              ∎
```

2.8 ▪ Functions

There are two ways to define your own mathematical functions.

If a function can be described by a single formula, then it can be entered directly at the command prompt using *anonymous function* syntax, which is signaled by the "at" (@) operator.

Example 2.18. Let $f(x) = 2x + 1$. Construct this function in the MATLAB environment and then evaluate $f(4)$.

Solution.

```
>> f = @(x) 2*x+1;
>> f(4)
ans =
     9                                                              ∎
```

In the example, the code segment @(x) 2*x+1 should be read, "A function with input x and output $2x + 1$." By itself, this is an anonymous function. The statement f = @(x) 2*x+1 assigns the name f to this function.

More complicated or more permanent functions can be saved to separate files, as illustrated by the next example.

Example 2.19. Write a MATLAB function to compute the positive part of a given number.

Solution. The positive part of a number is naturally expressed piecewise:

$$f(x) = \begin{cases} x & \text{if } x \geq 0, \\ 0 & \text{if } x < 0. \end{cases}$$

We will implement this as a MATLAB function and store our code in a single-purpose file pospart.m. To begin, type

```
>> edit pospart.m
```

You will likely be asked to create the file. Reply yes. Then enter the following code:

```
function y = pospart(x)
%POSPART   The positive part of a real number.
if x>=0
  y = x;
else
  y = 0;
end
```

(The text after the % character above is a comment intended for human readers and is ignored by the MATLAB interpreter.) Save the file, and try it out:

```
>> pospart(5)
ans =
     5
>> pospart(-2)
ans =
     0
```

The function works as intended. ∎

A file containing MATLAB code should end in extension .m and is called an *M-file*. The file pospart.m from the example is more specifically called a *function M-file* because it defines a function.

2.9 ▪ Plots

The MATLAB graphics library supports two fundamentally different types of plots of a single variable. One is for a sequence of discrete points, and the other is for a function of a continuous variable.

The plot command plots a sequence of points.

Example 2.20. Plot the sequence of points (k, k^2), $k = 0, 1, 2, \ldots, 5$, marking each by an asterisk.

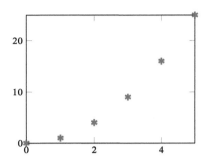

Figure 2.1. *A plot of a sequence of points.*

Solution. The x-values for the points are given in the problem.

```
>> x = 0:5
x =
     0     1     2     3     4     5
```

The MATLAB function `arrayfun`, which evaluates a function at every entry in an array, allows us to compute the corresponding y-values with a single line of code.

```
>> y = arrayfun(@(x) x^2,x)
y =
     0     1     4     9    16    25
```

The `newfig` function provided with this book clears any existing graph and sets some formatting options to match the figures in this book.

```
>> newfig;
```

The MATLAB built-in function `plot` creates the desired plot.

```
>> plot(x,y,'*');
```

The final argument `'*'` to `plot` selects asterisks for the point markers. See Figure 2.1. On your computer, this figure is likely in a separate window; make sure that you can see it. ∎

The `plotfun` command provided with this book is for plotting a function of a continuous variable.

Example 2.21. Plot the graph of $f(x) = \sin x, 0 \le x \le 4\pi$.

Solution. This requires just three lines of code.

```
>> newfig;
>> plotfun(@(x) sin(x),[0 4*pi],'displayname','sin');
>> xlabel('x');
```

See Figure 2.2. ∎

Note that the `displayname` option to `plot` or `plotfun` specifies a legend entry, and `xlabel` adds a label to the horizontal axis.

2.10 ▪ Scripts

In addition to function M-files, there are also *script M-files*. These also end in .m but take no input. They are useful for collecting sequences of statements that accomplish a single goal.

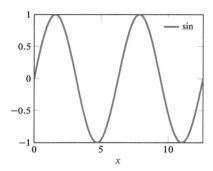

Figure 2.2. *A plot of a continuous function.*

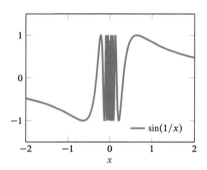

Figure 2.3. *A function with a wild discontinuity.*

Example 2.22. Create a script M-file that plots the function $f(x) = \sin(1/x)$, $-2 \le x \le 2$. Set the range on the vertical axis using `ylim`, label the horizontal axis using `xlabel`, and identify the function with a legend entry. Then modify the script to add horizontal lines at $y = \pm 1$.

Solution. Create and open a file for editing:

```
>> edit plotsinrecip.m
```

Enter the following code into the new file:

```
newfig;
plotfun(@(x) sin(1/x),[-2 2],'displayname','sin(1/x)');
ylim([-1.5 1.5]);
xlabel('x');
```

Save the file, and then run it:

```
>> plotsinrecip
```

You should see the curious function in Figure 2.3.

Now modify `plotsinrecip.m` to read as follows:

```
newfig;
plotfun(@(x) sin(1/x),[-2 2],'displayname','sin(1/x)');
plotfun(@(x) 1,[-2 2],'displayname','max');
plotfun(@(x) -1,[-2 2],'displayname','min');
ylim([-1.5 1.5]);
xlabel('x');
```

Rerun the script to verify that horizontal lines now mark the maximum and minimum values. ∎

Because we used a script M-file in the example, we were able to make changes without re-entering every line of code from scratch. (An aside: A reader with prior MATLAB experience may have noticed the absence of a hold on command in the previous example. The newfig command provided with this book executes hold on behind the scenes for every pair of axes it creates.)

2.11 ▪ Loops

Many computations require repetition. A for *loop* executes a sequence of statements repeatedly over a sequence of indices. The syntax is illustrated by the next example.

Example 2.23. Compute the first four positive square numbers using a for loop.

Solution. The numbers will be stored in a 1-by-4 array (one row and four columns). Initially, every entry of the array is set equal to NaN, a special value representing "Not a Number," to indicate that the values have not yet been computed.

```
>> squares = nan(1,4)
squares =
   NaN    NaN    NaN    NaN
```

The entries of the array are set one at a time using a for loop.

```
>> for i = 1:4
       squares(i) = i^2
   end
squares =
     1    NaN    NaN    NaN
squares =
     1      4    NaN    NaN
squares =
     1      4      9    NaN
squares =
     1      4      9     16
```

The final instantiation of squares contains all of the desired numbers. The display of the intermediate values can be suppressed by terminating the line squares(i) = i^2 with a semicolon. ∎

Note that the previous problem can be solved more simply using arrayfun.

Example 2.24. *Horner's rule* is a method for evaluating a polynomial. Given a polynomial in standard form,

$$p(x) = c_0 + c_1 x + c_2 x^2 + c_3 x^3 + \cdots + c_{n-1} x^{n-1} + c_n x^n,$$

common factors of x are factored out, resulting in the following expression:

$$p(x) = c_0 + x(c_1 + x(c_2 + x(c_3 + \cdots + x(c_{n-1} + x c_n)))).$$

Note that this can be evaluated without exponentiation. Write a MATLAB function for evaluating a polynomial with Horner's rule and demonstrate its use.

Solution. Our solution is polynomialeval in Listing 2.1.

The if statement tests whether any coefficients are provided. If they are not, it terminates the computation early with return. The expression n:-1:1 in the for loop counts down from n to 1 in steps of -1, i.e., through the sequence $n, n-1, n-2, \ldots, 2, 1$.

Listing 2.1. Evaluate a polynomial.

```
function y = polynomialeval(c,x)

if isempty(c), y = 0; return; end
n = length(c)-1;
y = c(n+1);
for j = n:-1:1
  y = c(j)+x*y;
end
```

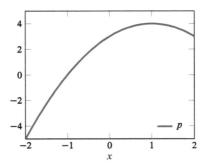

Figure 2.4. *The polynomial* $p(x) = 3 + 2x - x^2$ *evaluated with Horner's rule.*

Also, note the unfortunate indexing: the mathematical expressions refer to coefficients $c_0, \ldots, c_n$, but the entries of the MATLAB vector are identified by c(1), $\ldots$, c(n+1).

To illustrate the use of polynomialeval, consider the polynomial $p(x) = 3 + 2x - x^2$.

```
>> c = [ 3   2  -1 ];
```

Its value at $x = 1.5$ is computed as follows:

```
>> polynomialeval(c,1.5)
ans =
    3.750000000000000
```

The polynomial can also be plotted using plotfun:

```
>> newfig;
>> plotfun(@(x) polynomialeval(c,x),[-2 2],'displayname','p');
>> ylim([-5 5]);
>> xlabel('x');
```

See Figure 2.4. ■

Beware of the indexing issue mentioned in the previous example—whether we count from 0 or 1. This is a persistent source of mistakes for beginners and experts alike.

2.12 ▪ Recursion

A *recursive* definition is one that refers to itself. Classic examples are the factorial function

$$n! = \begin{cases} 1 & \text{if } n = 0, \\ n(n-1)! & \text{if } n \geq 1 \end{cases}$$

and the Fibonacci sequence

$$F_n = \begin{cases} 1 & \text{if } n = 1 \text{ or } n = 2, \\ F_{n-2} + F_{n-1} & \text{if } n \geq 3. \end{cases} \tag{2.3}$$

Note that to avoid infinite descent, one or more *base cases* are necessary. For the factorial function, the base case is $n = 0$, and for the Fibonacci sequence, the base cases are $n = 1$ and $n = 2$.

Recursion is also a useful computational tool, as the next example demonstrates.

Example 2.25. *Binary search* is an algorithm for finding an element x in an ordered list. First, x is compared against an element at the middle of the list called the *pivot*. If they are equal, then the search is complete. If x is less than the pivot, then the pivot and all greater elements can be discarded; if x is greater than the pivot, then the pivot and all lesser elements may be discarded. The process repeats, approximately halving the list each time, until either x is found or there are no more elements to inspect. Implement binary search in MATLAB code.

Solution. Our solution binarysearch is below.

```
function k = binarysearch(x,ys,a,b)
if nargin<4, a = 1; b = length(ys); end
if b<a, k = []; return; end
pivot = a+ceil((b-a)/2);
if x==ys(pivot)
  k = pivot;
elseif x<ys(pivot)
  k = binarysearch(x,ys,a,pivot-1);
else
  k = binarysearch(x,ys,pivot+1,b);
end
```

The user specifies x, the item to find; ys, the list to search; and optionally a and b, indices within which to restrict the search. If the user does not provide values for a and b, then the number nargin of input arguments is less than four, and the implementation defaults to searching over the entire list (a = 1 and b = length(ys)).

The base case is the empty list, in which case the routine returns [] to signal that x is not present.

The recursive step locates a pivot entry at the middle of the list, compares the desired element to the pivot, and then repeats the process on just half of the remaining list.

The following code demonstrates the use of binarysearch:

```
>> ys = [ 1 3 8 11 14 17 ];
>> binarysearch(8,ys)
ans =
     3
>> binarysearch(9,ys)
ans =
     []
```

It finds that 8 is the third entry of list ys and that 9 is not present. ∎

At any given stage in a recursive computation, the *recursion depth* is the number of recursive calls currently pending. For example, if at some point in binary search the list has been split in half and split in half again, then the recursion depth is 2. In any recursive algorithm, a base case or termination criterion is required to prevent infinite descent.

2.13 ▪ Getting help

The `help` command provides brief summaries of MATLAB functions. For example, if you forget the order of the arguments to `linspace`, try this:

```
>> help linspace
```

Notes

The original version of MATLAB was written by Cleve Moler in the late 1970's [56]. The system is currently developed by The MathWorks.

Exercises

Exercises 1–4: Create a function M-file satisfying the given specification. Include a demonstration of your function.

1. input: r; output: area of a circle of radius r
2. input: v, d; output: the time required to drive d miles at a constant speed of v miles per hour
3. input: m_1, m_2, r; output: the force of gravity between objects of masses m_1 and m_2 (in kilograms) separated by a distance r (in meters)
4. input: x; output: $|x|$. (Do not use the built-in function abs.)

Exercises 5–15: Create a function satisfying the given specification using one or more for loops. Include a demonstration of your function. If there is a built-in MATLAB function that satisfies the specification, create your own rather than using the built-in function.

5. input: n; output: $n!$
6. input: n; output: nth Fibonacci number
7. input: real numbers a, b and integer n; output: n-by-1 vector $(a, a + h, a + 2h, \ldots, b - h, b)$ of uniform steps
8. input: n-by-1 vector $(x_1, x_2, \ldots, x_n)$; output: vector of differences $(x_2 - x_1, x_3 - x_2, \ldots, x_n - x_{n-1})$
9. input: n; output: the n-by-n identity matrix
10. input: n-by-1 vector $(d_1, d_2, \ldots, d_n)$; output: diagonal matrix with diagonal entries $d_1, d_2, \ldots, d_n$
11. input: an m-by-n matrix; output: the 1-by-n row vector of column sums
12. input: m-by-n matrix; output: 1-by-n vector with maximum entry of each column
13. input: a matrix; output: transpose of the matrix
14. input: m-by-n matrix $A = [a_{ij}]$; output: m-by-n matrix $C = [c_{ij}]$ of cumulative column sums $c_{ij} = \sum_{k=1}^{i} a_{kj}$
15. input: matrices $\mathbf{A}$, $\mathbf{B}$; output: matrix product $\mathbf{AB}$

16. Let F_n be the nth Fibonacci number, starting with $F_1 = F_2 = 1$. Prove that

$$\left[\begin{array}{c} F_n \\ F_{n+1} \end{array} \right] = \left[\begin{array}{cc} 0 & 1 \\ 1 & 1 \end{array} \right]^{n-1} \left[\begin{array}{c} 1 \\ 1 \end{array} \right]$$

for $n = 1, 2, 3, \ldots$.

17. Use the result of Exercise 16 to compute the 50th Fibonacci number.

18. The nth Fibonacci number equals $(\phi^n - \psi^n)/\sqrt{5}$, in which $\phi = (1 + \sqrt{5})/2$ and $\psi = (1 - \sqrt{5})/2$. Use this formula to compute the 50th Fibonacci number.

19. Create a plot with the first 10 Fibonacci numbers represented by dots and the function $(1/\sqrt{5})\phi^x$ represented by a curve, in which $\phi = (1 + \sqrt{5})/2$. Label the x-axis and include a legend.

20. Consider the continued fraction

$$a_0 + \cfrac{1}{a_1 + \cfrac{1}{a_2 + \cfrac{1}{\ddots + \frac{1}{a_n}}}},$$

also denoted $[a_0; a_1, a_2, \ldots, a_n]$. Define p_k and q_k by

$$p_k = a_k p_{k-1} + p_{k-2}, \quad p_{-2} = 0, \quad p_{-1} = 1,$$
$$q_k = a_k q_{k-1} + q_{k-2}, \quad q_{-2} = 1, \quad q_{-1} = 0.$$

Then the continued fraction equals p_n/q_n. Create a MATLAB function for computing the continued fraction $[a_0; a_1, a_2, \ldots, a_n]$ using the above recurrence relations. Demonstrate the function's correctness on a small example.

Chapter 3

Experimental accuracy analysis

Because exact solution formulas do not exist for many problems, our numerical methods are typically indirect. One common approach is to express a desired quantity as a limit of a sequence: $x = \lim_{n \to \infty} x_n$. As the sequence is traversed, the terms get closer and closer to the desired answer. Rather than taking the limit $n \to \infty$, many of our numerical methods stop at some large n. The resulting error is called *truncation error*.

Commonly available computers store real numbers to about 15 or 16 digits of precision. For this reason, a sequence of approximations may settle on a fixed value rather than becoming forever more accurate. When this happens, we say that *numerical convergence* has occurred. This often indicates that high accuracy has been achieved.

This chapter presents a number of examples in which numerical convergence is sought and rate of convergence is measured empirically. These experimental techniques can justify high confidence in the accuracy of computations, and in later chapters, they are complemented by rigorous proofs.

3.1 ▪ Numerical convergence

When generating a sequence of approximations in the limited precision of a computer, numerical convergence occurs if the terms cease to change significantly. If this happens, then the computation can be concluded.

Example 3.1. While studying compound interest in 1683, Jakob Bernoulli investigated $(1 + 1/n)^n$, showing that the $n \to \infty$ limit is between 2 and 3. Nowadays, we know that the limit equals e, the base of the natural logarithm:

$$e = \lim_{n \to \infty} \left(1 + \tfrac{1}{n}\right)^n. \tag{3.1}$$

Approximate e by computing $(1 + 1/n)^n$ for increasing values of n. Does numerical convergence occur? When? To what value?

Solution. By trial and error, we have found that extremely large values of n are required to understand the behavior of the sequence. Below, the nth term is computed for $n = 1$, 100, 10^4, 10^6, ..., 10^{18}.

```
>> n = 10.^(0:2:18);
>> eapprox = nan(size(n));
```

```
>> for k = 1:length(n)
     eapprox(k) = (1+1/n(k))^n(k);
   end
```

The results are displayed in a table using a command `disptable` provided with this book. The columns of the table are specified one at a time, and each column specification consists of a column heading, a data vector, and an `sprintf` format string (as seen in Example 2.3).

```
>> disptable('n',n,'%5.0e','(1+1/n)^n',eapprox,'%17.15f');
    n          (1+1/n)^n
1e+00     2.000000000000000
1e+02     2.704813829421528
1e+04     2.718145926824926
1e+06     2.718280469095753
1e+08     2.718281798347358
1e+10     2.718282053234788
1e+12     2.718523496037238
1e+14     2.716110034087023
1e+16     1.000000000000000
1e+18     1.000000000000000
```

Technically, numerical convergence occurs around $n = 10^{16}$, but the final value according to the computer is 1. The computation is a failure. At first, the terms appear to be approaching a number around 2.718282, presumably a decent approximation for e, but then they veer away. An effective method would converge numerically to an accurate approximation. ∎

The sequence $(1 + 1/n)^n$ in the previous example converges very slowly. Before a highly accurate approximation for e is found, *roundoff error* overwhelms the computation. In the extreme case, $1/n$ becomes so small that $1 + 1/n$ is rounded to 1 by the computer, so that the formula $(1 + 1/n)^n$ is effectively replaced by $1^n = 1$. A progression of less and less accurate computations precedes this total collapse. Roundoff error is explored in more detail in Chapter 5.

The next example investigates a different method for computing e.

Example 3.2. About a half-century after Bernoulli, Leonhard Euler stated the following equation:

$$e = \sum_{i=0}^{\infty} \frac{1}{i!}. \tag{3.2}$$

Approximate e by truncating the series. Does numerical convergence occur? When? To what value?

Solution. Euler's series is also a limit formula because it can be rewritten as

$$e = \lim_{n \to \infty} \sum_{i=0}^{n} \frac{1}{i!}.$$

Experimentally, the sequence of partial sums

$$\sum_{i=0}^{n} \frac{1}{i!}, \quad n = 0, 1, 2, 3, \ldots,$$

is found to converge rapidly.

```
>> n = 0:2:20;
>> eapprox = nan(size(n));
>> for k = 1:length(n)
      eapprox(k) = sum(arrayfun(@(i) 1/factorial(i),0:n(k)));
   end
>> disptable('n',n,'%2d','partial sum',eapprox,'%17.15f');
 n         partial sum
 0    1.000000000000000
 2    2.500000000000000
 4    2.708333333333333
 6    2.718055555555555
 8    2.718278769841270
10    2.718281801146385
12    2.718281828286169
14    2.718281828458230
16    2.718281828459043
18    2.718281828459046
20    2.718281828459046
```

(The format string '%2d' above specifies an integer with a width of two digits.) Numerical convergence occurs by $n = 18$. We suspect that the final approximation

$$e \approx 2.718281828459046 \tag{3.3}$$

is highly accurate.   ■

Examples 3.1 and 3.2 teach an important lesson: A slow-converging method may be costly or unstable, while a fast-converging method may deliver an accurate solution in short time.

3.2 ▪ Rate of convergence

Associated to a sequence of approximations $x_1, x_2, x_3, \ldots$ converging to a limit x is the sequence of absolute errors $|x - x_n|$, $n = 1, 2, 3, \ldots$. Often, such a sequence resembles a power function n^{-p} or an exponential function r^{-n}. Figures 3.1 and 3.2 display examples of these and other familiar functions. We will compare our experimental data with these plots.

The plots on the left of Figures 3.1 and 3.2 use familiar linear scales, but the plots on the right may be less familiar. The plot on the right of Figure 3.1 is called a *log-log* plot because it uses logarithmic scales for the vertical and horizontal axes. Note that the value on each axis grows by a constant *factor* with each tick mark. (The notations $\cdot^{-15}$, $\cdot^{-10}$, and $\cdot^{-5}$ are short for 10^{-15}, 10^{-10}, and 10^{-5}, respectively.) The plot on the right of Figure 3.2 is called a *log-linear* plot because only the vertical axis is measured on a logarithmic scale.

A power function n^{-p} appears as a straight line on a log-log plot because $y = \log_{10} n^{-p}$ is related to $x = \log_{10} n$ by $y = -px$. An exponential function r^{-n} appears as a straight line on a log-linear plot because $y = \log_{10} r^{-n}$ is related to $x = n$ by $y = -(\log_{10} r)x$. Exponential functions r^{-n} converge to zero much more quickly than power functions n^{-p}.

In MATLAB code, xlog or ylog specifies a logarithmic scale for the horizontal or vertical axis, respectively.

Example 3.3. Our best approximation for e is (3.3). Using this value, measure the absolute errors $|e - (1 + 1/n)^n|$ associated with Bernoulli's limit formula. Plot the

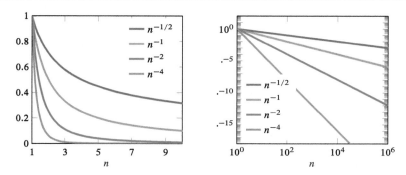

Figure 3.1. *Power functions on standard (left) and log-log (right) axes. In the plot on the right, the marks* $\cdot^{-5}$, $\cdot^{-10}$, *and* $\cdot^{-15}$ *are abbreviations for* 10^{-5}, 10^{-10}, *and* 10^{-15}.

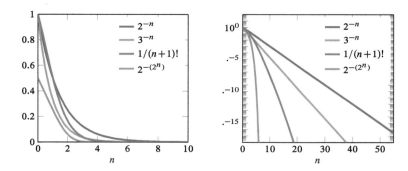

Figure 3.2. *Exponential functions (and more) on standard (left) and log-linear (right) axes.*

errors, and, if possible, find a power function or exponential function that converges to zero at the same rate.

Solution. The approximations are computed as in Example 3.1.

```
>> n = 10.^(0:2:18);
>> eapprox = nan(size(n));
>> for k = 1:length(n)
      eapprox(k) = (1+1/n(k))^n(k);
   end
```

We plan to make two graphs: one of the approximations themselves, and one of their absolute errors. An *m*-by-*n* grid of subfigures is created with ax = newfig(m,n).

```
>> ax = newfig(1,2);
```

The approximations $(1 + 1/n)^n$ are plotted in the left graph of Figure 3.3.

```
>> xlog;
>> plot(n,eapprox,'*','displayname','(1+1/n)^n');
>> ylim([0 3.5]);
>> xlabel('n');
```

At first, the computation makes progress toward a value around 2.7, but things go haywire sometime before $n = 10^{16}$.

The absolute errors are estimated by comparing with (3.3), the best available approximation.

```
>> err = abs(2.718281828459046-eapprox);
>> disptable('n'             ,n        ,'%5.0e'  ...
            ,'(1+1/n)^n' ,eapprox ,'%17.15f' ...
```

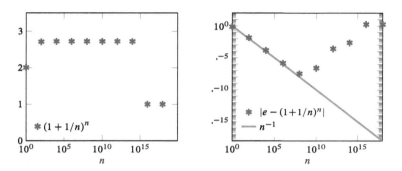

Figure 3.3. *Approximations for e from Bernoulli's limit (left) and their absolute errors (right).*

```
              ,'abs. error',err      ,'%17.15f');
    n            (1+1/n)^n              abs. error
1e+00    2.000000000000000      0.718281828459046
1e+02    2.704813829421528      0.013467999037517
1e+04    2.718145926824926      0.000135901634120
1e+06    2.718280469095753      0.000001359363293
1e+08    2.718281798347358      0.000000030111688
1e+10    2.718282053234788      0.000000224775742
1e+12    2.718523496037238      0.000241667578192
1e+14    2.716110034087023      0.002171794372023
1e+16    1.000000000000000      1.718281828459046
1e+18    1.000000000000000      1.718281828459046
```

The most accurate computation occurs around $n = 10^8$, with an absolute error of about 3×10^{-8}. The absolute errors are also plotted in the right graph of Figure 3.3. The invocation of subplot below selects the second subfigure before plotting.

```
>> subplot(ax(2));
>> xlog; ylog; ylim([1e-18 1e2]);
>> plot(n,err,'*','displayname','|e-(1+1/n)^n|');
```

The absolute errors are intended to converge to zero. At first, they appear to decrease at a predictable rate, but eventually they veer upward. By trial and error, the graph of n^{-1} is found to be a good fit for the initial trend. This function appears as a straight line on the log-log plot.

```
>> plotfun(@(n) n^(-1),[n(1) n(end)],'displayname','n^{-1}');
>> xlabel('n');
```

Mathematically, $|e - (1 + 1/n)^n|$ appears to converge to 0 at the same rate as n^{-1}, but evidently the computer is unable to calculate $(1 + 1/n)^n$ accurately for $n > 10^8$. ∎

In the previous example, the truncation error is comparable to a power function n^{-p}. Better numerical methods converge at least as fast as an exponential function r^{-n}. To emphasize such distinctions, some language is introduced.

Given sequences $a_1, a_2, a_3, \ldots$ and $b_1, b_2, b_3, \ldots$ of nonnegative real numbers that converge to zero, we say that a_n *converges as fast as* b_n if there exist a constant M and a positive integer N such that $|a_n| \le Mb_n$ for all $n \ge N$. We say that a_n and b_n *converge at the same rate* if a_n converges as fast as b_n and vice versa. Consider a sequence $x_1, x_2, x_3, \ldots$ that converges to x.

- The convergence is *algebraic* if $|x - x_n|$ converges to zero as fast as n^{-p} for some $p > 0$. On a log-log plot, the absolute errors eventually lie on or below a line with slope $-p$.

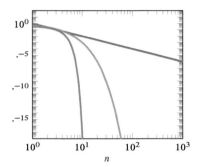 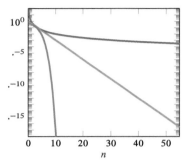

Figure 3.4. *Rates of convergence: algebraic (blue), geometric (orange), and supergeometric (green), on log-log axes (left) and log-linear axes (right).*

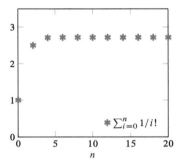 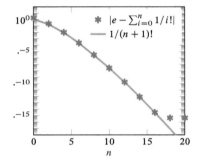

Figure 3.5. *Approximations for e from Euler's series (left) and their absolute errors (right).*

- The convergence is *geometric* if $|x - x_n|$ converges to zero as fast as r^{-n} for some $r > 1$. On a log-linear plot, the absolute errors eventually lie on or below a line with slope $-\log_{10} r$.

- The convergence is *supergeometric* if $|x - x_n|$ converges to zero as fast as r^{-n} for *any* $r > 1$. On a log-linear plot, the absolute errors fall toward zero faster than any straight line.

These classes are illustrated in Figure 3.4.

Bernoulli's formula for e converges algebraically because the associated truncation errors converge to zero, discounting roundoff error, as fast as the power function n^{-1}. Graphically, the truncation errors converge to zero as fast as a straight line in a log-log plot.

Example 3.4. How fast does Euler's series $\sum_{i=0}^{n} 1/i!$ converge? Classify the rate as algebraic, geometric, or supergeometric.

Solution. Several partial sums are computed as in Example 3.2.

```
>> n = 0:2:20;
>> eapprox = nan(size(n));
>> for k = 1:length(n)
      eapprox(k) = sum(arrayfun(@(i) 1/factorial(i),0:n(k)));
   end
```

The approximations are plotted in Figure 3.5.

```
>> ax = newfig(1,2);
>> plot(n,eapprox,'*','displayname','\Sigma_{i=0}^n 1/i!');
>> ylim([0 3.5]);
>> xlabel('n');
```

In place of the true absolute error $|e - \sum_{i=0}^{n} 1/i!|$, we consider the difference

$$\left| 2.718281828459046 - \sum_{i=0}^{n} \frac{1}{i!} \right|.$$

```
>> err = abs(2.718281828459046-eapprox);
>> disptable('n'              ,n          ,'%2d'       ...
             ,'partial sum',eapprox,'%17.15f' ...
             ,'abs. error' ,err        ,'%17.15f');
     n         partial sum             abs. error
     0    1.000000000000000      1.718281828459046
     2    2.500000000000000      0.218281828459046
     4    2.708333333333333      0.009948495125713
     6    2.718055555555555      0.000226272903491
     8    2.718278769841270      0.000003058617776
    10    2.718281801146385      0.000000027312661
    12    2.718281828286169      0.000000000172877
    14    2.718281828458230      0.000000000000816
    16    2.718281828459043      0.000000000000003
    18    2.718281828459046      0.000000000000000
    20    2.718281828459046      0.000000000000000
```

The absolute errors are added to the figure.

```
>> subplot(ax(2));
>> ylog; ylim([1e-18 1e2]);
>> plot(n,err,'*','displayname','|e-\Sigma_{i=0}^n 1/i!|');
```

Note that the error plot is on log-linear axes, in contrast to the log-log axes of Example 3.3. Because the errors fall toward zero even faster than a straight line on a log-linear plot (until numerical convergence occurs at $n = 18$), the convergence is supergeometric. In fact, the absolute error is close to $1/(n + 1)!$, as seen in the plot.

```
>> plot(n,1./factorial(n+1),'displayname','1/(n+1)!');
>> xlabel('n');                                                                        ∎
```

Bernoulli's sequence $(1 + 1/n)^n$ converges algebraically, whereas Euler's series $\sum_{i=0}^{n} 1/i!$ converges supergeometrically. Euler's series begets a superior numerical method.

So far, we have seen examples of algebraic and supergeometric convergence. The next example illustrates a geometric rate of convergence.

Example 3.5. The quantity $\sqrt{2}$ is the limit of the sequence $x_0, x_1, x_2, \ldots$ defined by $x_0 = 1$ and

$$x_{n+1} = \begin{cases} x_n + 2^{-n-1} & \text{if } (x_n + 2^{-n-1})^2 \leq 2, \\ x_n & \text{otherwise.} \end{cases}$$

The recursion starts with the low estimate of 1 and adds powers of two as long as the sum does not exceed the desired quantity. Here are the first few terms:

$$
\begin{aligned}
x_0 &   & &= 1   & &= 1, \\
x_1 &= x_0   & &= 1   & &= 1, \\
x_2 &= x_1 + 2^{-2} = 1 + 1/4   & &= 5/4 & &= 1.25, \\
x_3 &= x_2 + 2^{-3} = 5/4 + 1/8   & &= 11/8 & &= 1.375, \\
x_4 &= x_3   & &= 11/8 & &= 1.375, \\
x_5 &= x_4 + 2^{-5} = 11/8 + 1/32 = 45/32 & &= 45/32 & &= 1.40625.
\end{aligned}
$$

Classify the rate of convergence as algebraic, geometric, or supergeometric.

Solution. The sequence of approximations $x_0, x_1, x_2, \ldots, x_{60}$ is computed.

```
>> n = 0:60;
>> x = nan(size(n));
>> x(1) = 1;
>> for k = 1:length(n)-1
      if (x(k)+2^(-n(k)-1))^2<=2
        x(k+1) = x(k)+2^(-n(k)-1);
      else
        x(k+1) = x(k);
      end
    end
```

The sequence of approximations is plotted in the left graph of Figure 3.6.

```
>> ax = newfig(1,2);
>> plot(n,x,'*','displayname','x_n');
>> ylim([0 2]);
>> xlabel('n');
```

The final approximation for $\sqrt{2}$ is the following:

```
>> x(end)
ans =
    1.414213562373095
```

The absolute errors $|\sqrt{2} - x_n|$ are measured using the built-in function sqrt. The errors are shown in the right graph of Figure 3.6.

```
>> subplot(ax(2));
>> ylog; ylim([1e-18 1e2]);
>> plot(n,abs(sqrt(2)-x),'*','displayname','|sqrt(2)-x_n|');
```

The errors appear to lie below a straight line on a log-linear plot. This is evidence of geometric convergence. More specifically, the errors appear to converge at the same rate as 2^{-n}.

```
>> plotfun(@(n) 2^(-n),[n(1) n(end)],'displayname','2^{-n}');
>> xlabel('n');
```
∎

3.3 ▪ Residuals

A major difficulty in measuring error is a chicken-and-egg problem: to measure the error $|x - \hat{x}|$ directly, the exact value x is required, but the whole reason for computing the approximation $\hat{x}$ is that x is unknown.

When a computed quantity $\hat{x}$ is an approximate solution to an equation $f(x) = b$, the corresponding *residual* is $b - f(\hat{x})$, and the *absolute residual* is $|b - f(\hat{x})|$. A small residual indicates that the computed value nearly solves the equation. Unlike absolute or

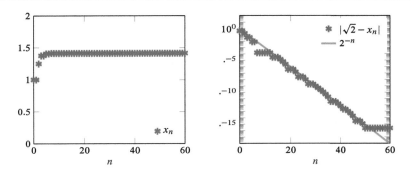

Figure 3.6. *Approximations for $\sqrt{2}$ (left) and their absolute errors (right).*

relative error, a residual can be measured without knowing the true value of the quantity
of interest.

Example 3.6. Approximate $\sqrt{2}$ by x_n as in Example 3.5. Because $\sqrt{2}$ is a solution of
$x^2 = 2$, an appropriate residual for evaluating x_n is $2 - x_n^2$. Plot the absolute residuals
$|2 - x_n^2|$ and measure the rate of convergence.

Solution. The approximations are computed again.

```
>> n = 0:60;
>> x = nan(size(n));
>> x(1) = 1;
>> for k = 1:length(n)-1
     if (x(k)+2^(-n(k)-1))^2<=2
       x(k+1) = x(k)+2^(-n(k)-1);
     else
       x(k+1) = x(k);
     end
   end
```

The absolute residuals $|2 - x_n^2|$ are measured and displayed in Figure 3.7.

```
>> ax = newfig;
>> ylog; ylim([1e-18 1e2]);
>> plot(n,abs(2-x.^2),'*','displayname','|2-x_n^2|');
```

The residuals appear to converge at the rate 2^{-n}, the same rate seen for the absolute
errors in Example 3.5.

```
>> plotfun(@(n) 2^(-n),[n(1) n(end)],'displayname','2^{-n}');
>> xlabel('n');
```
■

In Example 3.6, the residuals are measured without knowledge of the true value of
$\sqrt{2}$, and they suggest the same geometric rate of convergence as the absolute errors in
Example 3.5. Clearly, residuals are useful. However, the connection between error and
residual is subtle, and for some problems, a small residual may not indicate a small
error. The full story must wait for now.

Notes

The notion that a_n converges to zero as fast as b_n can be expressed using the popular "big-O"
notation as $a_n = O(b_n), n \to \infty$.

Havil's book *The Irrationals* provides additional historical background on the number e [41].

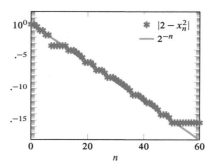

Figure 3.7. *Absolute residuals for the approximations of $\sqrt{2}$ in Figure 3.6.*

Exercises

Exercises 1–9: A mathematical constant is expressed as a limit of a sequence. Approximate the constant by truncating the sequence and seek numerical convergence. Either report the value of n at which the sequence converges numerically, or, if numerical convergence is not achieved with reasonable resources, find approximately the value of n at which the absolute error falls below 10^{-4}. Either way, provide evidence. You may compare with the known values $\pi = 3.141592653589793\ldots$, $\gamma = 0.577215664901533\ldots$, and $e = 2.718281828459046\ldots$.

1. $\pi = \lim_{n\to\infty} 2 \prod_{k=1}^{n} \frac{(2k)^2}{(2k-1)(2k+1)}$ (Wallis's product)

2. $\pi = \lim_{n\to\infty} (2 + \frac{1}{2n}) \prod_{k=1}^{n} \frac{(2k)^2}{(2k-1)(2k+1)}$

3. $\pi = \lim_{n\to\infty} 4 \sum_{i=0}^{n} \frac{(-1)^i}{2i+1}$ (Gregory series)

4. $\pi = \lim_{n\to\infty} \frac{(-1)^{n+1}}{n} + 4 \sum_{i=0}^{n} \frac{(-1)^i}{2i+1}$

5. $\pi = \lim_{n\to\infty} 2\sqrt{3} \sum_{i=0}^{n} \frac{(-1)^i}{(2i+1)3^i}$ (Sharp series)

6. $\pi = \lim_{n\to\infty} \frac{16}{5} \sum_{i=0}^{n} \frac{(-1)^i}{(2i+1)5^{2i}} - \frac{4}{239} \sum_{i=0}^{n} \frac{(-1)^i}{(2i+1)239^{2i}}$ (Machin's formula)

7. $\pi = \lim_{n\to\infty} \left(6 \sum_{k=1}^{n} \frac{1}{k^2} \right)^{1/2}$ (the Basel problem)

8. $\gamma = \lim_{n\to\infty} \sum_{k=1}^{n} \frac{1}{k} - \log n$ (Euler–Mascheroni constant)

9. $e = \lim_{n\to\infty} n \prod_{k=1}^{n} k^{-1/n}$ (from Stirling's approximation)

Exercises 10–17: Refer to the specified exercise to find a mathematical constant expressed as a limit of a sequence. Classify the rate of convergence as algebraic, geometric, or supergeometric. If algebraic or geometric, find a function of the form n^{-p} or r^{-n} that converges at approximately the same rate as the absolute error.

10. Exercise 1
11. Exercise 2
12. Exercise 3
13. Exercise 4
14. Exercise 5
15. Exercise 6

16. Exercise 7

17. Exercise 8

18. Modify Example 3.5 to compute $\sqrt{3}$. Run the method until it converges numerically. Report the approximations $x_0, x_1, x_2, \ldots$ and the corresponding sequence of absolute residuals $|3 - x_n^2|$, $n = 0, 1, 2, \ldots$.

19. Modify Example 3.5 to compute $\sqrt[3]{2}$. Run the method until it converges numerically. Report the approximations $x_0, x_1, x_2, \ldots$ and the corresponding absolute residuals $|2 - x_n^3|$, $n = 0, 1, 2, \ldots$.

20. Compute $\sqrt{2}$ using the Babylonian method of Example 1.1, starting from the guess $x_0 = 1.5$. Graph the absolute residuals $|2 - x_n^2|$ and classify the rate of convergence as algebraic, geometric, or supergeometric.

21. Suppose the numerical value of π is unknown and an approximation $\hat{x}$ is computed. Two possible residuals are $\cos \hat{x} - (-1)$ and $\sin \hat{x} - 0$. As $\hat{x}$ approaches π, one of these residuals converges to 0 considerably faster than the other. Which one, and why? Which do you think provides a better measure of the approximation's accuracy? Explain.

Exercises 22–24: The continued fraction $[a_0; a_1, a_2, a_3, \ldots, a_n]$ and an approach to its computation are introduced in Exercise 20 in Chapter 2. The infinite continued fraction $[a_0; a_1, a_2, a_3, \ldots]$ is defined by

$$[a_0; a_1, a_2, a_3, \ldots] = \lim_{n \to \infty} [a_0; a_1, a_2, a_3, \ldots, a_n].$$

For each of the equations below, approximate the mathematical constant on the left-hand side by truncating and evaluating the continued fraction on the right-hand side. Does numerical convergence occur with a reasonable expenditure of computer time? If so, how many terms are necessary, and what is the final approximation?

22. golden ratio $\phi = [1; 1, 1, 1, 1, \ldots]$

23. $\sqrt{2} = [1; 2, 2, 2, 2, \ldots]$

24. $e = [2; 1, 2, 1, 1, 4, 1, 1, 6, 1, 1, 8, 1, \ldots]$

25. The golden ratio ϕ satisfies the equation $\phi^2 - \phi - 1 = 0$. Let x_n be the truncated continued fraction

$$x_n = [1; \underbrace{1, 1, 1, \ldots, 1}_{n}].$$

Compare the rates of convergence for the absolute error $|\phi - x_n|$ and the absolute residual $|x_n^2 - x_n - 1|$. Do the two measures converge at the same rate, in the technical sense introduced in this chapter, or does one converge faster than the other?

Chapter 4

Experimental accuracy analysis for functions

This chapter continues the study of experimental error analysis, but the focus moves from scalar quantities to functions.

4.1 ▪ Error functions

Suppose that a function $f(x)$ is approximated by another function $\hat{f}(x)$. A plot of the *absolute error function* $|f(x) - \hat{f}(x)|$ can reveal the magnitude of the error and how it varies.

Our first example involves a Taylor polynomial. Recall that the degree-m Taylor polynomial for a function $f(x)$, centered at $x = x_0$, is

$$p(x) = f(x_0) + f'(x_0)(x - x_0) + \frac{f''(x_0)}{2!}(x - x_0)^2 + \cdots + \frac{f^{(m)}(x_0)}{m!}(x - x_0)^m,$$

i.e.,

$$p(x) = \sum_{i=0}^{m} \frac{f^{(i)}(x_0)}{i!}(x - x_0)^i,$$

when the function $f(x)$ is m-times differentiable at $x = x_0$. The polynomial p and its first m derivatives are constructed to agree with f and its first m derivatives at the central point. Therefore, in a neighborhood of the central point, the polynomial may strongly resemble the original function.

Example 4.1. Let $p(x)$ be the quadratic Taylor polynomial for e^x about $x = 0$. Plot $p(x)$ and its absolute error function over $-1 \le x \le 1$. Discuss.

Solution. The ith derivative of e^x is $\frac{d^i}{dx^i}[e^x] = e^x$, whose value at $x = 0$ is $e^0 = 1$. Hence, the Taylor polynomial is

$$p(x) = 1 + x + \frac{1}{2}x^2.$$

We construct this in code.

```
>> p = @(x) 1+x+(1/2)*x^2;
```

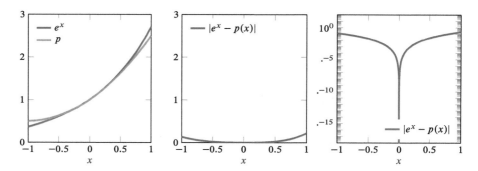

Figure 4.1. *A Taylor polynomial for e^x: the approximating polynomial (left), the absolute error on linear axes (middle), and the absolute error on log-linear axes (right).*

The Taylor polynomial is graphed in the left plot of Figure 4.1 along with `exp` of MATLAB for comparison.

```
>> ax = newfig(1,3);
>> plotfun(@(x) exp(x),[-1 1],'displayname','e^x');
>> plotfun(p,[-1 1],'displayname','p');
>> xlabel('x');
```

The absolute error function for $p(x)$ is $|e^x - p(x)|$. The error is measured by comparing with the MATLAB `exp` routine, and it is plotted in the middle graph of Figure 4.1.

```
>> subplot(ax(2));
>> ylim([0 3]);
>> plotfun(@(x) abs(exp(x)-p(x)),[-1 1],'displayname','|e^x-p(x)|');
>> xlabel('x');
```

The error is smallest near the expansion point $x = 0$ as expected. The absolute error function is shown again but on log-linear axes in the final plot.

```
>> subplot(ax(3));
>> ylog; ylim([1e-18 1e2]);
>> plotfun(@(x) abs(exp(x)-p(x)),[-1 1],'displayname','|e^x-p(x)|');
>> xlabel('x');
```

The vertical asymptote in the log-linear plot indicates that $e^x - p(x) = 0$ at $x = 0$. In other words, the Taylor polynomial equals the original function at the central point $x = 0$, as expected. Moving away from this central point, the absolute error grows, reaching a magnitude a little over $10^{-1} = 0.1$ near the ends of the interval. ∎

The concept of the residual also extends to functions, as the next example demonstrates.

Example 4.2. In Example 3.5, $\sqrt{2}$ was computed by repeatedly adding powers of 2 to an initial underestimate. To extend this method to the square root function over $1 \le x < 4$, let $f_0(x) = 1$ and define, for each $x \in [1, 4)$,

$$f_{n+1}(x) = \begin{cases} f_n(x) + 2^{-n-1} & \text{if } (f_n(x) + 2^{-n-1})^2 \le x, \\ f_n(x) & \text{otherwise.} \end{cases}$$

In the limit, $f_n(x)$ approaches $\sqrt{x}$ as $n \to \infty$ (Exercise 9).

Write a MATLAB function to implement this method. Plot the approximation $f_4(x)$, its absolute error $|\sqrt{x} - f_4(x)|$, and its absolute residual $|x - (f_4(x))^2|$.

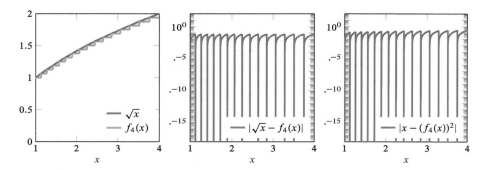

Figure 4.2. *Computing the square root function: an approximation (left), its absolute error (middle), and its absolute residual (right).*

Solution. Create a new M-file, enter the following code, and save as sqrtbitbybit.m.

```
function y = sqrtbitbybit(n,x)
y = 1;
for i = 0:n-1
  if (y+2^(-i-1))^2<=x
    y = y+2^(-i-1);
  end
end
```

The left plot in Figure 4.2 shows the approximation for the square root function.

```
>> n = 4;
>> ax = newfig(1,3);
>> ylim([0 2]);
>> plotfun(@(x) sqrt(x),[1 4],'displayname','sqrt(x)');
>> plotfun(@(x) sqrtbitbybit(n,x),[1 4],'displayname','f_4(x)');
>> xlabel('x');
```

The middle plot shows the absolute error.

```
>> subplot(ax(2));
>> ylog; ylim([1e-18 1e2]);
>> plotfun(@(x) abs(sqrt(x)-sqrtbitbybit(n,x)),[1 4] ...
        ,'displayname','|sqrt(x)-f_4(x)|');
>> xlabel('x');
```

And the right plot shows the absolute residual.

```
>> subplot(ax(3));
>> ylog; ylim([1e-18 1e2]);
>> plotfun(@(x) abs(x-sqrtbitbybit(n,x)^2),[1 4] ...
        ,'displayname','|x-(f_4(x))^2|');
>> xlabel('x');
```

The graph of $f_4(x)$ has a staircase shape, and it regularly intersects the graph of $\sqrt{x}$. Hence, the error and residual hit zero repeatedly. ∎

4.2 ▪ Infinity norm

A useful one-number summary of an error function is its *infinity norm*. Before defining the infinity norm, we need a few facts about sets of real numbers.

Let S be a set of real numbers. A number M is an upper bound for the set if every $x \in S$ satisfies $x \leq M$. If an upper bound exists, then S is said to be bounded above.

Whenever a set of real numbers is bounded above, a least upper bound exists, and it is called the *supremum* of the set. In particular, the supremum of a closed interval $[c, d]$ is d, and the supremum of the open interval (c, d) is also d.

Now let $g(x)$ be a function on an interval $a \leq x \leq b$. Its infinity norm is defined to be

$$\|g\|_\infty = \sup_{a \leq x \leq b} |g(x)| \tag{4.1}$$

if the supremum exists or $\|g\|_\infty = \infty$ otherwise. This is also called the *supremum norm*. If $g(x)$ is continuous on the closed interval $a \leq x \leq b$, then $|g(x)|$ is guaranteed to attain a maximum. In this common scenario, $\|g\|_\infty$ equals the global maximum value of $|g(x)|$.

For an error function $f(x) - \hat{f}(x)$, the infinity norm $\|f - \hat{f}\|_\infty$ provides a measure of the greatest error over the whole domain.

This book provides a routine `infnorm` for estimating the infinity norm. A call to the routine takes the form `infnorm(g,[a b],N)`, in which `g` is the function, `[a b]` is its domain (a closed interval), and `N` is the number of points at which to start the search for the supremum. If there is reason to believe that `infnorm` does not find the extreme point of a particular function, a larger value of `N` can be used to perform a finer search. The next example illustrates `infnorm`.

Example 4.3. Let $p(x)$ be the degree-2 Taylor polynomial for $f(x) = e^x$ about $x = 0$. Considering the middle plot in Figure 4.1, the absolute error $|f(x) - p(x)|$ on $[-1, 1]$ appears to be around 0.2 at its greatest. Verify this using `infnorm`.

Solution. The infinity norm of the error is $\|f - p\|_\infty$. This is estimated as follows:

```
>> p = @(x) 1+x+(1/2)*x^2;
>> M = infnorm(@(x) exp(x)-p(x),[-1 1],100)
M =
    0.218281828459046
```

As expected, the absolute error is around 0.2 at its greatest.

The infinity norm of the error is illustrated graphically in Figure 4.3.

```
>> newfig;
>> ylog; ylim([1e-18 1e2]);
>> plotfun(@(x) abs(exp(x)-p(x)),[-1 1],'displayname','|f(x)-p(x)|');
>> plotfun(@(x) M,[-1 1],'displayname','||f-p||');
>> xlabel('x');
```

The horizontal line at height $M = \|f - p\|_\infty$ touches the graph of the absolute error at its highest point. ∎

The infinity norm is particularly helpful when testing for numerical convergence or measuring rate of convergence.

Example 4.4. Let $p_m(x)$ be the degree-m Taylor polynomial for $f(x) = e^x$ about $x = 0$. Plot the infinity norm of the error on $-1 \leq x \leq 1$ against m. Determine whether numerical convergence occurs and, if so, at what value of m. Classify the convergence as algebraic, geometric, or supergeometric.

Solution. As mentioned in Example 4.1, $\frac{d^i}{dx^i}[e^x]|_{x=0} = 1$ for all i. Hence, the degree-m Taylor polynomial about $x = 0$ is

$$p_m(x) = \sum_{i=0}^{m} \frac{1}{i!} x^i.$$

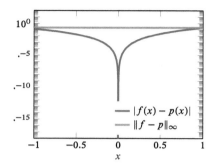

Figure 4.3. *The absolute error function for a Taylor polynomial and its infinity norm.*

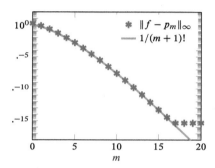

Figure 4.4. *Rate of convergence for a Taylor polynomial.*

The infinity norm of the error is calculated for $m = 0, 1, 2, \ldots, 20$ as follows:

```
>> m = 0:20;
>> c = arrayfun(@(i) 1/factorial(i),0:20);
>> err = nan(size(m));
>> for k = 1:length(m)
     p = @(x) polynomialeval(c(1:m(k)+1),x);
     err(k) = infnorm(@(x) exp(x)-p(x),[-1 1],100);
   end
```

The computed errors are plotted against m in Figure 4.4.

```
>> newfig;
>> ylog; ylim([1e-18 1e2]);
>> plot(m,err,'*','displayname','||f-p_m||');
```

Numerical convergence occurs at $m = 17$.

Because the errors approach zero faster than a straight line on a log-linear plot, the rate of convergence is supergeometric. In fact, the rate of convergence appears to match $1/(m + 1)!$.

```
>> plot(0:20,arrayfun(@(m) 1/factorial(m+1),0:20) ...
     ,'displayname','1/(m+1)!');
>> xlabel('n');
```
∎

Example 4.5. Using the infinity norm, measure the absolute error and absolute residual for sqrtbitbybit from Example 4.2. Plot the results against n. Classify the rate of convergence as algebraic, geometric, or supergeometric, and, if possible, find a power function n^{-p} or exponential function r^{-n} that converges to zero at the same rate.

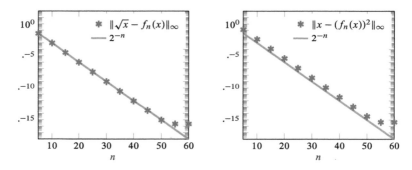

Figure 4.5. *Rate of convergence for an approximation to the square root function: absolute error (left) and absolute residual (right).*

Solution. The infinity norms for the residuals and errors are computed.

```
>> n = 5:5:60;
>> err = nan(size(n));
>> res = nan(size(n));
>> for k = 1:length(n)
      f = @(x) sqrtbitbybit(n(k),x);
      err(k) = infnorm(@(x) sqrt(x)-f(x),[1 4],100);
      res(k) = infnorm(@(x) x-f(x)^2,[1 4],100);
   end
```

They are plotted in Figure 4.5.

```
>> ax = newfig(1,2);
>> ylog; ylim([1e-18 1e2]);
>> plot(n,err,'*','displayname','||sqrt(x)-f_n(x)||');
>> xlabel('n');
>> subplot(ax(2));
>> ylog; ylim([1e-18 1e2]);
>> plot(n,res,'*','displayname','||x-(f_n(x))^2||');
>> xlabel('n');
```

The straight-line graph of 2^{-n} appears to capture the rate of convergence for both measures.

```
>> subplot(ax(1));
>> plotfun(@(n) 2^(-n),[n(1) n(end)],'displayname','2^{-n}');
>> subplot(ax(2));
>> plotfun(@(n) 2^(-n),[n(1) n(end)],'displayname','2^{-n}');
```

Because the rate of convergence is matched by an exponential function, the convergence is geometric. ∎

4.3 ▪ The shape of roundoff error

Roundoff error appears to be random. Recognizing the shape of roundoff error can provide further evidence of numerical convergence.

Example 4.6. In Example 4.4, we concluded that the sequence of Taylor polynomials for e^x about $x = 0$ converges numerically at $m = 17$, when considered on the domain $-1 \leq x \leq 1$. Plot the degree-17 Taylor polynomial and its absolute error. Comment on the absolute error plot.

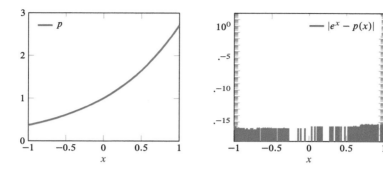

Figure 4.6. *A Taylor polynomial for e^x (left) and its absolute error (right).*

Solution. The Taylor polynomial is plotted on the left of Figure 4.6. It clearly resembles the natural exponential function.

```
>> m = 17;
>> c = arrayfun(@(i) 1/factorial(i),0:m);
>> p = @(x) polynomialeval(c,x);
>> ax = newfig(1,2);
>> plotfun(p,[-1 1],'displayname','p');
>> xlabel('x');
```

The absolute error is plotted on the right.

```
>> subplot(ax(2));
>> ylog; ylim([1e-18 1e2]);
>> plotfun(@(x) abs(exp(x)-p(x)),[-1 1],'displayname','|e^x-p(x)|');
>> xlabel('x');
```

The absolute error is clearly very small, on the order of 10^{-15}. In addition, the herky-jerky movement of the absolute error function is a hallmark of roundoff error and its effectively random perturbations. It suggests that the computation has reached the limits of hardware precision. ∎

Notes

A reader who is familiar with MATLAB may notice that `sqrtbitbybit` and some other functions in this book are not "vectorized." While a vectorized routine accepts vector arguments, often applying an identical operation to each entry, `sqrtbitbybit` requires scalar arguments n and x. Whether vectorization is helpful or harmful depends on the specific configuration of programming language, libraries, and hardware. For this reason, we do not insist on vectorizing code in this book and prioritize convenience instead.

Exercises

Exercises 1–8: A function $f(x)$, a central point x_0, and an interval are given. Let $p_m(x)$ be the Taylor polynomial of degree m for the function about $x = x_0$.

(a) Plot $f(x)$ and $p_4(x)$ on the same axes over the given interval.

(b) Plot $\|f - p_m\|_\infty$ against m. Does p_m converge numerically as m increases? If so, at what value of m?

(c) If the error appears to converge to zero, infer the rate of convergence. Is it algebraic, geometric, or supergeometric? If possible, find a function of the form m^{-s} or ρ^{-m} that decreases at approximately the same rate as the error.

1. $\log x$; $x_0 = 1$; $1/2 \le x \le 3/2$
2. $\cos x$; $x_0 = 0$; $-\pi/2 \le x \le \pi/2$
3. $\sin x$; $x_0 = 0$; $-\pi/2 \le x \le \pi/2$
4. $\arcsin x$; $x_0 = 0$; $-1/2 \le x \le 1/2$. *Hint.* Express the Taylor polynomial as

$$p_m(x) = x \sum_{i=0}^{\lceil m/2 \rceil - 1} \frac{1}{2i+1} \left(\prod_{j=1}^{i} \left(1 - \frac{1}{2j} \right) \right) x^{2i},$$

in which $\lceil \cdot \rceil$ denotes the ceiling function.
5. $\arcsin x$; $x_0 = 0$; $-1 \le x \le 1$. *Hint.* Use the expression for the Taylor polynomial in the previous exercise.
6. $\arctan x$; $x_0 = 0$; $-1/2 \le x \le 1/2$
7. $\arctan x$; $x_0 = 0$; $-1 \le x \le 1$
8. $\arctan x$; $x_0 = 0$; $-2 \le x \le 2$

9. Let $f_n(x)$ be the approximation to the square root function defined in Example 4.2. Using induction, prove that $0 \le \sqrt{x} - f_n(x) < 2^{-n}$ for all $x \in [1, 4)$ and all $n \ge 0$.

10. Consider the degree-30 Taylor polynomial $p(x)$ for $\cos x$ about $x = 0$. Plot the absolute error function $|\cos x - p(x)|$ and the relative error function $|\cos x - p(x)|/|\cos x|$ over $0 \le x \le \pi$ by comparing with the built-in cos function. Describe the difference between the error graphs and explain the reason for the difference.

11. Let $p(x)$ be the degree-60 Taylor polynomial for e^x about $x = 0$, and consider $p(-x^2)$ as an approximation for e^{-x^2}. Plot the absolute error $|e^{-x^2} - p(-x^2)|$ and the relative error $|e^{-x^2} - p(-x^2)|/|e^{-x^2}|$ over $-4 \le x \le 4$. Is the absolute error significantly larger than the relative error, or vice versa? Provide evidence and discuss.

12. The natural logarithm function satisfies $e^{\log x} = x$. Thus, if $f(x)$ is an approximation for $\log x$, the residual $x - e^{f(x)}$ provides a measure of accuracy. Let $p(x)$ be the degree-10 Taylor polynomial for $\log x$ about $x = 1$. Plot the absolute residual $|x - e^{p(x)}|$ on $1/2 \le x \le 3/2$. How does the absolute residual compare with the absolute error $|\log x - p(x)|$ on the same interval? You may use the built-in exp function to compute the residual, and you may compare with the built-in log function to measure the absolute error.

13. Suppose $f(x)$ is $(m+1)$-times differentiable on (a, b) and $f^{(m+1)}(x)$ is continuous on the interval. Let $p(x)$ be the degree-m Taylor polynomial for $f(x)$ about $x = x_0$ for some $x_0 \in (a, b)$. Prove that

$$f(x) = p(x) + \frac{1}{m!} \int_{x_0}^{x} f^{(m+1)}(t)(x-t)^m \, dt$$

for all $x \in (a, b)$. *Hint.* Use induction, justifying the base case $f(x) = f(x_0) + \int_{x_0}^{x} f'(t) \, dt$ and then using integration by parts for the induction step.

14. A version of the mean-value theorem says that if (1) $g(x)$ is continuous on $[a, b]$, (2) $h(x) \geq 0$ on $[a, b]$, and (3) $\int_a^b h(x)\, dx$ exists, then $\int_a^b g(x) h(x)\, dx = g(\eta) \int_a^b h(x)\, dx$ for some $\eta \in [a, b]$. Using this fact, prove the following corollary of Exercise 13:

$$f(x) = f(x_0) + f'(x_0)(x - x_0) + \frac{f''(x_0)}{2!}(x - x_0)^2 + \cdots + \frac{f^{(m)}(x_0)}{m!}(x - x_0)^m + R_m$$

with

$$R_m = \frac{f^{(m+1)}(\eta)}{(m+1)!}(x - x_0)^{m+1}$$

for some η between x_0 and x. The error term R_m is known as the *Lagrange remainder* for the Taylor polynomial expansion.

15. Let $p(x)$ be the Taylor polynomial for $f(x) = e^x$ centered at $x_0 = 0$. Use Exercise 14 to prove $\|f - p\|_\infty \leq e/(m+1)!$, in which the infinity norm is measured over the interval $-1 \leq x \leq 1$. Compare with Example 4.4.

16. Let $p(x)$ be the Taylor polynomial for $f(x) = e^x$ centered at $x_0 = 0$. Use the previous exercise to find a degree m for which the Taylor polynomial error is guaranteed to be below 10^{-15} for all x in the interval $-1 \leq x \leq 1$. Compare with Example 4.4.

Chapter 5

Floating-point arithmetic

Computers contain special hardware for real-number arithmetic. As seen in earlier experiments, about 15 or 16 digits of precision are available. In this chapter, the level of precision is made precise, and laws governing arithmetic with these limited-precision numbers are stated.

The main goal is to be aware of the limitations of finite-precision arithmetic, so that we can recognize roundoff error when it appears. Fortunately, roundoff error is negligible for many of our methods and need not consume our thoughts.

5.1 ▪ Utility of finite precision

Computers nowadays are very powerful. Wouldn't it be convenient if they simply kept all of the digits, so we wouldn't have to worry about roundoff at all? While tempting, this approach is impractical.

Example 5.1. Compute the quantity

$$\sum_{i=1}^{n} \frac{1}{i^3}$$

for $n = 0, 1, 2, \ldots, 10$, once exactly and once using the computer's finite precision. Compare and contrast.

Solution. The desired quantity is a rational number and can therefore be represented exactly as a fraction.

n	$\sum_{i=1}^{n} i^{-3}$
1	1
2	9/8
3	251/216
4	2035/1728
5	256103/216000
6	28567/24000
7	9822481/8232000
8	78708473/65856000
9	19148110939/16003008000
10	19164113947/16003008000

Even so, computer hardware produces a numerical expansion rather than an exact fraction:

```
>> n = (1:10)';
>> x = nan(size(n));
>> for k = 1:length(n)
      x(k) = sum(arrayfun(@(i) i^(-3),1:n(k)));
   end
>> x
x =
    1.000000000000000
    1.125000000000000
    1.162037037037037
    1.177662037037037
    1.185662037037037
    1.190291666666667
    1.193207118561710
    1.195160243561710
    1.196531985674193
    1.197531985674193
```

Although the fractions have the advantage of being exact, they are prohibitively costly. With each additional arithmetic operation, the number of digits in the numerator and denominator can grow significantly, steadily increasing storage and computation cost. In contrast, the storage and computation cost for the numerical expansions is capped. ∎

Exact computation is impractical for two reasons: First, irrational numbers, which are difficult or impossible to represent exactly, are frequently encountered. Second, operations like addition of fractions and exponentiation tend to make the number of digits grow with every arithmetic step. Hence, roundoff error is accepted as a fact of life.

5.2 ▪ Binary notation

We often say informally that an approximation is accurate to a certain number of decimal places or digits. In reality, though, computers store numbers in *binary* instead of decimal. In binary notation, numbers are expressed in powers of 2 rather than powers of 10. To indicate that a number is expressed in binary, a subscript of 2 can be appended.

Example 5.2. The number

$$11010_2$$

is expressed in binary notation. Convert it to decimal.

Solution. The 0's and 1's in 11010_2 are coefficients for powers of 2:

$$1 \times 2^4 + 1 \times 2^3 + 0 \times 2^2 + 1 \times 2^1 + 0 \times 2^0 = 16 + 8 + 2 = 26.$$ ∎

A binary number has *bits* instead of digits. The allowable values for a bit are just 0 and 1.

Rather than a decimal point, a number in binary may have a *binary point*, and the place value of any bit to the right of the binary point is a negative power of 2.

Example 5.3. Convert the binary representation

$$101.011_2$$

to decimal.

Solution. The value is

$$1 \times 2^2 + 0 \times 2^1 + 1 \times 2^0 + 0 \times 2^{-1} + 1 \times 2^{-2} + 1 \times 2^{-3}$$
$$= 4 + 1 + \frac{1}{4} + \frac{1}{8} = \frac{43}{8} = 5.375. \quad \blacksquare$$

Converting from decimal to binary is perhaps a little more difficult. To do so, find the largest power of 2 that is no greater than the given number, subtract it, and repeat.

Example 5.4. Express 87 in binary.

Solution. Successively remove powers of 2, starting with the largest possible:

$$87 = 2^6 + 23$$
$$= 2^6 + 2^4 + 7$$
$$= 2^6 + 2^4 + 2^2 + 3$$
$$= 2^6 + 2^4 + 2^2 + 2^1 + 1$$
$$= 2^6 + 2^4 + 2^2 + 2^1 + 2^0$$
$$= 1 \times 2^6 + 0 \times 2^5 + 1 \times 2^4 + 0 \times 2^3 + 1 \times 2^2 + 1 \times 2^1 + 1 \times 2^0.$$

Hence,

$$87 = 1010111_2. \quad \blacksquare$$

To express a fraction in binary, use the same procedure, but include negative powers of 2.

Example 5.5. Express the fraction $23/32$ in binary.

Solution. Successively remove powers of 2, starting with the largest possible:

$$\frac{23}{32} = \frac{1}{2} + \frac{7}{32}$$
$$= \frac{1}{2} + \frac{1}{8} + \frac{3}{32}$$
$$= \frac{1}{2} + \frac{1}{8} + \frac{1}{16} + \frac{1}{32}$$
$$= 2^{-1} + 2^{-3} + 2^{-4} + 2^{-5}.$$

Hence,

$$\frac{23}{32} = 0.10111_2. \quad \blacksquare$$

Many rational numbers require a repeating binary expansion.

Example 5.6. Evaluate the repeating binary number

$$0.0\overline{0011}_2 = 0.0001100110011\ldots_2$$

as a fraction.

Solution. Note that the given number is a geometric series:

$$0.0\overline{0011}_2 = 0.0011_2 \times 2^{-1} + 0.0011 \times 2^{-5} + 0.0011 \times 2^{-9} + \cdots$$

$$= \frac{3}{16} \times 2^{-1} \times (1 + 2^{-4} + 2^{-8} + \cdots)$$

$$= \frac{3}{32} \sum_{k=0}^{\infty} \left(\frac{1}{16}\right)^k.$$

Using the fact that $\sum_{k=0}^{\infty} r^k = 1/(1-r)$, the value is found to be

$$0.0\overline{0011}_2 = \frac{3/32}{1 - (1/16)} = \left(\frac{3}{32}\right)\left(\frac{16}{15}\right) = \frac{1}{10}. \qquad \blacksquare$$

It is remarkable that a number as basic as $1/10$ in our normal counting system is so complicated in binary. Perhaps this shouldn't be surprising though—in our standard decimal system, $1/3 = 0.\overline{3}$ isn't very pretty, and $1/7$ is downright ugly: $1/7 = 0.\overline{142857}$.

5.3 ▪ IEEE 754 representation

The nearly universal computer number system is defined by the *IEEE 754 standard*. It specifies rules for *floating-point* number representation and arithmetic, inspired by scientific notation. Numbers are represented in the form

$$(-1)^s \times f \times 2^\beta. \tag{5.1}$$

The number s is either 0 or 1 and determines the sign. The *significand* f is a number in $[1, 2)$, so its most significant bit is always immediately to the left of the binary point. The *exponent* β then moves the binary point to the appropriate position.

Example 5.7. Express 1.5 in the form (5.1).

Solution. We find

$$1.5 = (-1)^0 \times 1.1_2 \times 2^0. \qquad \blacksquare$$

Example 5.8. Express -3.125 in the form (5.1).

Solution. We find

$$-3.125 = (-1)^1 \times 1.1001_2 \times 2^1$$

because $(-1)^1 \times 1.1001_2 \times 2^1 = -1 \times 11.001_2 = -3.125.$ $\blacksquare$

Representation (5.1) is the basic idea underlying any floating-point number system. IEEE 754 specifies further technical details. Its *double-precision* format is most common. It specifies that a number should be stored in 64 bits, partitioned into three groups:

s eeeeeeeeeee ffffffffffffffffffffffffffffffffffffffffffffffffffff

Referring again to (5.1), these bits are filled as follows:

- If $s = 0$, then s is filled with 0. If $s = 1$, then s is filled with 1.

- The exponent β must be an integer between -1022 and 1023, inclusive. The number e is defined to be $\beta + 1023$, so that e is an integer between 1 and 2046. The binary representation of this number fills the bits labeled e.

- The significand f is any number in $[1, 2)$ that can be expressed in binary with 52 bits to the right of the binary point. Because the bit immediately to the left of the binary point is always 1, it does not need to be stored electronically. The bits labeled f above are thus filled with the 52 bits of f to the right of the binary point.

- The number zero receives a special representation, consisting of 64 consecutive 0's.

- The special "numbers" +Inf (positive infinity), -Inf (negative infinity), NaN ("not a number") and certain exceptionally tiny numbers called "denormalized" numbers are signaled with a special code of 0 or 2047 in place of the exponent e. The details are omitted here.

The set of nonzero numbers that can be represented as above (not including +Inf, -Inf, NaN or the denormalized numbers) will be denoted $\mathbb{F}$.

The library routine ieee754 reveals the internal representation of a floating-point number.

Example 5.9. Express 1.5 in the IEEE 754 format. Check against ieee754.

Solution. First, express the number in the form (5.1):

$$1.5 = (-1)^0 \times 1.1_2 \times 2^0.$$

Hence, $s = 0$, $\beta = 0$, and $f = 1.1_2$.

The 64-bit representation

s eeeeeeeeeee ffffffffffffffffffffffffffffffffffffffffffffffffffff
0 01111111111 1000000000000000000000000000000000000000000000000000

is determined as follows:

- The first bit 0 encodes s, fixing the positive sign.

- The next 11 bits 01111111111 are the binary representation of $e = \beta + 1023 = 1023$.

- The final 52 bits 1000...0 are the bits to the right of the binary point in $f = 1.1000\ldots0_2$.

Let's check this.
```
>> [sbin,ebin,fbin] = ieee754(1.5)
sbin =
     '0'
ebin =
     '01111111111'
fbin =
     '1000000000000000000000000000000000000000000000000000'                    ∎
```

Example 5.10. Express -3.125 in the IEEE 754 representation.

Solution. From
$$-3.125 = (-1)^1 \times 1.1001_2 \times 2^1,$$

we find $s = 1$, $\beta = 1$ (and thus $e = 1024 = 10000000000_2$), and $f = 1.1001_2$. The IEEE 754 representation is

```
s eeeeeeeeeee ffffffffffffffffffffffffffffffffffffffffffffffffffffff
1 10000000000 1001000000000000000000000000000000000000000000000000000
```

This checks out:
```
>> [sbin,ebin,fbin] = ieee754(-3.125)
sbin =
     '1'
ebin =
     '10000000000'
fbin =
     '1001000000000000000000000000000000000000000000000000'                    ∎
```

The IEEE 754 system is designed to represent many numbers with small relative error. The following theorem makes this precise. The floating-point representative $\text{fl}(x)$ for a real number x is defined to be a number in $\mathbb{F}$ that is nearest to x. When x is equally close to two representatives, the standard permits several different tie-breakers. In this book, we round away from zero for simplicity.

Theorem 5.11. *If*
$$2^{-1022} \le |x| < (2 - 2^{-52}) \times 2^{1023}, \tag{5.2}$$

then the IEEE 754 double-precision representative $\text{fl}(x) \in \mathbb{F}$ *satisfies*

$$\text{fl}(x) = x(1 + \delta) \text{ for some } \delta \text{ with } |\delta| \le 2^{-53}. \tag{5.3}$$

Proof. Assume without loss of generality that $x > 0$. Let $f \times 2^\beta$ be the greatest floating-point number that is no greater than x. The next greater floating-point number is $(f + 2^{-52}) \times 2^\beta$. The gap is $2^{-52+\beta}$, and the distance from x to the nearer of these two is at most half the gap size. In other words, $|x - \text{fl}(x)| \le 2^{-53+\beta}$. For the relative error, note that $x \ge 2^\beta$, so $|x - \text{fl}(x)|/|x| \le 2^{-53+\beta-\beta} = 2^{-53}$.   □

Note that $2^{-53} \approx 1.1 \times 10^{-16}$. When we speak of "about 15 or 16 digits of precision," we are really referring to the representation error of $\le 2^{-53}$. This bound on representation error is called *unit roundoff*.

Example 5.12. Measure the distance from $1/10$ to the nearest floating-point number. Compare with the theorem.

Solution. In binary,

$$1/10 = 0.0001\overline{1100}_2 = 1.1001\overline{1100}_2 \times 2^{-4}.$$

The nearest floating-point number is

$$\text{fl}(1/10) = 1.1001100110011001100110011001100110011001100110011010_2 \times 2^{-4}.$$

The difference is

$$1/10 - \text{fl}(1/10) = -1.1001\overline{1100}_2 \times 2^{-58}.$$

The relative error is

$$\frac{|1/10 - \text{fl}(1/10)|}{|1/10|} = \frac{1.1001\overline{1100}_2 \times 2^{-58}}{1.1001\overline{1100}_2 \times 2^{-4}} = 2^{-54}.$$

This is no greater than 2^{-53}, as guaranteed by the previous theorem. ∎

5.4 ▪ Floating-point arithmetic

Simply storing a number in floating-point incurs a small roundoff error. Arithmetic incurs additional roundoff error.

A computer satisfying the IEEE 754 standard provides addition, subtraction, multiplication, and division instructions that guarantee small relative error for inputs in $\mathbb{F}$. To distinguish the inexact instructions of the computer from the corresponding exact mathematical operators, the symbols $\oplus$, $\ominus$, $\otimes$, and $\oslash$ are used in place of $+$, $-$, $\times$, and $/$. The following operations and error terms are guaranteed for $x, y \in \mathbb{F}$. In each case, δ is a real number that is guaranteed to exist.

$$x \oplus y = (x + y)(1 + \delta), \quad |\delta| \leq 2^{-53}, \tag{5.4}$$
$$x \ominus y = (x - y)(1 + \delta), \quad |\delta| \leq 2^{-53}, \tag{5.5}$$
$$x \otimes y = (xy)(1 + \delta), \quad |\delta| \leq 2^{-53}, \tag{5.6}$$
$$x \oslash y = (x/y)(1 + \delta), \quad |\delta| \leq 2^{-53}. \tag{5.7}$$

The next example utilizes the guarantees of IEEE 754 arithmetic and the *triangle inequality*

$$|x + y| \leq |x| + |y|. \tag{5.8}$$

The triangle inequality can quantify the following intuition: If x is close to y and y is close to z, then x must be close to z. Specifically, we find

$$|x - z| = |(x - y) + (y - z)| \leq |x - y| + |y - z|; \tag{5.9}$$

the distance from x to z is no larger than the sum of the distances from x to y and from y to z.

Example 5.13. The following computation of $(1/3) + (1/5)$ accumulates error at each step:

```
>> onethird = 1/3;
>> onefifth = 1/5;
>> onethird+onefifth
ans =
    0.533333333333333
```

Find an upper bound for the error on $|(8/15) - (\text{fl}(1/3) \oplus \text{fl}(1/5))|$ using (5.4)–(5.7) and the triangle inequality.

Solution. Each individual error is bounded as follows:

$$|\varepsilon_1| = |(1/3) - \text{fl}(1/3)| \le 2^{-53}(1/3) < 3.8 \times 10^{-17},$$
$$|\varepsilon_2| = |(1/5) - \text{fl}(1/5)| \le 2^{-53}(1/5) < 2.3 \times 10^{-17},$$
$$|\varepsilon_3| = |(\text{fl}(1/3) + \text{fl}(1/5)) - (\text{fl}(1/3) \oplus \text{fl}(1/5))|$$
$$\le 2^{-53}(\text{fl}(1/3) + \text{fl}(1/5)) < 6.0 \times 10^{-17}.$$

By the triangle inequality, the absolute error in the final answer is no greater than the sum of the three absolute errors, which is

$$|\varepsilon_1| + |\varepsilon_2| + |\varepsilon_3| < 1.3 \times 10^{-16}. \qquad \blacksquare$$

In the previous example, three separate errors accumulate to produce an error that is slightly larger than any of the individual errors. More worrisome are diabolical combinations of errors such as the phenomenon of *catastrophic cancellation* in the next example.

Example 5.14. The following code computes $c = (3.000001)^2 - 3^2$ in floating-point:

```
>> a = 3.000001;
>> b = 3;
>> capprox = a*a-b*b
capprox =
    6.000001000927568e-06
```

Measure the absolute and relative errors. Discuss.

Solution. The exact answer is $c = 6.000001 \times 10^{-6}$, so the final six digits of `capprox` are garbage.

The absolute error is

$$|6.000001 \times 10^{-6} - 6.000001000927568 \times 10^{-6}| \approx 9.3 \times 10^{-16}.$$

The relative error is

$$\frac{|6.000001 \times 10^{-6} - 6.000001000927568 \times 10^{-6}|}{|6.000001 \times 10^{-6}|} \approx \frac{9.3 \times 10^{-16}}{6 \times 10^{-6}} \approx 1.5 \times 10^{-10}.$$

The loss of six digits of accuracy is revealed by the relative error—note that it is on the order of 10^{-10} rather than 10^{-16}.

What is happening? First, note that all of the absolute errors are as small as we could hope. The exact value of a^2 is 9.000006000001, and the computed quantity is off by only about 10^{-15}:

```
>> a*a
ans =
    9.000006000001001
```

The exact value of b^2 is 9, and this is computed perfectly:

```
>> b*b
ans =
    9
```

Even the difference $a^2 - b^2$ looks fine if we focus on absolute error; as already observed, the absolute error is about 10^{-15}.

Nothing looks amiss until we look at relative error. Although the errors on a*a and b*b are tiny relative to $a^2 = 9.000006000001$ and $b^2 = 9$, the errors are not so tiny compared to the ultimate answer $a^2 - b^2 = 0.000006000001$.

Although each individual computation is very accurate, their combined relative error is not so tiny.   ■

In catastrophic cancellation, the subtraction of two nearly equal quantities leads to large relative error. Essentially, the signal is removed to expose the noise. Catastrophic cancellation will not occur very often in this book, but it cannot be totally ignored.

Notes

The original IEEE 754 standard was published in 1985 [45]. It was updated in 2008 [46].

Exercises

Exercises 1–4: Convert binary to decimal by hand.

1. 1011_2
2. 10101_2
3. 0.011_2
4. 11.0001_2

Exercises 5–8: Convert decimal to binary by hand. Express as a repeating binary expansion if necessary.

5. 17
6. 72
7. 0.6875
8. 6.4

Exercises 9–12: Convert the IEEE 754 representation to decimal.

9. 0 10000000000 0010000000000000000000000000000000000000000000000000
10. 1 01111111110 1110000000000000000000000000000000000000000000000000
11. 1 01111111111 0101010101010101010101010101010101010101010101010101
12. 0 10000000111 1111110000000000000000000000000000000000000000000000

Exercises 13–16: Represent the number in the IEEE 754 format.

13. 2.625

14. −0.005859375

15. −1.01

16. 15.1

Exercises 17–20: Suppose the given number is represented by the closest number in the IEEE 754 set. Give upper bounds on the absolute and relative representation error using Theorem 5.11. Report the bounds in decimal notation with two significant digits.

17. 3.87

18. 199.414

19. e

20. $1/\sqrt{1000}$

21. The value $\pi + e$ is computed with the following code:
    ```
    pi = 3.1415926535897932;
    e = 2.7182818284590452;
    pi+e
    ```
    The values for `pi` and `e` are correct to 17 digits and therefore are rounded to the nearest floating-point representatives. Using (5.3) and (5.4)–(5.7), find upper bounds on the absolute and relative errors for the final computed value.

22. The value $\sqrt{3} - \sqrt{2}$ is computed with the following code:
    ```
    sqrt3 = 1.7320508075688772;
    sqrt2 = 1.4142135623730950;
    sqrt3-sqrt2
    ```
    The values for `sqrt3` and `sqrt2` are correct to 17 digits and therefore are rounded to the nearest floating-point representatives. Using (5.3) and (5.4)–(5.7), find upper bounds on the absolute and relative errors for the final computed value.

23. Find a computation involving addition and/or subtraction (and no other operations) for which the relative roundoff error equals 1 when computed at the MATLAB prompt. *Hint.* Find an expression that would evaluate to 1 in exact arithmetic but that evaluates to 0 in finite precision.

24. Find a computation involving addition and/or subtraction (and no other operations) for which the relative roundoff error is about $1/2$ when computed at the MATLAB prompt.

25. Routine `sqrtbitbybit` from Example 4.2 can compute the square root of any number in the interval $[1, 4)$. In this exercise, you will compute $\sqrt{x}$ for an $x \notin [1, 4)$ by leveraging floating-point representation. To have a concrete example, take $x = 31$. Express this x in the form $f \times 2^\beta$ with $f \in [1, 4)$ and β an even integer. Then compute $\sqrt{x} = \sqrt{f \times 2^\beta}$ using algebra and `sqrtbitbybit`, but without using the built-in `sqrt` function or raising any number to a noninteger power. Achieve full accuracy.

26. Repeat the previous exercise with $x = 57.5$.

27. Let $x = f \times 2^\beta$ with $f \in [1, 2)$ and β an integer. Prove

$$\log x = \log(f/2) - (\beta + 1)\log(1/2).$$

    Then use this equation and a Taylor polynomial for $\log t$ about $t = 1$ to compute $\log 11$ to 15-digit accuracy.

28. Demonstrate that the MATLAB expression `exp(x)-1` fails to compute $f(x) = e^x - 1$ to high relative accuracy for some values of x, and explain why the relative error is large.

29. Design and implement a method to compute $f(x) = e^x - 1$ with relative error on the order of 10^{-15} for $-1 \le x \le 1$. (According to the previous exercise, the naive approach doesn't work.)

30. Demonstrate that the MATLAB expression `log(1+x)` fails to compute $f(x) = \log(1 + x)$ to high relative accuracy for some values of x, and explain why the relative error is large.

31. Design and implement a method to compute $f(x) = \log(1 + x)$ with relative error on the order of 10^{-15} for $-1/2 \le x \le 1/2$. (According to the previous exercise, the naive approach doesn't work.)

Part II

Interpolation

Chapter 6

Polynomial interpolation

Because polynomials are relatively well understood, a common first step in a numerical method is to replace a given function by a close-fitting polynomial. Our primary approach to this substitution is *interpolation*. Several points on the graph of the original function are sampled, and the polynomial surrogate is constructed to pass through the same finite set of points. The approximating polynomial is called an *interpolant* or *interpolating polynomial*. Later, the interpolant can be used to compute integrals, find zeros, solve differential equations, and more.

6.1 ▪ Existence and uniqueness

Definition 6.1. *Given points (x_i, y_i), $i = 0, \ldots, m$, with distinct $x_0, \ldots, x_m$, a Lagrange interpolating polynomial is a polynomial of degree at most m whose graph passes through the points.*

The numbers $x_0, \ldots, x_m$ are interpolation *nodes*, and together they form a *grid* of *degree m*. Throughout the book, the nodes in a grid are required to be distinct, unless stated otherwise.

Often, the points (x_i, y_i) come from the graph of an existing function. Matching a function by polynomial interpolation is illustrated in Figure 6.1.

An interpolating polynomial of degree at most m can be constructed with the help of the *Lagrange basis polynomials* of degree m,

$$l_i(x) = \prod_{\substack{k=0,\ldots,m \\ k \neq i}} (x - x_k), \quad i = 0, \ldots, m. \tag{6.1}$$

Because the degree m is omitted in the notation l_i, we should be careful to make it clear from context. Note that there are $m + 1$ Lagrange basis polynomials $l_0, \ldots, l_m$ of degree m. In particular, there is exactly one Lagrange basis polynomial of degree zero: $l_0(x) = 1$.

The ith Lagrange basis polynomial is constructed to equal zero at every node except x_i. That is, $l_i(x_j) = 0$ for all $j \neq i$. This is useful for solving the Lagrange interpolation problem, as seen in the proof of the following theorem.

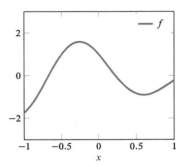
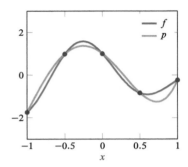

Figure 6.1. *A function f (left) and a degree-4 polynomial interpolant p (right).*

Theorem 6.2. *For a given set of points (x_i, y_i), $i = 0, \ldots, m$, with distinct $x_0, \ldots, x_m$, there exists a unique Lagrange interpolating polynomial, and it is*

$$p(x) = \sum_{i=0}^{m} \frac{y_i}{l_i(x_i)} l_i(x). \tag{6.2}$$

Proof. We first prove that (6.2) solves the interpolation problem. Function $p(x)$ is a polynomial of degree at most m because it is a linear combination of $l_0(x), \ldots, l_m(x)$, which are themselves polynomials of degree m. At each node x_j, we have $l_i(x_j) = 0$ for $i \neq j$, and so

$$p(x_j) = \frac{y_j}{l_j(x_j)} l_j(x_j) + \sum_{i \neq j} \frac{y_i}{l_i(x_i)} l_i(x_j) = y_j + 0 = y_j.$$

To prove uniqueness, suppose that $q(x)$ is also a Lagrange interpolating polynomial for the given points. We want to show that q necessarily equals p. Let $r(x) = p(x) - q(x)$. On one hand, $r(x)$ is a polynomial of degree at most m. On the other hand, $r(x)$ is divisible by the degree-$(m + 1)$ polynomial $\prod_{i=0}^{m}(x - x_i)$ because it has a zero at each x_i: $r(x_i) = p(x_i) - q(x_i) = y_i - y_i = 0$. The only resolution is that $r(x) = 0$ everywhere. □

According to the theorem, a degree-m polynomial is determined by $m + 1$ points. This should seem reasonable—in particular, a line is determined by two points and a parabola is determined by three.

Example 6.3. Construct the quadratic polynomial whose graph intersects the points $(0, 2)$, $(1, 5)$, and $(2, 4)$.

Solution. The Lagrange basis polynomials for the nodes $x_0 = 0$, $x_1 = 1$, and $x_2 = 2$ are

$$l_0(x) = (x - x_1)(x - x_2) = (x - 1)(x - 2),$$
$$l_1(x) = (x - x_0)(x - x_2) = (x - 0)(x - 2),$$
$$l_2(x) = (x - x_0)(x - x_1) = (x - 0)(x - 1).$$

The denominators in (6.2) are $l_0(x_0) = l_0(0) = 2$, $l_1(x_1) = l_1(1) = -1$, and $l_2(x_2) = l_2(2) = 2$. Thus the interpolating polynomial is constructed from the scaled basis

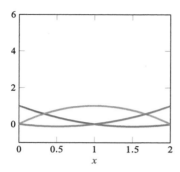 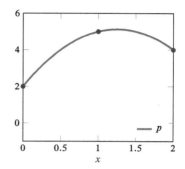

Figure 6.2. *The solution to a Lagrange interpolation problem: three scaled basis polynomials (left) and a linear combination that solves the interpolation problem (right).*

polynomials

$$\frac{l_0(x)}{l_0(x_0)} = \frac{(x-1)(x-2)}{2},$$
$$\frac{l_1(x)}{l_1(x_1)} = \frac{(x-0)(x-2)}{-1},$$
$$\frac{l_2(x)}{l_2(x_2)} = \frac{(x-0)(x-1)}{2}.$$

These are plotted in the left graph of Figure 6.2.

```
>> ax = newfig(1,2); legend hide;
>> plotfun(@(x) (1/2)*(x-1)*(x-2),[0 2]);
>> plotfun(@(x)     -1*(x-0)*(x-2),[0 2]);
>> plotfun(@(x) (1/2)*(x-0)*(x-1),[0 2]);
>> ylim([-1 6]);
>> xlabel('x');
```

Notice that $l_i(x)/l_i(x_i)$ equals 1 at the ith node and 0 at every other node.

The interpolating polynomial is formed from a linear combination of the scaled basis polynomials. The coefficients are the y-values of the points $(0,2)$, $(1,5)$, and $(2,4)$ given in the problem:

$$p(x) = 2\frac{(x-1)(x-2)}{2} + 5\frac{(x-0)(x-2)}{-1} + 4\frac{(x-0)(x-1)}{2}.$$

The interpolating polynomial p is plotted in the right graph.

```
>> p = @(x) (x-1)*(x-2)-5*(x-0)*(x-2)+2*(x-0)*(x-1);
>> subplot(ax(2));
>> plotfun(p,[0 2],'displayname','p');
>> ylim([-1 6]);
>> xlabel('x');
```

6.2 ▪ Lagrange and barycentric forms

The right-hand side of (6.2) is known as the *Lagrange form* of the interpolating polynomial.

The Lagrange interpolating polynomial can be evaluated more efficiently by first manipulating it algebraically. The result is (6.4) in the following theorem, and it is called the *barycentric form*. The formula involves a sequence of numbers $w_0, \ldots, w_m$ associated with the grid called *barycentric weights*. The sequence is required to satisfy

$$w_i = \frac{C}{l_i(x_i)}, \quad i = 0, \ldots, m, \tag{6.3}$$

for any nonzero number C that is constant with respect to i.

Theorem 6.4. *Let (x_i, y_i), $i = 0, \ldots, m$, be a sequence of points with distinct $x_0, \ldots, x_m$. Define*

$$p(x) = \frac{\displaystyle\sum_{i=0}^{m} \frac{w_i y_i}{x - x_i}}{\displaystyle\sum_{i=0}^{m} \frac{w_i}{x - x_i}} \tag{6.4}$$

for any x that is not equal to any of the nodes $x_0, \ldots, x_m$. At the nodes, define

$$p(x_i) = y_i, \quad i = 0, \ldots, m. \tag{6.5}$$

Then p equals the Lagrange interpolating polynomial for the points (x_i, y_i).

According to the proof below, the barycentric form (6.4) really does represent a polynomial!

Proof. Starting from the Lagrange form and using (6.3), we find

$$p(x) = \sum_{i=0}^{m} \frac{y_i}{l_i(x_i)} l_i(x) = \sum_{i=0}^{m} \frac{y_i}{C/w_i} \prod_{k \neq i} (x - x_k) = \sum_{i=0}^{m} \frac{w_i y_i}{C} \frac{\prod_{k=0}^{m}(x - x_k)}{x - x_i}$$

for any $x \notin \{x_0, \ldots, x_m\}$, and so

$$p(x) = \prod_{k=0}^{m} (x - x_k) \times \sum_{i=0}^{m} \frac{w_i y_i}{C(x - x_i)}.$$

The trick for eliminating $\prod_k (x - x_k)$ is to interpolate by the same process the constant function that is equal to 1 everywhere:

$$1 = \prod_{k=0}^{m} (x - x_k) \times \sum_{i=0}^{m} \frac{w_i \times 1}{C(x - x_i)}.$$

Dividing gives

$$\frac{p(x)}{1} = \frac{\prod_{k=0}^{m}(x - x_k) \times \sum_{i=0}^{m} \frac{w_i y_i}{C(x-x_i)}}{\prod_{k=0}^{m}(x - x_k) \times \sum_{i=0}^{m} \frac{w_i}{C(x-x_i)}} = \frac{\sum_{i=0}^{m} \frac{w_i y_i}{x-x_i}}{\sum_{i=0}^{m} \frac{w_i}{x-x_i}}. \qquad \square$$

Unlike the factors $l_i(x)$ in the original Lagrange form, the barycentric weights w_i do not depend on x. Hence, they can often be computed in advance and retrieved when needed. The constant C in (6.3) can take any nonzero value because it is factored out of the numerator and denominator of (6.4) and canceled in the last step of the proof.

One immediate benefit of the barycentric form over the Lagrange form is a significantly reduced operation count (Exercises 17–18).

Routine `baryweights` of Listing 6.1 computes a sequence of barycentric weights for a grid, and routine `interpgen` of Listing 6.2 implements polynomial interpolation by applying the barycentric formula. The suffix gen in the name `interpgen` indicates that the routine is useful for *general* grids, in contrast to more specialized routines in the upcoming chapters.

Listing 6.1. Compute barycentric weights.

```
function ws = baryweights(xs)

m = length(xs)-1;
ws = nan(m+1,1);
for i = 0:m
  k = [0:i-1 i+1:m];
  ws(i+1) = 1/prod(xs(i+1)-xs(k+1));
end
```

Listing 6.2. Construct a polynomial interpolant.

```
function p = interpgen(xs,ps)

ws = baryweights(xs);
p = @(x) interp_(xs,ws,ps,x);
```

```
function px = interp_(xs,ws,ys,x)

px = nan(size(x));
for k = 1:numel(x)
  % evaluate barycentric formula
  zs = ws./(x(k)-xs);
  px(k) = sum(ys.*zs)/sum(zs);
  % fix if at node
  if isnan(px(k))
    i = find(~isfinite(zs),1);
    px(k) = ys(i);
  end
end
end
```

The real work of barycentric interpolation is completed by the helper routine `interp_`, also shown in Listing 6.2. (An underscore at the end of a function name indicates an internal routine that is typically not invoked by an end user.) This routine first applies the barycentric form (6.4). If it happens that an x-value $x(k)$ coincides with one of the interpolation nodes, then the barycentric formula is undefined because of division by zero. This situation is detected in the computer code by the test `isnan(px(k))`, and then (6.5) is used instead.

6.3 ▪ Interpolating a function

A function f may be replaced by a polynomial interpolant p in two steps: (1) evaluate f at the interpolation nodes, and (2) construct the interpolating polynomial from the collected data. For the sake of concise notation, we place the interpolation nodes into a vector:

$$\mathbf{x} = \left[\begin{array}{c} x_0 \\ \vdots \\ x_m \end{array} \right].$$

Then

$$f|_\mathbf{x} = \left[\begin{array}{c} f(x_0) \\ \vdots \\ f(x_m) \end{array} \right]$$

is a *sample* of f that contains all of the data required to construct an interpolating polynomial p.

The polynomial p may be called an interpolating polynomial for the function f, although it would be more proper to say that it is an interpolating polynomial for the sample $f|_\mathbf{x}$.

Example 6.5. Let $p(x)$ be the interpolating polynomial for $f(x) = \sin x$ on the grid with nodes $x_0 = 0$, $x_1 = \pi/4$, and $x_2 = \pi/2$. Evaluate $p(\pi/3)$ and plot p over $[0, \pi/2]$.

Solution. The grid $\mathbf{x}$ is constructed.

```
>> xs = [0; pi/4; pi/2]
xs =
                 0
   0.785398163397448
   1.570796326794897
```

(Read xs as "x's"; rather than a single x, there are multiple x's in the vector.) Next, the sample $\mathbf{p} = f|_\mathbf{x}$ is drawn.

```
>> ps = arrayfun(@(x) sin(x),xs)
ps =
                 0
   0.707106781186547
   1.000000000000000
```

The interpolating polynomial is constructed with interpgen.

```
>> p = interpgen(xs,ps);
```

The interpolant is evaluated at $x = \pi/3$ as requested.

```
>> p(pi/3)
ans =
   0.850761583276931
```

This is reasonably close to $\sin(\pi/3) = \sqrt{3}/2 = 0.87\ldots$.

The graph of the interpolant is shown in Figure 6.3. It is produced by the following code:

```
>> newfig;
>> plotfun(@(x) sin(x),[0 pi/2],'displayname','sin');
>> plotfun(p,[0 pi/2],'displayname','p');
>> plotsample(xs,ps);
>> xlabel('x');
```

Note the usage of the new function plotsample for plotting a sample of a function. ∎

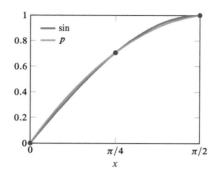

Figure 6.3. *A portion of the sine function and a quadratic interpolant p.*

6.4 ▪ Interpolation accuracy

When a function $f(x)$ is approximated by an interpolant $p(x)$, the interpolation error function is $f(x) - p(x)$. The next theorem provides an upper bound on the magnitude of this error.

The theorem applies to functions that are sufficiently smooth, in a sense of possessing sufficiently high derivatives. To indicate that a function $f(x)$ has a kth derivative $f^{(k)}(x)$ that is continuous on $a \leq x \leq b$, we write $f \in C^k[a, b]$. (At each endpoint of the interval, the derivative is the appropriate one-sided derivative.)

Theorem 6.6. *Let $f \in C^{m+1}[a, b]$, and let p be the interpolating polynomial for f on nodes $x_0, \ldots, x_m \in [a, b]$. Then, for every $x \in [a, b]$, we have*

$$f(x) - p(x) = \frac{f^{(m+1)}(\eta_x)}{(m+1)!} \prod_{i=0}^{m} (x - x_i) \tag{6.6}$$

for some $\eta_x \in [a, b]$.

The proof is at the end of this section.

The theorem implies one of our first instances of an *error bound*:

$$|f(x) - p(x)| \leq \frac{\|f^{(m+1)}\|_\infty}{(m+1)!} \prod_{i=0}^{m} |x - x_i|, \tag{6.7}$$

in which $\|f^{(m+1)}\|_\infty$ is taken over the interval $[a, b]$. The left-hand side of the inequality is the absolute error, measuring the quality of our approximation. The right-hand side is an upper bound on the error. If the right-hand side can be shown to be small, then the interpolant value $p(x)$ is guaranteed to be close to the original function value $f(x)$. In general, the right-hand side of an error bound may reveal the accuracy and efficiency of a numerical method and may indicate necessary conditions for the numerical method to work to its full potential.

Regarding the particular error bound (6.7), our plan is to illustrate and prove the bound in this chapter, and then to derive rates of convergence for common scenarios in the next few chapters.

Before beginning the work of proving the theorem, we consider an example.

Example 6.7. Let $f(x) = \sin x$, and let $p(x)$ be the interpolating polynomial of degree 2 from Example 6.5. Use Theorem 6.6 to find an upper bound on $\|f - p\|_\infty$. Verify the bound experimentally.

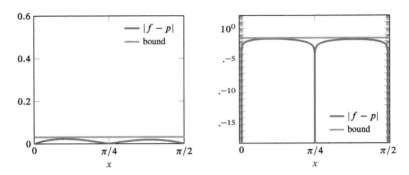

Figure 6.4. *Interpolation error and an upper bound on standard (left) and log-linear (right) axes.*

Solution. In this example, the grid is of degree $m = 2$. The derivative in (6.6) is

$$f^{(m+1)}(x) = f'''(x) = -\cos x,$$

which satisfies $|f'''(x)| \leq 1$ for all x. The final factor in (6.6) is

$$\prod_{i=0}^{m}(x - x_i) = \left(x - 0\right)\left(x - \frac{\pi}{4}\right)\left(x - \frac{\pi}{2}\right).$$

Using basic calculus, we can show that this cubic polynomial has global extreme values of $\pm\pi^3/(96\sqrt{3})$. Thus,

$$|f(x) - p(x)| \leq \frac{\|f^{(m+1)}\|_\infty}{(m+1)!}\prod_{i=0}^{m}|x - x_i| \leq \left(\frac{1}{3!}\right)\left(\frac{\pi^3}{96\sqrt{3}}\right) = \frac{\pi^3}{576\sqrt{3}}$$

for all x, and hence

$$\|f - p\|_\infty \leq \frac{\pi^3}{576\sqrt{3}} = 0.031\ldots.$$

The actual interpolation error is measured experimentally and plotted in Figure 6.4 on standard and log-linear axes.

```
>> xs = [0; pi/4; pi/2];
>> ps = arrayfun(@(x) sin(x),xs);
>> p = interpgen(xs,ps);
>> ax = newfig(1,2);
>> plotfun(@(x) abs(sin(x)-p(x)),[0 pi/2],'displayname','|f-p|');
>> ylim([0 0.6]);
>> xlabel('x');
>> subplot(ax(2));
>> ylog; ylim([1e-18 1e2]);
>> plotfun(@(x) abs(sin(x)-p(x)),[0 pi/2],'displayname','|f-p|');
>> xlabel('x');
```

Note that the absolute error equals zero at the interpolation nodes $x_0 = 0$, $x_1 = \pi/4$, and $x_2 = \pi/2$. In the log-linear plot, the zeros appear as vertical asymptotes.

Finally, the error bound is graphed as a horizontal line on each plot.

```
>> subplot(ax(1));
>> plotfun(@(x) pi^3/(576*sqrt(3)),[0 pi/2],'displayname','bound');
>> subplot(ax(2));
>> plotfun(@(x) pi^3/(576*sqrt(3)),[0 pi/2],'displayname','bound');
```

As expected, the absolute error never exceeds the bound. ∎

A key tool in the proof of Theorem 6.6 is Rolle's theorem. The simplest version of the theorem is this: Between any two real zeros of a smooth function must lie a stationary point. In more formal language, if $g \in C^1[a, b]$ and $g(a) = g(b) = 0$, then there exists a point $\eta \in (a, b)$ at which $g'(\eta) = 0$. The following generalization is also true.

Theorem 6.8. *If $g \in C^r[a, b]$ has $r + 1$ zeros, then $g^{(r)}$ has at least one zero in (a, b).*

This is a special case of Theorem A.6 in Appendix A.

The following lemma gets us most of the way toward proving Theorem 6.6.

Lemma 6.9. *Suppose $g \in C^{m+1}[a, b]$ has zeros $\zeta_0, \ldots, \zeta_m$. Then, for every $x \in [a, b]$, we have*

$$g(x) = \frac{g^{(m+1)}(\eta_x)}{(m+1)!} \prod_{i=0}^{m} (x - \zeta_i)$$

for some $\eta_x \in [a, b]$.

Proof. Fix an $x \in [a, b]$. If x is one of the zeros ζ_i, then the conclusion follows immediately with, say, $\eta_x = a$, so assume that x is not equal to any of the ζ_i. Let

$$p_x(\eta) = \frac{g(x)}{\prod_{i=0}^{m}(x - \zeta_i)} \prod_{i=0}^{m} (\eta - \zeta_i).$$

Notice that $p_x(\eta)$ is a degree-$(m + 1)$ polynomial in η that shares zeros $\zeta_0, \ldots, \zeta_m$ with $g(\eta)$ and that satisfies $p_x(x) = g(x)$. Let $E_x(\eta) = g(\eta) - p_x(\eta)$, which has zeros $\zeta_0, \ldots, \zeta_m$ and x. Take $m + 1$ derivatives:

$$E_x^{(m+1)}(\eta) = g^{(m+1)}(\eta) - \frac{g(x)}{\prod_{i=0}^{m}(x - \zeta_i)} \frac{\mathrm{d}^{m+1}}{\mathrm{d}\eta^{m+1}} \left[\prod_{i=0}^{m} (\eta - \zeta_i) \right]$$

$$= g^{(m+1)}(\eta) - \frac{g(x)}{\prod_{i=0}^{m}(x - \zeta_i)} (m+1)!.$$

$E_x(\eta)$ belongs to $C^{m+1}[a, b]$ and has at least $m + 2$ zeros, so by Theorem 6.8, there exists an η_x for which $E_x^{(m+1)}(\eta_x) = 0$. Therefore,

$$0 = g^{(m+1)}(\eta_x) - \frac{g(x)}{\prod_{i=0}^{m}(x - \zeta_i)} (m+1)!,$$

or equivalently,

$$g(x) = \frac{g^{(m+1)}(\eta_x)}{(m+1)!} \prod_{i=0}^{m} (x - \zeta_i). \qquad \square$$

Now we are ready to return to polynomial interpolation. Applying the lemma to the error function $f(x) - p(x)$ gives our desired result.

Proof of Theorem 6.6. Let $E(x) = f(x) - p(x)$. Note that $x_0, \ldots, x_m$ are zeros of $E(x)$. Fix an $x \in [a, b]$. By Lemma 6.9,

$$E(x) = \frac{f^{(m+1)}(\eta_x) - p^{(m+1)}(\eta_x)}{(m+1)!} \prod_{i=0}^{m} (x - x_i)$$

for some η_x. Also, $p^{(m+1)}(x)$ is zero everywhere because $p(x)$ is a degree-m polynomial. Thus,

$$f(x) - p(x) = \frac{f^{(m+1)}(\eta_x)}{(m+1)!} \prod_{i=0}^{m} (x - x_i). \qquad \square$$

Remark 6.10. There is a more general notion of interpolation in which a node may be repeated. Theorem 6.6 holds in this more general setting; see Appendix A.

6.5 ▪ The monomial basis

Central to many of our numerical methods are interpolating polynomials, which are identified by points on their graphs and which are expressed naturally in the Lagrange and barycentric forms. Although it is possible to reexpress an interpolating polynomial in the more familiar form

$$p(x) = c_0 + c_1 x + c_2 x^2 + \cdots + c_m x^m \tag{6.8}$$

(in the *monomial basis*), it is important *not* to do this! Evaluating a polynomial expressed in the monomial basis can produce inaccurate results (Exercise 19).

At first, you may feel that you don't "have" the polynomial because you don't know its coefficients. Over time, though, you will come to appreciate that the coefficients $c_0, \ldots, c_m$ are no more fundamental than the sample values $y_0, \ldots, y_m$ that the polynomial interpolates. Each is a sequence of $m + 1$ numbers, and each uniquely identifies a polynomial of degree at most m.

6.6 ▪ Inexact data

Given a function $f(x)$ and a grid $x_0, \ldots, x_m$ of distinct nodes, an interpolating polynomial $p(x)$ satisfies $p(x_i) = f(x_i)$, $i = 0, \ldots, m$. Attaining true equality is usually unrealistic on a finite-precision computer. A more realistic model is a polynomial $\hat{p}(x)$ satisfying $\hat{p}(x_i) \approx f(x_i)$, $i = 0, \ldots, m$. Is this enough to keep the polynomial $\hat{p}(x)$ close to the original function $f(x)$ over the entire domain of interest?

Associated to the sequence of distinct nodes $x_0, \ldots, x_m$ in an interval $[a, b]$ are the *Lebesgue function*

$$\lambda(x) = \sum_{i=0}^{m} \left| \frac{l_i(x)}{l_i(x_i)} \right| \tag{6.9}$$

and the *Lebesgue constant*

$$\Lambda = \max_{a \le x \le b} \lambda(x). \tag{6.10}$$

(The maximum is defined because $\lambda(x)$ is continuous and $[a, b]$ is closed and bounded.) The following theorem states that the distance between two interpolating polynomials is proportional to the distance between the data that define them, and that the constant of proportionality is the Lebesgue constant.

Theorem 6.11. *Let $x_0, \ldots, x_m$ be distinct nodes, and suppose $p(x)$ and $\hat{p}(x)$ are polynomials of degree at most m satisfying $p(x_i) = y_i$ and $\hat{p}(x_i) = \hat{y}_i$, $i = 0, \ldots, m$. If*

$$|y_i - \hat{y}_i| \le \delta, \quad i = 0, \ldots, m,$$

then

$$\|p - \hat{p}\|_\infty \le \Lambda\delta. \tag{6.11}$$

In addition, for any fixed p, there exists a $\hat{p}$ for which $\|p - \hat{p}\|_\infty = \Lambda\delta$.

Proof. For the inequality, note that

$$p(x) - \hat{p}(x) = \sum_{i=0}^{m} \frac{y_i}{l_i(x_i)} l_i(x) - \sum_{i=0}^{m} \frac{\hat{y}_i}{l_i(x_i)} l_i(x) = \sum_{i=0}^{m} (y_i - \hat{y}_i) \frac{l_i(x)}{l_i(x_i)},$$

so

$$|p(x) - \hat{p}(x)| \le \sum_{i=0}^{m} \left| (y_i - \hat{y}_i) \frac{l_i(x)}{l_i(x_i)} \right| \le \sum_{i=0}^{m} \delta \left| \frac{l_i(x)}{l_i(x_i)} \right| = \delta\lambda(x) \le \Lambda\delta.$$

For the second half of the theorem, let $\eta \in [a, b]$ satisfy $\lambda(\eta) = \Lambda$ and then define

$$\omega_i = \text{sign}\left(\frac{l_i(\eta)}{l_i(x_i)} \right), \quad \hat{y}_i = y_i - \delta\omega_i$$

for $i = 0, \ldots, m$. (The *sign function* $\text{sign}(x)$ is defined to equal 1 for positive x, -1 for negative x, and 0 for $x = 0$. Note that $\text{sign}(x)x = |x|$.) Then the interpolating polynomial $\hat{p}(x)$ determined by $\hat{p}(x_i) = \hat{y}_i$ satisfies

$$p(\eta) - \hat{p}(\eta) = \sum_{i=0}^{m} (y_i - \hat{y}_i) \frac{l_i(\eta)}{l_i(x_i)} = \sum_{i=0}^{m} \delta\omega_i \frac{l_i(\eta)}{l_i(x_i)} = \delta \sum_{i=0}^{m} \left| \frac{l_i(\eta)}{l_i(x_i)} \right| = \delta\lambda(\eta) = \Lambda\delta.$$

Therefore, $\|p - \hat{p}\|_\infty \ge \Lambda\delta$. But $\|p - \hat{p}\|_\infty$ cannot be any greater than $\Lambda\delta$ without violating (6.11), so we must have $\|p - \hat{p}\|_\infty = \Lambda\delta$. $\square$

The theorem is mostly encouraging. Think of p as an interpolating polynomial for some function f on the grid $x_0, \ldots, x_m$, and think of $\hat{p}$ as a polynomial that differs slightly from f at the nodes because of measurement error or roundoff error. The theorem says that the sampling errors at the nodes perturb the entire interpolant by a proportional amount. However, the way in which the grid determines the constant of proportionality Λ is a subtle matter, one that will have to wait for now.

Notes

Meijering surveys the history of interpolation [53]. The reference by Davis contains many of the important results [21].

The "Lagrange form" of the interpolating polynomial was originally reported by Waring in 1779 [81]. Lagrange published the result in 1795 [48].

A 2004 article by Berrut and Trefethen [6] has increased the visibility of the barycentric form. According to this article, the barycentric form likely originated in a 1945 paper by Taylor [70]. The numerical stability of the barycentric form was proved in 2004 by Higham [44].

The interpolation error theorem is due to Cauchy in 1840 [12].

Exercises

Exercises 1–4: Construct the interpolating polynomial through the given points by hand. Express your answer in Lagrange form, and check it by graphing.

1. $(0,0)$, $(\pi/6, 1/2)$, $(\pi/4, 1/\sqrt{2})$
2. $(-1, 1/e)$, $(0, 1)$, $(1, e)$
3. $(0,0)$, $(\pi/6, 1/2)$, $(\pi/4, 1/\sqrt{2})$, $(\pi/3, \sqrt{3}/2)$
4. $(1,0)$, $(\sqrt{e}, 1/2)$, $(e, 1)$, $(e^2, 2)$

Exercises 5–8: Compute barycentric weights by hand for the given grid.

5. 0, $1/4$, 1
6. 0, $\pi/6$, $\pi/4$
7. -1, $-1/3$, $1/3$, 1
8. 0, $\pi/6$, $\pi/4$, $\pi/3$

Exercises 9–14: Construct the polynomial that interpolates the given function at the given nodes using `interpgen`. On one graph, plot the original function and the interpolant, marking the interpolation points, and on a second graph plot the absolute error function on log-linear axes.

9. $1/x$; 0.5, 1, 1.5, 2, 2.5
10. $(x+2)/(x^2 + 2x + 2)$; -1, -0.5, 0, 0.5, 1
11. $\log x$; 1, 1.5, 2
12. $\exp(-x^2)$; 0, 0.25, 0.5, 0.75, 1
13. $\cot x$; $\pi/12$, $\pi/6$, $\pi/3$, $\pi/2$, $2\pi/3$, $5\pi/6$, $11\pi/12$
14. $\arctan x$; -2, -1, 0, 1, 2

15. Suppose $f(x) = e^{2x}$ is interpolated on the grid with nodes $x_0 = -1/2$, $x_1 = 0$, and $x_2 = 1/2$. Derive an upper bound on the interpolation error $\|f - p\|_\infty$ using (6.7). Find the tightest possible bound that can be derived in this manner. Graph the absolute error and the bound and verify that the bound is correct.

16. Suppose $f(x) = \cos(x/2)$ is interpolated on the grid with nodes $x_0 = 0$, $x_1 = \pi/3$, and $x_2 = \pi/2$. Derive an upper bound on the interpolation error $\|f - p\|_\infty$ using (6.7). Find the tightest possible bound that can be derived in this manner. Graph the absolute error and the bound and verify that the bound is correct.

17. In the Lagrange form (6.2), the factors $y_i / l_i(x_i)$ do not depend on x and therefore can be computed once for a given interpolant rather than at each evaluation of $p(x)$. Assuming these factors are precomputed, argue that a direct evaluation of (6.2) requires $m^2 + 2m$ additions/subtractions and $m^2 + m$ multiplications.

18. Assuming that the barycentric weights are precomputed, argue that an evaluation of the barycentric form (6.4) requires $3m + 1$ additions/subtractions and $2m + 3$ multiplications/divisions.

19. In finite precision, the barycentric form (6.4) and the monomial form (6.8) may produce different values. To illustrate this, consider the *Chebyshev polynomial*

$T_{25}(x)$. This is a polynomial of degree 25 whose graph intersects the points

$$\left(\cos\left(\frac{(2i+1)\pi}{50}\right), 0\right), \quad i = 0, \ldots, 24, \quad \text{and} \quad (1,1).$$

Construct the interpolating polynomial through these points using interpgen and evaluate it at $x = \cos(\pi/25)$. The same polynomial can be expressed in the monomial basis as

$$T_{25}(x) = 25x - 2600x^3 + 80\,080x^5 - 1\,144\,000x^7 + 9\,152\,000x^9 - 45\,260\,800x^{11}$$
$$+ 146\,227\,200x^{13} - 317\,521\,920x^{15} + 466\,944\,000x^{17} - 458\,752\,000x^{19}$$
$$+ 288\,358\,400x^{21} - 104\,857\,600x^{23} + 16\,777\,216x^{25}.$$

Evaluate this expression at $x = \cos(\pi/25)$ as well. How close are the computed values? Which do you suspect is more accurate, and why?

20. Prove the triangle inequality for the infinity norm: $\|f + g\|_\infty \leq \|f\|_\infty + \|g\|_\infty$. Also prove the following corollary: $\|f - h\|_\infty \leq \|f - g\|_\infty + \|g - h\|_\infty$ for any function g.

21. Let $x_0, \ldots, x_m \in [a, b]$ and $f \in C^{m+1}[a, b]$. Suppose that $\hat{p}(x)$ is a polynomial of degree at most m satisfying $|f(x_i) - \hat{p}(x_i)| < \delta$ for $i = 0, \ldots, m$. Prove that

$$|f(x) - \hat{p}(x)| \leq \frac{\|f^{(m+1)}\|_\infty}{(m+1)!} \prod_{i=0}^{m} |x - x_i| + \Lambda\delta$$

for all $x \in [a, b]$, in which Λ is the Lebesgue constant.

22. Graph the Lebesgue function for the grid with nodes $x_0 = 0$, $x_1 = \pi/4$, and $x_2 = \pi/2$ over $[0, \pi/2]$. Then find the Lebesgue constant for the grid.

23. Let $p(x)$ be the quadratic polynomial whose graph intersects the points $(0, 0)$, $(\pi/4, 0.71)$, and $(\pi/2, 1)$. (Note that these points lie close to the graph of $\sin x$.) Find a bound on $\|\sin x - p(x)\|_\infty$, in which the infinity norm is measured over $[0, \pi/2]$, using Example 6.7, Theorem 6.11, and Exercise 22.

24. Let x_0, x_1 and y_0, y_1, y_0', y_1' be real numbers. Find a polynomial $p(x)$ of degree at most three for which $p(x_i) = y_i$ and $p'(x_i) = y_i'$ for $i = 0, 1$. Prove it. *Hint.* The polynomial can be expressed in the form $p(x) = a(x - x_1)^2 + b(x - x_0)^2 + c(x - x_0)(x - x_1)^2 + d(x - x_0)^2(x - x_1)$ for constant coefficients a, b, c, d.

25. The *Newton basis polynomials* for distinct nodes $x_0, \ldots, x_m$ are

$$n_i(x) = \prod_{k=0}^{i-1} (x - x_k), \quad i = 0, \ldots, m,$$

and a polynomial is in *Newton form* when it is written as $p(x) = c_0 n_0(x) + c_1 n_1(x) + \cdots + c_m n_m(x)$. Prove that the system of equations

$$\sum_{k=0}^{i} c_k n_k(x_i) = y_i, \quad i = 0, \ldots, m,$$

has a unique solution $c_0, c_1, \ldots, c_m$ and that the polynomial with these coefficients is the Lagrange interpolating polynomial for (x_i, y_i), $i = 0, \ldots, m$.

26. The Newton form of an interpolating polynomial is defined in Exercise 25. Express the interpolating polynomial for the points $(1, 4), (2, 3), (3, 5)$ in Newton form.

Chapter 7

Interpolation on a piecewise-uniform grid

In *piecewise-polynomial interpolation*, the graph of a function is broken into small segments, and each is interpolated by a low-degree polynomial. Figure 7.1 contains an example of a piecewise-linear interpolant. As the graph suggests, even straight-line segments can provide a reasonably good fit if enough segments are used.

7.1 ▪ Piecewise-uniform grids

A natural grid for piecewise-polynomial interpolation is a *piecewise-uniform grid*, which is really a succession of grids on short subintervals. Given an interval $[a, b]$ and a positive integer n, the interval is first chopped into n subintervals of common width $l = (b - a)/n$:

$$[a_j, b_j] = [a + (j - 1)l, a + jl], \quad j = 1, \dots, n. \tag{7.1}$$

This sequence of subintervals is known as a *partition* of the original interval. (Generally, a partition is a sequence of contiguous subintervals whose union is the original interval. The partition above is more specifically a *uniform partition* because the subintervals are of equal width.) On the jth subinterval is placed a degree-m grid of equally spaced nodes,

$$\mathbf{x}_j = \begin{bmatrix} x_{0j} \\ x_{1j} \\ \vdots \\ x_{mj} \end{bmatrix}, \tag{7.2}$$

defined by

$$x_{ij} = \begin{cases} a_j + ih & \text{if } m \geq 1, \\ (a_j + b_j)/2 & \text{if } m = 0, \end{cases} \tag{7.3}$$

in which $h = l/m$. The piecewise-defined grid can be represented by the matrix

$$\begin{bmatrix} \mathbf{x}_1 & \mathbf{x}_2 & \cdots & \mathbf{x}_n \end{bmatrix} = \begin{bmatrix} x_{01} & x_{02} & \cdots & x_{0n} \\ x_{11} & x_{12} & \cdots & x_{1n} \\ \vdots & \vdots & \ddots & \vdots \\ x_{m1} & x_{m2} & \cdots & x_{mn} \end{bmatrix}, \tag{7.4}$$

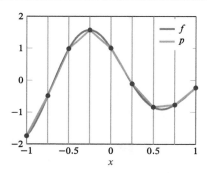

Figure 7.1. *A smooth function and a piecewise-linear interpolant.*

and it is called an m-by-n piecewise-uniform grid for short. If $n = 1$, then the grid is simply called a *uniform grid* of degree m.

A piecewise-uniform grid is computed by routine griduni, shown in Listing 7.1. The code also computes the associated barycentric weights, which are derived later in this chapter.

Listing 7.1. Construct a piecewise-uniform grid.

```
function [xs,ws] = griduni(ab,m,n)

if nargin<3, n = 1; end
a = ab(1);
b = ab(2);
if m==0
  % compute nodes and barycentric weights for degree 0
  l = (b-a)/n;
  xs = linspace(a+l/2,b-l/2,n);
  ws = [ 1 ];
else
  % compute nodes for degree >= 1
  xs = linspace(a,b,m*n+1);
  xs = [ reshape(xs(1:end-1),m,n); xs(m+1:m:end) ];
  % compute barycentric weights for degree >= 1
  ws = nan(m+1,1);
  ws(1) = 1;
  for i = 1:floor(m/2)
    ws(i+1) = -ws(i)*(m-i+1)/i;
  end
  k = ceil(m/2);
  ws(end:-1:end-k+1) = (-1)^m*ws(1:k);
end
```

Example 7.1. Compute and graph piecewise-uniform grids on $[0, 1]$ with $m = 0, 1, 2, 3$ and $n = 4$.

Solution. In each case, the interval is divided into $n = 4$ parts: $[0, 0.25]$, $[0.25, 0.5]$, $[0.5, 0.75]$, and $[0.75, 1]$.

When $m = 0$, a single node is placed at the center of each subinterval.

```
>> format short;
>> griduni([0 1],0,4)
ans =
    0.1250    0.3750    0.6250    0.8750
```

(The command `format short` causes subsequent output to be displayed with fewer digits to save space.)

When $m = 1$, nodes are placed at the ends of each subinterval.

```
>> griduni([0 1],1,4)
ans =
         0    0.2500    0.5000    0.7500
    0.2500    0.5000    0.7500    1.0000
```

In the matrix, each column contains the nodes for a single subinterval. Note that the right node on one subinterval coincides with the left node on the next subinterval. See also the following graphic:

When $m = 2$, there are three nodes on each subinterval.

```
>> griduni([0 1],2,4)
ans =
         0    0.2500    0.5000    0.7500
    0.1250    0.3750    0.6250    0.8750
    0.2500    0.5000    0.7500    1.0000
```

Again, a node lying at the intersection of two subintervals belongs to both incident subgrids.

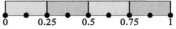

Finally, when $m = 3$, there are four nodes per subinterval.

```
>> griduni([0 1],3,4)
ans =
         0    0.2500    0.5000    0.7500
    0.0833    0.3333    0.5833    0.8333
    0.1667    0.4167    0.6667    0.9167
    0.2500    0.5000    0.7500    1.0000
```

```
>> format long;
```

∎

Theorem 7.2. *Let*

$$w_i = (-1)^i \binom{m}{i}, \quad i = 0, \ldots, m, \tag{7.5}$$

in which $\binom{m}{i}$ denotes the binomial coefficient $m!/(i!(m-i)!)$. Then $w_0, \ldots, w_m$ is a sequence of barycentric weights for a degree-m uniform grid, regardless of the interval on which it is placed.

Proof. If $m = 0$, then any number is a valid barycentric weight because of the arbitrary constant C in the definition (6.3). Assume for the rest of the proof that $m \geq 1$.

Suppose the grid is placed on an interval $[a, b]$, and let $h = (b - a)/m$. We have $x_i = a + ih$ and therefore $x_i - x_k = (a + ih) - (a + kh) = h(i - k)$. With $l_i(x)$

denoting the ith Lagrange basis polynomial,

$$l_i(x_i) = \prod_{k \neq i}(x_i - x_k) = h^m \prod_{k \neq i}(i - k) = h^m \left(\prod_{k=0}^{i-1}(i - k) \prod_{k=i+1}^{m}(i - k) \right)$$

$$= h^m \left(\prod_{s=1}^{i} s \prod_{s=1}^{m-i}(-s) \right) = (-1)^{m-i} h^m i!(m - i)!.$$

Let $C = (-1)^m h^m m!$. Then a valid sequence of barycentric weights is given by

$$\frac{C}{l_i(x_i)} = \frac{(-1)^m h^m m!}{(-1)^{m-i} h^m i!(m - i)!} = (-1)^i \frac{m!}{i!(m - i)!}. \qquad \Box$$

Routine griduni computes the barycentric weights using the recurrence relation

$$\binom{m}{i} = \frac{m - i + 1}{i} \binom{m}{i - 1}, \quad i = 1, \ldots, m,$$

and the symmetry relation

$$\binom{m}{m - i} = \binom{m}{i}, \quad i = 0, \ldots, m.$$

7.2 ▪ Piecewise-polynomial interpolation

Once a piecewise-uniform grid is in place, interpolation can begin. The interpolant should come from a class of piecewise-polynomial functions.

Definition 7.3. *Let $[a_j, b_j]$, $j = 1, \ldots, n$, form a uniform partition of $[a, b]$ as in (7.1). For $m \geq 1$, a function p is of class $P_{mn}[a, b]$ if*

1. *p is continuous on $[a, b]$ and*

2. *p equals a polynomial of degree at most m on each subinterval $[a_j, b_j]$.*

Also, p is of class $P_{0n}[a, b]$ if p is constant on each subinterval $[a_j, b_j)$, $j = 1, \ldots, n - 1$, and on $[a_n, b_n]$.

To indicate that p is of class $P_{mn}[a, b]$, we write $p \in P_{mn}[a, b]$.

In the definition, there is no deep reason to require that a piecewise-constant function be continuous from the right at $x = a_j$; it is just a convenient convention.

A piecewise-polynomial interpolant $p \in P_{mn}[a, b]$ is determined by its values on an m-by-n piecewise-uniform grid. Often these values are chosen to match an existing function f. On the subgrid $\mathbf{x}_j$ of (7.2), a sample of f is

$$f|_{\mathbf{x}_j} = \begin{bmatrix} f(x_{0j}) \\ f(x_{1j}) \\ \vdots \\ f(x_{mj}) \end{bmatrix}. \tag{7.6}$$

Combining the subintervals gives a sample on the entire piecewise-uniform grid:

$$
\begin{bmatrix}
f|_{\mathbf{x}_1} & f|_{\mathbf{x}_2} & \cdots & f|_{\mathbf{x}_n}
\end{bmatrix}
=
\begin{bmatrix}
f(x_{01}) & f(x_{02}) & \cdots & f(x_{0n}) \\
f(x_{11}) & f(x_{12}) & \cdots & f(x_{1n}) \\
\vdots & \vdots & \ddots & \vdots \\
f(x_{m1}) & f(x_{m2}) & \cdots & f(x_{mn})
\end{bmatrix}. \tag{7.7}
$$

Routine `sampleuni` is for sampling on an m-by-n piecewise-uniform grid. See Listing 7.2.

Listing 7.2. Sample a function on a piecewise-uniform grid.

```
function ys = sampleuni(f,ab,m,n)

if nargin<4, n = 1; end
a = ab(1);
b = ab(2);
if m==0
  l = (b-a)/n;
  ys = arrayfun(f,linspace(a+l/2,b-l/2,n));
else
  ys = arrayfun(f,linspace(a,b,m*n+1));
  ys = [ reshape(ys(1:end-1),m,n); ys(m+1:m:end) ];
end
```

Given such a sample, routine `interpuni` in Listing 7.3 below, with its helper routine `interpuni_`, constructs a piecewise-polynomial interpolant that passes through the given points and that belongs to $P_{mn}[a,b]$. The piecewise-linear function in Figure 7.1 is an example of such a piecewise-polynomial interpolant. As seen in the implementation of `interpuni`, evaluation at a given x is a two-step process. First, a subinterval containing x is identified by its index j. Second, the polynomial interpolant on that subinterval is evaluated using `interp_`, which applies the barycentric formula.

Example 7.4. Interpolate $f(x) = \sin x$ on the 2-by-4 piecewise-uniform grid over $[0, 2\pi]$. Evaluate the interpolant at $x = 5\pi/6$.

Solution. The original function is sampled.

```
>> f = @(x) sin(x);
>> a = 0;
>> b = 2*pi;
>> m = 2;
>> n = 4;
>> format short;
>> ps = sampleuni(f,[a b],m,n)
ps =
         0    1.0000    0.0000   -1.0000
    0.7071    0.7071   -0.7071   -0.7071
    1.0000    0.0000   -1.0000   -0.0000
>> format long;
```

Reading the sample values from top to bottom then left to right, the shape of the sine function is evident.

The interpolant p is constructed.

```
>> p = interpuni(ps,[a b]);
```

Listing 7.3. Construct a piecewise-polynomial interpolant on a piecewise-uniform grid.

```
function p = interpuni(ps,ab)

% order subintervals from left to right on the real line
if ab(1)>ab(2)
  ps = fliplr(flipud(ps));
  ab = ab(end:-1:1);
end
% construct piecewise-polynomial interpolant
a = ab(1);
b = ab(2);
m = size(ps,1)-1;
n = size(ps,2);
l = (b-a)/n;
[xs,ws] = griduni([0 1],m,1);
p = @(x) interpuni_(xs,ws,ps,a,b,l,n,x);
```

```
function y = interpuni_(xs,ws,ps,a,b,l,n,x)

y = nan(size(x));
for k = 1:length(x)
  if x(k)>=a&&x(k)<=b
    % find subinterval
    j = min(floor((x(k)-a)/l)+1,n);
    % evaluate polynomial piece
    t = (x(k)-(a+(j-1)*l))/l;
    y(k) = interp_(xs,ws,ps(:,j),t);
  end
end
```

At the requested point $x = 5\pi/6$, the interpolant takes the following value:
```
>> p(5*pi/6)
ans =
    0.517428249943598
```
Considering the coarseness of the grid, this is reasonably close to the original value $\sin(5\pi/6) = 0.5$. ∎

An interpolant can be graphed with the familiar command plotfun. Two additional routines, plotsample and plotpartition, help to visualize the piecewise-uniform grid on which the interpolant is constructed. The following example illustrates their usage.

Example 7.5. Plot the interpolant of the previous example, marking each sample point with a dot and delineating the polynomial pieces by vertical lines. Then plot the absolute error function.

Solution. The interpolant is constructed as before.
```
>> f = @(x) sin(x);
>> a = 0;
>> b = 2*pi;
>> m = 2;
>> n = 4;
>> ps = sampleuni(f,[a b],m,n);
>> p = interpuni(ps,[a b]);
```
The original function and the interpolant are plotted in the left graph of Figure 7.2.

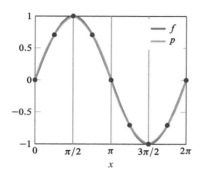
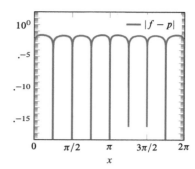

Figure 7.2. *A piecewise-polynomial interpolant on a piecewise-uniform grid (left) and its absolute error (right).*

Notice that the interpolant's graph is parabolic on each subinterval.

```
>> ax = newfig(1,2);
>> plotfun(f,[a b],'displayname','f');
>> plotpartition([a b],n);
>> xs = griduni([a b],m,n);
>> plotsample(xs,ps);
>> plotfun(p,[a b],'displayname','p');
>> ylim([-1 1]);
>> xlabel('x');
```

The absolute error function is plotted in the right graph.

```
>> subplot(ax(2));
>> ylog; ylim([1e-18 1e2]);
>> plotfun(@(x) abs(f(x)-p(x)),[a b],'displayname','|f-p|');
>> xlabel('x');
```

Note that the error equals zero at every interpolation node. ∎

7.3 ▪ A first look at efficiency

Intuitively, finer grids should lead to better fits. An m-by-n grid can be refined by increasing either the number of subintervals n or the degree m. The following example illustrates the effects of these refinements.

Example 7.6. Interpolate $f(x) = e^{2x}$, $-1 \leq x \leq 1$, using m-by-n piecewise-uniform grids with every combination of $m = 0, 1, 2, 3$ and $n = 1, 2, 3$.

Solution. Each plot in Figure 7.3 shows the original function f and a piecewise-polynomial interpolant. From top to bottom, the degree is $m = 0, 1, 2, 3$, and from left to right, the number of subintervals is $n = 1, 2, 3$.

```
>> f = @(x) exp(2*x);
>> a = -1;
>> b = 1;
>> m = 0:3;
>> n = 1:3;
>> ax = newfig(length(m),length(n));
>> for k = 1:length(m)
      for l = 1:length(n)
         xs = griduni([a b],m(k),n(l));
```

```
    ps = sampleuni(f,[a b],m(k),n(l));
    p = interpuni(ps,[a b]);
    subplot(ax(k,l));
    legend hide;
    plotfun(f,[a b]);
    plotpartition([a b],n(l));
    plotsample(xs,ps);
    plotfun(p,[a b]);
    ylim([-2 10]);
  end
end
```

In this example, higher-degree polynomials and narrower subintervals produce better fits.　∎

In the next chapter, we fix a small degree m and increase the number of subintervals n to improve accuracy. Beginning in Chapter 9, we investigate high-degree polynomial interpolation and find that a good deal of care is required.

Notes

Meijering's survey [53] addresses the history of piecewise-polynomial interpolation.

Historically, interpolation was presented more often through tables than graphics. For example, the classic *Handbook of Mathematical Functions with Formulas, Graphs, and Mathematical Tables* [1] records selected values of elementary and special functions, and the reader can estimate other values using low-degree polynomial interpolation.

Exercises

Exercises 1–6: Interpolate the function on an m-by-n piecewise-uniform grid using `interpuni`. Plot the original function and the interpolant in one graph and the absolute error in a second graph.

1. $\sqrt{x}$, $1/4 \le x \le 1$; $m = 1, n = 14$
2. $(x + 1)/(x^2 + 4)$, $-3 \le x \le 3$; $m = 3, n = 4$
3. $\tan x$, $-\pi/3 \le x \le \pi/3$; $m = 0, n = 6$
4. $\arcsin x$, $0 \le x \le 1$; $m = 2, n = 3$
5. $\arctan x$, $-\sqrt{3} \le x \le \sqrt{3}$; $m = 1, n = 20$
6. $|x|$, $-1.25 \le x \le 1.25$; $m = 1, n = 5$

Exercises 7–12: A function and a degree m are given. Experimentally, determine the number of subintervals n in an m-by-n piecewise-uniform grid required to interpolate the function with an absolute error below 10^{-6}. Report the value of n to two significant digits and plot the absolute error on log-linear axes.

7. e^x, $-1 \le x \le 1$; $m = 1$
8. $\log x$, $1/2 \le x \le 1$; $m = 3$
9. $\sqrt{x}$, $1/4 \le x \le 1$; $m = 3$
10. $x^{3/2}$, $0 \le x \le 1$; $m = 2$
11. $\arctan x$, $-2 \le x \le 2$; $m = 1$

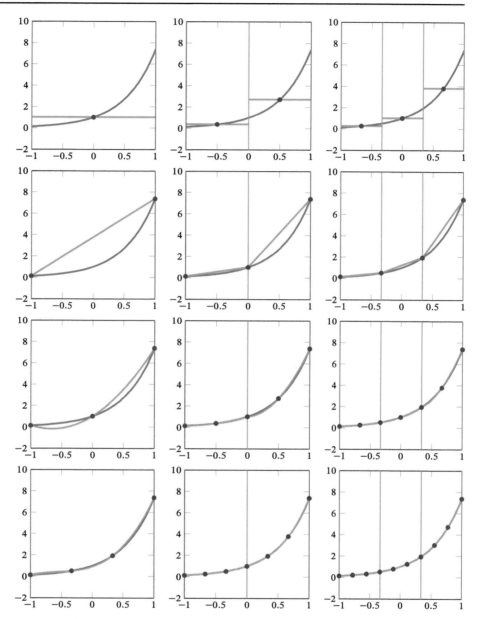

Figure 7.3. *Piecewise-polynomial interpolation of e^{2x}. From left to right, the inter-polants are constructed from $n = 1$, $n = 2$, and $n = 3$ pieces. From top to bottom, the polynomial pieces have degrees $m = 0$, $m = 1$, $m = 2$, and $m = 3$.*

12. $(x^2 + x + 1)/(x^2 - x + 1)$, $-5 \leq x \leq 5$; $m = 2$

13. Compute barycentric weights for a uniform grid of degree $m = 4$ on $[6, 8]$ using griduni and then using baryweights. Verify that they produce identical answers up to a scalar factor.

14. Consider on one hand the set $P_{0n}[a, b]$ of piecewise-constant functions. Consider on the other hand the set of row vectors with n entries. Prove that the function that

maps $p \in P_{0n}[a, b]$ to its sample

$$\begin{bmatrix} p(x_{01}) & p(x_{02}) & \cdots & p(x_{0n}) \end{bmatrix}$$

is a bijection from $P_{0n}[a, b]$ to the set of row vectors.

15. Let $m \geq 1$. Consider on one hand the set $P_{mn}[a, b]$ of piecewise-polynomial functions. Consider on the other hand the set of $(m + 1)$-by-n matrices in which the last entry in column j equals the first entry in column $j+1$ for $j = 1, \ldots, n-1$. Prove that the function that maps $p \in P_{mn}[a, b]$ to its sample

$$\begin{bmatrix} p(x_{01}) & p(x_{02}) & \cdots & p(x_{0n}) \\ p(x_{11}) & p(x_{12}) & \cdots & p(x_{1n}) \\ \vdots & \vdots & \ddots & \vdots \\ p(x_{m1}) & p(x_{m2}) & \cdots & p(x_{mn}) \end{bmatrix}$$

is a bijection from $P_{mn}[a, b]$ to the set of matrices described above.

16. Write a MATLAB routine to sample a function at the left endpoint of each subinterval in a uniform partition. Demonstrate your routine on the function $f(x) = 1/x$ with a partition of $[a, b] = [1, 4]$ into $n = 6$ subintervals of equal width. Plot f and the piecewise-constant interpolant $p \in P_{0n}[a, b]$ that matches f at the left endpoint of each subinterval.

17. Let $[a, b]$ be an interval of the real line and n a positive integer. Partition the interval into contiguous subintervals $[a_j, b_j]$, $j = 1, \ldots, n$, of equal width. Let p be the function that is linear on each subinterval and satisfies $p((a_j + b_j)/2) = y_j$, $p'((a_j + b_j)/2) = y'_j$ for given numbers y_j, y'_j, $j = 1, \ldots, n$. Write a MATLAB function analogous to `interpuni` for constructing a piecewise-linear function of this form given a, b, n, y_j, and y'_j, $j = 1, \ldots, n$. Demonstrate it.

18. Let $[a_j, b_j]$, $j = 1, \ldots, n$, be a partition of an interval $[a, b]$. The subintervals are not required to have the same width. For any nonnegative integer m, there is a piecewise-defined grid consisting of a simple uniform grid of degree m on each subinterval. Write a MATLAB routine for constructing this piecewise-defined grid, given the sequence of subintervals and the degree m. Demonstrate your routine.

19. Write a MATLAB routine for sampling a function on a grid of the form described in Exercise 18. Demonstrate your routine.

20. Write a MATLAB routine, analogous to `interpuni`, for constructing a piecewise-polynomial interpolant from a sample on a grid as described in Exercise 18. Demonstrate your routine.

Chapter 8

Accuracy of interpolation on a piecewise-uniform grid

In this chapter, we analyze interpolation error on a piecewise-uniform grid.

When designing a numerical method, the goal is high accuracy at low cost. In the case of interpolation, the primary measure of cost in this book is *sampling cost*. The sampling cost for a grid is the number of function evaluations required to produce a sample on the grid. Let

$$N = \begin{cases} n & \text{if } m = 0, \\ mn & \text{if } m \geq 1. \end{cases}$$

The sampling cost for an m-by-n piecewise-uniform grid is N if $m = 0$ or $N + 1$ if $m \geq 1$. (If $m = 0$, then there are n nodes, one per subinterval. If $m \geq 1$, then there are $(m - 1)n + (n + 1) = mn + 1$ nodes: $m - 1$ nodes in the interior of each of n subintervals plus $n + 1$ distinct nodes at the endpoints.) Note that N can be expressed more briefly as $N = (m \vee 1)n$, in which $a \vee b$ denotes the larger of a or b. We would like to show that the interpolation error approaches zero as N approaches infinity, and to derive a guarantee on the rate of convergence.

8.1 ▪ Rate of convergence

Our error bound is a consequence of Theorem 6.6.

Theorem 8.1. *Let $f \in C^{m+1}[a, b]$ and $p \in P_{mn}[a, b]$, and suppose that p interpolates f on the m-by-n piecewise-uniform grid over $[a, b]$. Then*

$$\|f - p\|_\infty \leq CN^{-(m+1)}, \tag{8.1}$$

in which $C = \frac{1}{m+1}\|f^{(m+1)}\|_\infty(b - a)^{m+1}$ and $N = (m \vee 1)n$.

Proof. Denote the nodes by x_{ij} as in (7.4).

First, consider $m > 0$. Let $x \in [a, b]$. Then x must lie at a node or between two consecutive nodes. Either way, $x_{kj} \leq x \leq x_{k+1,j}$ for some pair of nodes $x_{kj}, x_{k+1,j}$. We apply (6.7) to the jth subinterval of the grid. The final factor in the error bound splits into two parts, one involving nodes to the left of x and the other involving nodes to the right:

$$\prod_{i=0}^{m}|x - x_{ij}| = \prod_{i=1}^{k+1}|x - x_{k+1-i,j}| \prod_{i=1}^{m-k}|x - x_{k+i,j}|.$$

Let $h = (b-a)/(mn)$, the distance between consecutive nodes. For any $1 \le i \le k+1$, we have $x_{k+1-i,j} \le x_{kj} \le x \le x_{k+1,j}$ and therefore $|x - x_{k+1-i,j}| \le x_{k+1,j} - x_{k+1-i,j} = ih$. For any $1 \le i \le m-k$, we have $x_{kj} \le x \le x_{k+1,j} \le x_{k+i,j}$ and therefore $|x - x_{k+i,j}| \le x_{k+i,j} - x_{kj} = ih$. Hence,

$$\prod_{i=0}^{m}|x - x_{ij}| \le \left(\prod_{i=1}^{k+1} ih\right)\left(\prod_{i=1}^{m-k} ih\right) = \left(\prod_{i=1}^{k+1} i\right)\left(\prod_{i=2}^{m-k} i\right)h^{m+1}$$

$$\le \left(\prod_{i=1}^{k+1} i\right)\left(\prod_{i=k+2}^{m} i\right)h^{m+1} = m!\,h^{m+1}. \quad (8.2)$$

By (6.7),

$$|f(x) - p(x)| \le \frac{\|f^{(m+1)}\|_\infty}{(m+1)!}m!\,h^{m+1} = \frac{\|f^{(m+1)}\|_\infty (b-a)^{m+1}}{m+1}(mn)^{-(m+1)}.$$

Next, consider $m = 0$. Let $x \in [a,b]$. Then x must lie in some subinterval $[a_j, b_j] = [a + (j-1)\frac{b-a}{n}, a + j\frac{b-a}{n}]$. We apply (6.7) to the jth subinterval. The final factor in the error bound is bounded as follows:

$$\prod_{i=0}^{0}|x - x_{ij}| = |x - x_{0j}| \le \frac{h}{2}, \quad (8.3)$$

in which $h = (b-a)/n$. Hence,

$$|f(x) - p(x)| \le \frac{\|f'\|_\infty}{1}\frac{h}{2} = \frac{\|f'\|_\infty (b-a)}{2}n^{-1}.$$

Whether $m > 0$ or $m = 0$,

$$|f(x) - p(x)| \le \frac{\|f^{(m+1)}\|_\infty (b-a)^{m+1}}{m+1}N^{-(m+1)}$$

for all $x \in [a,b]$, so the infinity norm is bounded by the same quantity.   □

In many computations, the degree m is held fixed while the number of subintervals n is increased until a desired accuracy is achieved. According to Theorem 8.1, as the grid is refined, the error approaches zero as fast as $N^{-(m+1)}$.

Example 8.2. Predict the rate of convergence of piecewise-uniform interpolation on $f(x) = e^{2x}$, $-1 \le x \le 1$. Verify experimentally for $m = 0, 1, 2, 3$.

Solution. Because e^{2x} is infinitely differentiable, the function belongs to $C^{m+1}[a,b]$ for any m. The rate of convergence is predicted to be $N^{-(m+1)}$.

This prediction is tested experimentally. First, the function and interval are defined.

```
>> f = @(x) exp(2*x);
>> a = -1;
>> b = 1;
```

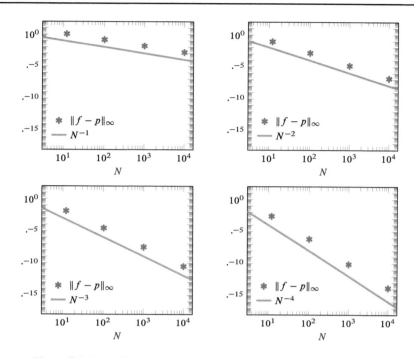

Figure 8.1. *Rate of convergence for interpolation on a piecewise-uniform grid: $m = 0$ (top-left), $m = 1$ (top-right), $m = 2$ (bottom-left), and $m = 3$ (bottom-right).*

We use sequences of grids with increasing N to analyze the rate of convergence. The values $N = 12, 102, 1002, 10002$ are convenient because they are divisible by 1, 2, and 3.

```
>> m = [0 1 2 3];
>> N = [12 102 1002 10002];
```

For each combination of parameters, we find an interpolant p and measure the error $\|f - p\|_\infty$.

```
>> err = nan(length(m),length(N));
>> for k = 1:length(m)
     for l = 1:length(N)
       n = N(l)/max(m(k),1);
       ps = sampleuni(f,[a b],m(k),n);
       p = interpuni(ps,[a b]);
       err(k,l) = infnorm(@(x) f(x)-p(x),[a b],100);
     end
   end
```

The observed errors are represented by asterisks in Figure 8.1.

```
>> ax = newfig(2,2); ax = ax';
>> for k = 1:length(m)
     subplot(ax(k));
     xlog; ylog; ylim([1e-18 1e2]);
     plot(N,err(k,:),'*','displayname','||f-p||');
     xlabel('N');
   end
```

The predicted convergence rate is represented by a straight line in each subplot.

```
>> for k = 1:length(m)
     subplot(ax(k));
```

```
plotfun(@(N) N^(-(m(k)+1)),[3 17000] ...
        ,'displayname',sprintf('N^{%d}',-(m(k)+1)));
end
```

The observed error seems to converge to zero at the predicted rate in each case. ∎

8.2 ▪ Achieving an accuracy goal

Theorem 8.1 enables one to achieve accuracy goals. The interpolation error $f(x)-p(x)$ satisfies $\|f - p\|_\infty \leq CN^{-(m+1)}$. To achieve an error below some small number ε, solve $CN^{-(m+1)} < \varepsilon$ for N.

Example 8.3. Consider interpolating $f(x) = \cos(3x) - \sin(4x)$, $-1 \leq x \leq 1$, with a piecewise-uniform grid of degree $m = 3$. How many subintervals n are required to achieve absolute error below 10^{-12}?

Solution. The constant C in Theorem 8.1 includes the factor $\|f^{(m+1)}\|_\infty$. In this problem, $f^{(4)}(x) = 81\cos(3x)-256\sin(4x)$ and $\|f^{(4)}\|_\infty \leq 81+256 = 337$. Hence, the interpolation error for the specified piecewise-polynomial interpolant p satisfies

$$\|f - p\|_\infty \leq \frac{337}{4}(2^4)N^{-4} = 1348N^{-4}.$$

Set the upper bound equal to the desired level of 10^{-12} and solve for the integer N:

$$1348N^{-4} \leq 10^{-12} \iff N \geq 6060.$$

Equivalently, it should suffice to take $n = 6060/3 = 2020$.
   Let's try it.

```
>> f = @(x) cos(3*x)-sin(4*x);
>> a = -1;
>> b = 1;
>> ps = sampleuni(f,[a b],3,2020);
>> p = interpuni(ps,[a b]);
>> infnorm(@(x) f(x)-p(x),[a b],100)
ans =
      1.301181384860683e-13
```

The error is below 10^{-12} as desired. ∎

For many functions, deriving $\|f^{(m+1)}\|_\infty$ is prohibitively difficult. In this case, the constant C in (8.1) can be estimated from data.

Example 8.4. Consider interpolating $f(x) = (\sin x)/x$, $-\pi \leq x \leq \pi$, with a piecewise-uniform grid of degree $m = 3$. Find a sufficient number of subintervals n to achieve absolute error below 10^{-12}.

Solution. At first glance, the function looks dangerous because of possible division by zero. However, l'Hôpital's rule shows that $\lim_{x\to 0} f(x) = 1$. Setting $f(0) = 1$ produces a continuous function that is defined on the entire real line.

```
>> f = @(x) iif(x==0,@()1,@()sin(x)/x);
>> a = -pi;
>> b = pi;
```

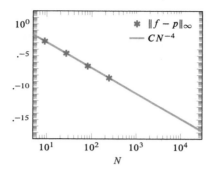

Figure 8.2. *Predicting interpolation error for finer piecewise-uniform grids.*

Error bound (8.1) includes $\|f^{(4)}\|_\infty$. We will not attempt to find this quantity by hand. Instead, we express the error bound in the form CN^{-4} and estimate C from an experiment.

```
>> m = 3;
>> n = 3.^(1:4);
>> err = nan(size(n));
>> for k = 1:length(n)
      ps = sampleuni(f,[a b],m,n(k));
      p = interpuni(ps,[a b]);
      err(k) = infnorm(@(x) f(x)-p(x),[a b],100);
   end
```

The asterisks in Figure 8.2 represent the interpolation error for four interpolants with increasing n.

```
>> N = max(m,1)*n;
>> newfig;
>> xlog; xlim([5 30000]); ylog; ylim([1e-18 1e2]);
>> plot(N,err,'*','displayname','||f-p||');
>> xlabel('N');
```

Based on Theorem 8.1, we know that $\|f - p\|_\infty \leq CN^{-4}$ for some C. We hypothesize that in fact $\|f - p\|_\infty \approx CN^{-4}$ and estimate C from the data.

```
>> C = err(end)*N(end)^4
C =
   12.749200286290723
>> plotfun(@(N) C*N^(-4),[5 30000],'displayname','C N^{-4}');
```

See the straight line in Figure 8.2. The fit looks good.

To keep the interpolation error below 10^{-12}, we solve for n:

$$CN^{-4} \leq 10^{-12} \iff C(3n)^{-4} \leq 10^{-12} \iff n \geq (10^{-12}/C)^{-1/4}/3.$$

```
>> nstar = ceil((1e-12/C)^(-1/4)/3)
nstar =
   630
```

Let's try it.

```
>> psstar = sampleuni(f,[a b],m,nstar);
>> pstar = interpuni(psstar,[a b]);
>> infnorm(@(x) f(x)-pstar(x),[a b],100)
ans =
     1.017186335161568e-12
```

The accuracy goal $\|f - p\|_\infty \leq 10^{-12}$ is nearly achieved at a sampling cost of $N = 3 \times 630 = 1890$ function evaluations.   ∎

8.3 ▪ Inexact data

A piecewise-polynomial interpolant is constructed from a sample of function values $f(x_{ij})$. If those values are known only approximately, then the interpolant incurs additional error.

The additional error can be bounded with the Lebesgue constant, defined in general by (6.10). For a degree-m uniform grid specifically, the Lebesgue constant Λ_m is independent of the interval on which the grid is placed (Exercise 18), and the values for $m = 0, 1, 2, 3$ are as follows (Exercises 19–21):

m	Λ_m
0	1
1	1
2	1.25
3	1.63...

Theorem 8.5. *Let $f \in C^{m+1}[a,b]$. If $\hat{p} \in P_{mn}[a,b]$ and $|f(x_{ij}) - \hat{p}(x_{ij})| \leq \delta$ at every node x_{ij} in the m-by-n piecewise-uniform grid on $[a,b]$, then*

$$\|f - \hat{p}\|_\infty \leq CN^{-(m+1)} + \Lambda_m\delta, \tag{8.4}$$

in which C is defined as in Theorem 8.1.

The proof is Exercise 17.

Note that the Lebesgue constant Λ_m in (8.4) depends only on m, not n. If the degree m is fixed while the number of subintervals n is increased, the error bound shrinks toward $\Lambda_m\delta$ but gets no smaller.

Exercises

Exercises 1–6: A function and a degree m are given. Consider interpolating the function on an m-by-n piecewise-uniform grid.

(a) Using the theory of this chapter, find a bound of the form CN^{-r} on the interpolation error.

(b) Using the bound from part (a), find a sufficient number n of subintervals to guarantee that the absolute interpolation error stays below 10^{-6}. Verify that this works with an experiment.

1. $e^{-x/2}, -1 \leq x \leq 1; m = 2$
2. $\sin(4x), 0 \leq x \leq \pi; m = 1$
3. $\cos(x) + 3\sin(x), -2\pi \leq x \leq 2\pi; m = 3$
4. $100 \log x, 0.5 \leq x \leq 1; m = 3$

5. $(x + 1)/(x - 1)$, $1.5 \le x \le 4$; $m = 2$

6. $x^2 \log x$, $0.1 \le x \le e$, $m = 1$

Exercises 7–12: A function and a degree m are given. Based on Theorem 8.1, the interpolation error associated with an m-by-n piecewise-uniform grid should be bounded by an expression of the form $CN^{-(m+1)}$. Estimate the value of C to one significant digit using experimental data.

7. $\arctan x$, $0 \le x \le 10$; $m = 3$

8. $\cot x$, $0.1 \le x \le 1$; $m = 2$

9. $(x^2 + x + 1)/(x^2 - 4)$, $-1 \le x \le 1$; $m = 3$

10. $\sqrt{1 + x^{-2}}$, $1 \le x \le 5$; $m = 2$

11. $x/\log x$, $2 \le x \le 4$; $m = 1$

12. $e^{1/x}$, $3 \le x \le 9$; $m = 2$

Exercises 13–16: Explain why Theorem 8.1 does not apply when the given function is interpolated on an m-by-n piecewise-uniform grid. Then measure the rate of convergence experimentally. You may need to use a very large number of sample points if measuring the error with infnorm.

13. $|x - \pi|$, $0 \le x \le 5$; $m = 0$

14. $\sqrt{|x - e|}$, $0 \le x \le 4$; $m = 0$

15. $|x - \sqrt{2}| \sin(x - \sqrt{2})$, $0 \le x \le 2\pi$; $m = 1$

16. $(|x - \sqrt{3}|)^{3/2}$, $0 \le x \le 3$; $m = 1$

17. Prove Theorem 8.5.

18. Prove that the Lebesgue constant for a degree-m uniform grid is

$$\Lambda_m = \max_{0 \le t \le m} \sum_{i=0}^{m} \frac{\prod_{k \ne i} |t - k|}{\prod_{k \ne i} |i - k|},$$

regardless of the interval on which the grid is placed.

19. Show that the Lebesgue constant for a degree-0 or degree-1 uniform grid equals 1.

20. Show that the Lebesgue constant for a degree-2 uniform grid equals $5/4$.

21. Show that the Lebesgue constant for a degree-3 uniform grid equals $(7/27)(1 + 2\sqrt{7})$.

22. When e^{-x} is computed with exp(-x), the absolute error is below 10^{-15} for any $x \in [0, 1]$. Suppose exp(-x) is sampled on a 3-by-n piecewise-uniform grid over $[0, 1]$, and then the sample is interpolated to find an approximating piecewise-polynomial $p(x)$ for e^{-x}. Find a bound on the interpolation error using Theorem 8.5. Provide a formula for the error bound and then graph it over $3 \le N \le 10^6$.

Chapter 9

Danger from uniform grids of high degree

We have seen that higher-degree polynomials often provide better fits than lower-degree polynomials. Why bother with *piecewise*-polynomial interpolation at all? Why not just fit a function with a single high-degree polynomial? Eventually we will do this, but there is danger ahead.

9.1 ▪ The Runge phenomenon

One danger of high-degree polynomial interpolation is exemplified by a function called *Runge's function*.

Example 9.1. Runge's function is

$$f(x) = \frac{1}{1 + 25x^2}.$$

Interpolate this function on a degree-m uniform grid over $-1 \le x \le 1$. Investigate the interpolation error as m increases.

Solution. In Figure 9.1, Runge's function is accompanied by polynomial interpolants of degrees $m = 5$, 10, and 20.

```
>> f = @(x) 1/(1+25*x^2);
>> a = -1;
>> b = 1;
>> m = [5 10 20];
>> ax = newfig(1,3);
>> for k = 1:length(m)
      subplot(ax(k));
      ps = sampleuni(f,[a b],m(k),1);
      p = interpuni(ps,[a b]);
      plotfun(f,[a b],'displayname','f');
      plotsample(griduni([a b],m(k),1),ps);
      plotfun(p,[a b],'displayname','p');
      ylim([-2 2]);
      xlabel('x');
   end
```

As the degree m increases, the quality of the interpolation worsens outside of a central interval, contrary to what we might expect. ▪

97

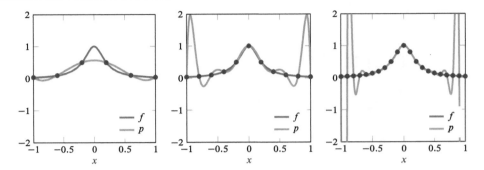

Figure 9.1. *Polynomial interpolants for Runge's function using uniform grids of degrees $m = 5$ (left), $m = 10$ (middle), and $m = 20$ (right).*

The example illustrates the *Runge phenomenon*, characterized by wild oscillations at the ends of an interpolant that become more severe as the degree is increased. More work leads to poorer results. This is unacceptable.

Why does the Runge phenomenon occur? It is important to realize that Runge's function is not as benign as it first seems. To see this requires a peek into the complex plane and a hunt for *singularities*, points where a function is undefined or not complex differentiable. (See Appendix B for a brief primer on complex functions.)

Example 9.2. Plot the real and imaginary parts of Runge's function $f(z) = 1/(1 + 25z^2)$ over the domain $z = x + yi$ with $-1 \le x \le 1$, $-0.4 \le y \le 0.4$. Identify any singularities.

Solution. The real and imaginary parts are graphed in Figure 9.2 using new routines `plotrealpart` and `plotimagpart`.

```
>> f = @(z) 1/(1+25*z^2);
>> ax = newfig(1,2);
>> legend hide;
>> plotrealpart(f,[-1 1 -0.4 0.4 -2 2]);
>> view(-25,30);
>> xlabel('x'); ylabel('y');
>> subplot(ax(2));
>> legend hide;
>> plotimagpart(f,[-1 1 -0.4 0.4 -2 2]);
>> view(-25,30);
>> xlabel('x'); ylabel('y');
```

In the above code, the vector `[-1 1 -0.4 0.4 -2 2]` sets the limits $[-1, 1]$ on the x-axis, $[-0.4, 0.4]$ on the y-axis, and $[-2, 2]$ on the vertical axis.

Note that the hump shape of Runge's function on the real line, familiar from Figure 9.1, is visible along the slice $y = 0$. This portion of the graph is drawn with a heavy pen for emphasis.

```
>> subplot(ax(1));
>> plotfun3(@(x) x,@(x) 0,@(x) real(f(x)),[-1 1],'k','linewidth',3);
>> subplot(ax(2));
>> plotfun3(@(x) x,@(x) 0,@(x) imag(f(x)),[-1 1],'k','linewidth',3);
```

Runge's function has exactly two singularities. Specifically, at $z = \pm(1/5)i$, the function has the form $f(\pm(1/5)i) = 1/0$, and in a neighborhood of either point the function is unbounded. The plots show infinite discontinuities called *poles* at these points. ■

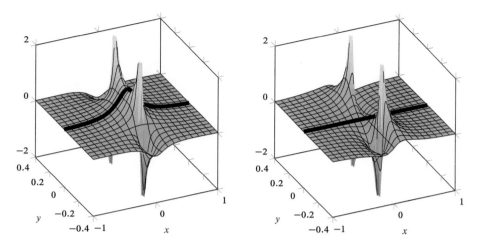

Figure 9.2. *The real (left) and imaginary (right) parts of Runge's function* $f(z) =$ $1/(1 + 25z^2)$ *in the complex plane.*

Although Runge's function looks well behaved on the real line, sea monsters lurk nearby in the complex plane. The function is undefined at $z = \pm(1/5)\mathrm{i}$, and its absolute value diverges to infinity as either of these points is approached. Perhaps a singularity near the interpolation interval can cause problems for polynomial interpolation, even if that singularity lies off the real line.

Indeed, the Runge phenomenon occurs when a singularity is too close to the interpolation interval (and a uniform grid of high degree is used). Exactly how close is too close is determined by a particular curve in the complex plane that we call the *Runge curve*. When there is a singularity enclosed by the Runge curve, the Runge phenomenon may occur. Although we do not have a simple formula for this curve, it can be drawn with the routine rungecurve demonstrated in the following example. The Runge curve does not depend on the degree m of the grid, and the curve for $[a, b]$ is a simple translation and rescaling of the one for $[-1, 1]$, so once you've seen a single Runge curve, you've essentially seen them all.

Example 9.3. Plot the Runge curve for the interval $[-1, 1]$ and the locations of the singularities of Runge's function. Comment on the implications for interpolation with a uniform grid of high degree.

Solution. The Runge curve is plotted in Figure 9.3.

```
>> a = -1;
>> b = 1;
>> newfig; legend hide;
>> rungecurve([a b]);
>> axis([-2 2 -2 2]);
>> axis square; grid on;
>> xlabel('real'); ylabel('imag');
```

Runge's function $f(z) = 1/(1 + 25z^2)$ has singularities at $\pm(1/5)\mathrm{i}$. These are marked by $\times$'s in the plot.

```
>> plot([0 0],[-1/5 1/5],'x');
```

Because the singularities lie within the Runge curve, interpolation of $f(x) = 1/(1 + 25x^2)$ is susceptible to the Runge phenomenon. As we saw in Example 9.1, the Runge phenomenon does indeed occur. ∎

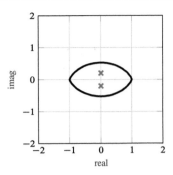

Figure 9.3. *The Runge curve for* $[-1, 1]$ *and the singularities of Runge's function.*

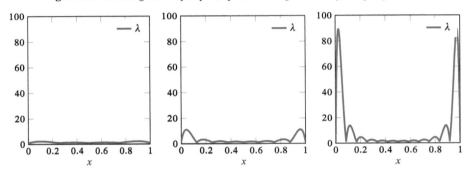

Figure 9.4. *Lebesgue function for a uniform grid of degree m:* $m = 4$ *(left),* $m = 8$ *(middle), and* $m = 12$ *(right).*

Note that the Runge curve for $[a, b]$ intersects the points a and b on the real line.

9.2 ▪ The Lebesgue constant for uniform grids

The Runge phenomenon isn't the end of the story. Just as damaging is the Lebesgue constant for a uniform grid of high degree. Recall that the Lebesgue constant was introduced in advance of Theorem 6.11 to quantify the effect of sampling errors on an interpolant. The conclusion was that sampling errors of size δ perturb the entire interpolant proportionally, up to $\Lambda\delta$ in the infinity norm, and the constant of proportionality Λ is the Lebesgue constant. Unfortunately, the Lebesgue constant for a high-degree uniform grid is huge, as proved below, so even tiny errors at the interpolation nodes can be blown into huge errors between the nodes.

The Lebesgue constant Λ for a grid is

$$\Lambda = \max_{a \leq x \leq b} \lambda(x),$$

in which $\lambda(x)$ is the Lebesgue function defined in (6.9). The Lebesgue functions for three uniform grids are plotted in Figure 9.4. The Lebesgue functions, and therefore the Lebesgue constants, appear to grow very quickly with m, a fact established by the following theorem.

Theorem 9.4. *The Lebesgue constant* Λ_m *for a uniform grid of degree* $m \geq 1$ *satisfies*

$$\Lambda_m \geq \frac{2^m}{4m^2}. \tag{9.1}$$

Proof. The Lebesgue constant for a uniform grid of degree m is independent of the interval on which the grid is placed (Exercise 18 in Chapter 8). For convenience, place the grid on $[a, b] = [0, m]$, so that the nodes are $x_i = i, i = 0, \dots, m$.

The ith Lagrange basis polynomial is, when $x \neq i$,

$$l_i(x) = \frac{\prod_{k=0}^{m}(x - x_k)}{x - x_i} = \frac{\prod_{k=0}^{m}(x - k)}{x - i},$$

and moreover,

$$l_i(x_i) = l_i(i) = \prod_{k=0}^{i-1}(i - k) \prod_{k=i+1}^{m}(i - k)$$

$$= i(i - 1) \cdots (1) \times (-1)(-2) \cdots (-(m - i)) = i! \times (-1)^{m-i}(m - i)!.$$

The Lebesgue constant is the maximum of the Lebesgue function on $[a, b] = [0, m]$. Figure 9.4 suggests an investigation near one of the endpoints. It will prove sufficient to concentrate on $x = 1/2$, halfway between the first two nodes. We find

$$l_i(1/2) = \frac{\left(\frac{1}{2} - 0\right)\left(\frac{1}{2} - 1\right)\prod_{k=2}^{m}\left(\frac{1}{2} - k\right)}{\frac{1}{2} - i} = (-1)^{m-1}\frac{\prod_{k=2}^{m}\left(k - \frac{1}{2}\right)}{4\left(i - \frac{1}{2}\right)}.$$

Therefore,

$$|l_i(1/2)| \geq \frac{\prod_{k=2}^{m}(k - 1)}{4m} = \frac{(m - 1)!}{4m} = \frac{m!}{4m^2}.$$

The Lebesgue function at $x = 1/2$ is bounded as follows:

$$\lambda(1/2) = \sum_{i=0}^{m}\left|\frac{l_i(1/2)}{l_i(x_i)}\right| \geq \sum_{i=0}^{m}\frac{m!/(4m^2)}{i!(m - i)!} = \frac{1}{4m^2}\sum_{i=0}^{m}\binom{m}{i},$$

in which $\binom{m}{i} = m!/(i!(m - i)!)$ is a binomial coefficient. The above sum of binomial coefficients equals 2^m. (The binomial coefficient $\binom{m}{i}$ counts the number of subsets of $\{1, \dots, m\}$ of size i. The sum therefore counts the number of subsets of any size, which equals 2^m.) Therefore,

$$\lambda(1/2) \geq \frac{2^m}{4m^2}.$$

The lower bound on $\lambda(1/2)$ provides a lower bound on the Lebesgue constant:

$$\Lambda_m = \max_{0 \leq x \leq m} \lambda(x) \geq \lambda(1/2) \geq \frac{2^m}{4m^2}. \qquad \square$$

The inequality $\Lambda_m \geq 2^m/(4m^2)$ shows that the Lebesgue constant grows very quickly—exponentially—with the degree of a uniform grid. For example, when $m = 64$, the Lebesgue constant is greater than 10^{15}.

Example 9.5. Interpolate $f(x) = e^{-x^2}$, $-6 \leq x \leq 6$, with uniform grids of increasing degree. What happens?

Solution. Three interpolants are computed and graphed in Figure 9.5.

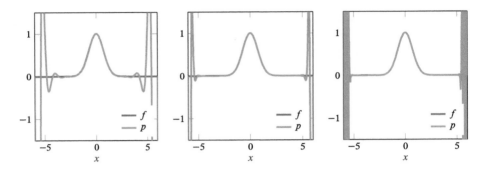

Figure 9.5. *Polynomial interpolant for e^{-x^2} on a uniform grid of degree m: $m = 20$ (left), $m = 40$ (middle), and $m = 80$ (right).*

```
>> f = @(x) exp(-x^2);
>> a = -6;
>> b = 6;
>> m = [20 40 80];
>> ax = newfig(1,3);
>> for k = 1:length(m)
     subplot(ax(k));
     ps = sampleuni(f,[a b],m(k),1);
     p = interpuni(ps,[a b]);
     plotfun(f,[a b],'displayname','f');
     plotfun(p,[a b],'displayname','p');
     xlabel('x');
     ylim([-1.5 1.5]);
   end
```

The graphed interpolants are highly inaccurate, and there is little hope that the accuracy would improve with higher degrees because the Lebesgue constant would only grow larger. ■

Note that the behavior in the previous example is not the Runge phenomenon. The Runge phenomenon occurs when a smooth interpolant tries to fit a singularity. When this happens, the polynomial interpolants diverge in theory, before roundoff error ever comes into play. In contrast, the function e^{-x^2} of the previous example is analytic (smooth, with no singularities) over the entire complex plane. In theory, the interpolants should converge to the original function, but in practice the sensitivity to tiny sampling errors quantified by the Lebesgue constant causes the numerical computations to become inaccurate.

9.3 ▪ Conclusion

The lesson of this chapter is that polynomial interpolation performs poorly on a uniform grid of high degree. This motivates the study of nonuniform grids beginning in the next chapter.

Notes

Runge described the phenomenon named after him in 1901 [64], using the function $1/(1+x^2)$ on $[-5, 5]$. Méray predicted the Runge phenomenon as early as 1884 [54] and considered Runge's function in 1896 [55]. The curve that we call the Runge curve is one of a family of level curves that Runge called *U-curves* in his original work. Epperson provides a summary of Runge's work as well as related original results [29]. The claim after Example 9.5 that the interpolants converge to $f(x) = e^{-x^2}$ is justified by the proposition on p. 331 of Epperson's article.

An article by Trefethen and Weideman discusses the history of the Lebesgue constant for a uniform grid [78]. The proof of Theorem 9.4 is based on a proof of a stronger result in Fornberg's book [32, pp. 171–172].

Exercises

Exercises 1–10: Interpolate the given function with uniform grids of increasing degree. Do the interpolants converge numerically to an accurate approximation? If so, provide evidence. If not, show what happens.

1. $1/x$, $1 \leq x \leq 5$
2. $1/(1 + 2x^2)$, $-1 \leq x \leq 1$
3. $1/(1 + x^4)$, $-5 \leq x \leq 5$
4. $\tanh x = (e^x - e^{-x})/(e^x + e^{-x})$, $-3 \leq x \leq 3$
5. $\tanh x$, $-5 \leq x \leq 5$
6. $\operatorname{erf} x$, $-4 \leq x \leq 4$
7. $\operatorname{erf} x$, $-6 \leq x \leq 6$
8. $\tan x$, $-\pi/3 \leq x \leq \pi/3$
9. $\arctan x$, $-2 \leq x \leq 2$
10. $\arctan x$, $-6 \leq x \leq 6$

Exercises 11–18: Locate the singularities, if any, of the given function in the complex plane.

11. $z/(z^2 - 2z + 2)$
12. $(z + 3)/(z^4 + 2)$
13. $\log z$
14. $\log(z^2 + 1)$
15. $\sqrt{z}$
16. $\sqrt{1 - z^2}$
17. e^z
18. e^{z^2}

Exercises 19–24: A function f and a rectangle $[a, b] \times [c, d]$ are given. Plot the real and imaginary parts of the function over $\{x + iy : a \leq x \leq b, c \leq y \leq d\}$ and locate any singularities in the rectangle.

19. $1/(z^3 - z^2 + z - 1)$, $[-2, 2] \times [-2, 2]$
20. $1/(z^4 - 2z^3 + 3z^2 - 2z + 2)$, $[-2, 2] \times [-2, 2]$
21. $\tan(\pi z)$, $[-2, 2] \times [-2, 2]$
22. $\tanh(\pi z)$, $[-2, 2] \times [-2, 2]$
23. $\arcsin$, $[-5, 5] \times [-2, 2]$
24. $\arctan$, $[-2, 2] \times [-5, 5]$

25. The Runge phenomenon is related to the notion of radius of convergence for Taylor series.

    (a) Find the radius of convergence for the Maclaurin series for $\arctan x$.

    (b) Find the smallest Runge curve centered at the origin that intersects singularities of $\arctan z$. Estimate the points where the Runge curve intersects the real axis to two significant digits.

    (c) If you attempted to approximate $\arctan x$ on larger and larger intervals $[-a, a]$ with a Taylor polynomial or by interpolation on a uniform grid, would divergence of the Taylor series or the Runge phenomenon become a problem first? *Note.* This says nothing of the Lebesgue constant, which is also a concern.

26. Let Λ_m be the Lebesgue constant for a uniform grid of degree m. Prove

$$\lim_{m \to \infty} \Lambda_m^{1/m} = 2.$$

    (Hence, Λ_m is on the order of 2^m for large m.) *Hint.* Find and prove an upper bound analogous to (9.1) and then show that the mth roots of the upper and lower bounds both converge to 2 as $m \to \infty$.

27. The gamma function $\Gamma(x)$ is an extension of the factorial function from nonnegative integers to real and complex numbers. Using the identity $\Gamma(x + 1) = x\Gamma(x)$, which holds for $x \neq 0, -1, -2, \ldots$, prove that the Lagrange basis polynomial in the proof of Theorem 9.4 satisfies

$$l_i(x) = (-1)^m \frac{x\Gamma(m + 1 - x)}{\Gamma(1 - x)} \frac{1}{x - i}$$

    for $x \neq i$.

28. This exercise considers the barycentric weights for a uniform grid.

    (a) For a given degree m, which barycentric weights have the largest magnitude? Which have the smallest magnitude? For a uniform grid of degree 20, what is the ratio between these weights?

    (b) Let $p(x)$ be an interpolating polynomial defined by $p(x_i) = y_i, i = 0, \ldots,$ 20. Suppose a single sample value y_i were perturbed to $\hat{y}_i$. Based on the barycentric weights, a perturbation in which sample value would have the most drastic effect on the entire interpolant? Which would produce the least change? Explain why, and then demonstrate the effect graphically.

Chapter 10

Chebyshev interpolation

We saw in the previous chapter that interpolation on a high-degree grid can go horribly wrong, with the interpolant experiencing wild oscillations near its ends. Fortunately, there is a surprisingly simple rehabilitation. The strategy is to pack extra nodes near the endpoints of the interpolation interval, clamping the polynomial extra firmly at its extremities.

10.1 ▪ Chebyshev grids

An especially effective grid on $[-1, 1]$ is the *Chebyshev grid*

$$x_i = -\cos\frac{i\pi}{m}, \quad i = 0, \ldots, m. \tag{10.1}$$

This grid can be generated by placing points uniformly along the upper half of the unit circle and then dropping the points to the horizontal axis. See Figure 10.1 and notice how the Chebyshev nodes are packed more tightly near the endpoints.

The effectiveness of the Chebyshev grid is thanks to complex analysis, specifically a mapping from the unit circle in the complex plane to the interval $[-1, 1]$ on the real line. Although the full story is outside the scope of this book, we will consider the complex plane in the next chapter.

On an arbitrary interval $[a, b]$, the Chebyshev grid of degree m is defined to be

$$x_i = \frac{a+b}{2} - \frac{b-a}{2}\cos\frac{i\pi}{m}, \quad i = 0, \ldots, m. \tag{10.2}$$

Note that (10.1) and (10.2) are identical when $[a, b] = [-1, 1]$.

The barycentric weights for a Chebyshev grid are remarkably simple.

Theorem 10.1. *A sequence of barycentric weights for a Chebyshev grid of degree m is given by*

$$w_i = \begin{cases} (-1)^i & \text{if } i = 1, \ldots, m-1, \\ (-1)^i/2 & \text{if } i = 0, m. \end{cases} \tag{10.3}$$

The proof is at the end of this chapter.

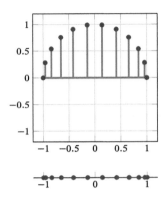

Figure 10.1. *A uniform grid on the upper semicircle (top) and a Chebyshev grid on the real line (bottom).*

The nodes and barycentric weights of a Chebyshev grid are computed by gridcheb, shown in Listing 10.1.

Listing 10.1. Construct a Chebyshev grid.

```
function [xs,ws] = gridcheb(ab,m)

a = ab(1);
b = ab(2);
xs = [ a; (a+b)/2-(b-a)/2*cos((1:m-1)'*pi/m); b ];
ws = [ 1/2; (-1).^(1:m-1)'; (1/2)*(-1)^m ];
```

10.2 ▪ Chebyshev interpolation

A polynomial interpolant on a Chebyshev grid is called a *Chebyshev interpolating polynomial* or *Chebyshev interpolant*.

Just as in Chapter 6, the interpolation nodes are conveniently placed in a vector **x**, and a sample of the function f is denoted $f|_\mathbf{x}$:

$$\mathbf{x} = \begin{bmatrix} x_0 \\ \vdots \\ x_m \end{bmatrix}, \quad f|_\mathbf{x} = \begin{bmatrix} f(x_0) \\ \vdots \\ f(x_m) \end{bmatrix}.$$

These data uniquely determine a polynomial interpolant of degree at most m. The difference from Chapter 6 is that we are now considering the more specific case in which **x** is a Chebyshev grid.

The routine samplecheb samples a function on a Chebyshev grid, and the routine interpcheb constructs a Chebyshev interpolant. See Listings 10.2 and 10.3.

Chebyshev interpolation can be highly accurate, as our next example suggests.

Example 10.2. Interpolate $f(x) = \sin x$, $0 \le x \le 2\pi$, using a Chebyshev grid of degree $m = 10$. Plot the interpolant p and its absolute error.

Solution. The interpolant is constructed.

Listing 10.2. Sample a function on a Chebyshev grid.

```
function ys = samplecheb(f,ab,m)

xs = gridcheb(ab,m);
ys = arrayfun(f,xs);
```

Listing 10.3. Construct a polynomial interpolant on a Chebyshev grid.

```
function p = interpcheb(ps,ab)

m = size(ps,1)-1;
[xs,ws] = gridcheb(ab,m);
p = @(x) interp_(xs,ws,ps,x);
```

```
>> f = @(x) sin(x);
>> a = 0;
>> b = 2*pi;
>> m = 10;
>> ps = samplecheb(f,[a b],m);
>> p = interpcheb(ps,[a b]);
```
The original function and the interpolant are plotted in Figure 10.2.
```
>> ax = newfig(1,2);
>> plotfun(f,[a b],'displayname','f');
>> xs = gridcheb([a b],m);
>> plotsample(xs,ps);
>> plotfun(p,[a b],'displayname','p');
>> xlabel('x');
```
Note how the sample points are denser near the endpoints $x = 0, 2\pi$. Also note that the fit is close enough that the original function disappears behind the interpolant.

An absolute error plot is also shown in Figure 10.2. The interpolant is quite accurate considering the small size of the sample.
```
>> subplot(ax(2));
>> ylog; ylim([1e-18 1e2]);
>> plotfun(@(x) abs(f(x)-p(x)),[a b],'displayname','|f-p|');
>> xlabel('x');                                                                  ∎
```

When we look at rate of convergence, the potential of Chebyshev interpolation is even more evident. Note that for a Chebyshev grid of degree m, the sampling cost is $m + 1$. In the following example, absolute error is plotted against the degree m to observe rate of convergence.

Example 10.3. Returning to $f(x) = \sin x$, $0 \le x \le 2\pi$, plot $\|f - p\|_\infty$ against the degree of the polynomial. Classify the rate of convergence as algebraic, geometric, or supergeometric if possible.

Solution. The function is defined.
```
>> f = @(x) sin(x);
>> a = 0;
>> b = 2*pi;
```
The errors are computed for $m = 2, 4, 6, \ldots, 26$.
```
>> m = 2:2:26;
```

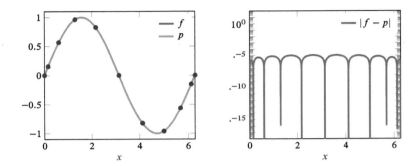

Figure 10.2. *A Chebyshev interpolant for* sin *x (left) and its absolute error (right).*

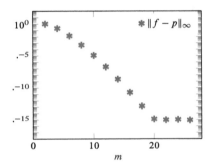

Figure 10.3. *Rate of convergence for Chebyshev interpolation of* sin x, $0 \leq x \leq 2\pi$.

```
>> err = nan(size(m));
>> for k = 1:length(m)
     ps = samplecheb(f,[a b],m(k));
     p = interpcheb(ps,[a b]);
     err(k) = infnorm(@(x) f(x)-p(x),[a b],100);
   end
```

The errors are plotted against m in Figure 10.3.

```
>> newfig;
>> ylog; ylim([1e-18 1e2]);
>> plot(m,err,'*','displayname','||f-p||');
>> xlim([0 28]);
>> xlabel('m');
```

Because the errors appear to approach zero faster than a straight line on a log-linear graph (before roundoff error becomes significant), the rate of convergence appears to be supergeometric. ■

Chebyshev interpolation commonly converges at a geometric or supergeometric rate. See Chapter 11.

10.3 ▪ Runge phenomenon

Chebyshev interpolation was motivated in part by the Runge phenomenon. The next two examples return to Runge's function.

Example 10.4. Apply Chebyshev interpolation to Runge's function with $m = 5, 10, 20$. Does the Runge phenomenon appear?

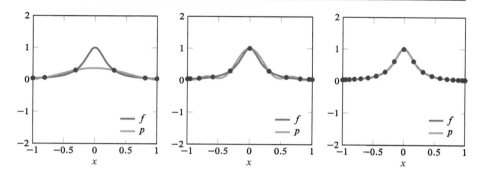

Figure 10.4. *Chebyshev interpolants for Runge's function: degree m = 5 (left), m = 10 (middle), and m = 20 (right).*

Solution. Runge's function is $f(x) = 1/(1 + 25x^2)$.

```
>> f = @(x) 1/(1+25*x^2);
>> a = -1;
>> b = 1;
```

The three requested Chebyshev interpolants are plotted in Figure 10.4.

```
>> m = [5 10 20];
>> ax = newfig(1,3);
>> for k = 1:length(m)
     ps = samplecheb(f,[a b],m(k));
     p = interpcheb(ps,[a b]);
     subplot(ax(k));
     plotfun(f,[a b],'displayname','f');
     xs = gridcheb([a b],m(k));
     plotsample(xs,ps);
     plotfun(p,[a b],'displayname','p');
     ylim([-2 2]);
     xlabel('x');
   end
```

There is no evidence of the Runge phenomenon. ∎

Example 10.5. Determine whether Chebyshev interpolation converges numerically on Runge's function. Classify the rate of convergence as algebraic, geometric, or super-geometric if possible.

Solution. The absolute interpolation error is plotted against m in Figure 10.5. Numerical convergence occurs around $m = 180$.

```
>> f = @(x) 1/(1+25*x^2);
>> a = -1;
>> b = 1;
>> m = 20:20:240;
>> err = nan(size(m));
>> for k = 1:length(m)
     ps = samplecheb(f,[a b],m(k));
     p = interpcheb(ps,[a b]);
     err(k) = infnorm(@(x) f(x)-p(x),[a b],100);
   end
>> newfig;
>> ylog; ylim([1e-18 1e2]);
>> plot(m,err,'*','displayname','||f-p||');
```

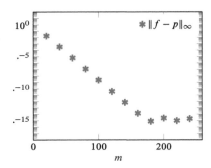

Figure 10.5. *Rate of convergence for Chebyshev interpolation on Runge's function.*

```
>> xlim([0 260]);
>> xlabel('m');
```
Because the error follows a straight line on a log-linear plot until numerical convergence, the rate of convergence appears to be geometric. ∎

Chebyshev interpolation succeeds on Runge's function! Chapter 11 argues that the Runge phenomenon is defeated more generally.

How can a Chebyshev grid succeed where a uniform grid fails? A partial explanation involves Theorem 6.6, the interpolation error theorem. Recall that if $p(x)$ is a polynomial interpolant for $f(x)$, then

$$f(x) - p(x) = \frac{f^{(m+1)}(\eta_x)}{(m+1)!} \prod_{i=0}^{m} (x - x_i) \tag{10.4}$$

for some $\eta_x \in [a, b]$. The component under control of the numerical analyst is the monic polynomial

$$l(x) = \prod_{i=0}^{m} (x - x_i), \tag{10.5}$$

which is determined by the choice of nodes $x_0, \ldots, x_m$. (A polynomial is monic if its leading coefficient equals 1.) For a uniform grid, $l(x)$ becomes relatively large in magnitude near the extreme nodes. For a Chebyshev grid, $l(x)$ stays uniformly small over the entire interval. The next example illustrates the difference between a uniform grid and a Chebyshev grid in this regard.

Example 10.6. Plot $l(x)$ for a uniform grid of degree 15 and then for a Chebyshev grid of the same degree. Compare.

Solution. The uniform grid and its monic polynomial are plotted in the left graph of Figure 10.6.
```
>> m = 15;
>> xsuni = griduni([-1 1],m,1);
>> ax = newfig(1,2);
>> legend hide;
>> plotsample(xsuni,zeros(size(xsuni)));
>> plotfun(@(x) prod(x-xsuni),[-1 1]);
>> axis([-1 1 -1.5e-3 1.5e-3]);
>> xlabel('x');
```

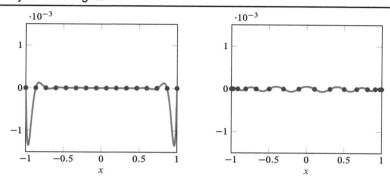

Figure 10.6. *The monic polynomial* $l(x) = \prod_{i=0}^{m}(x - x_i)$ *for a uniform grid (left) and a Chebyshev grid (right).*

The Chebyshev grid and its monic polynomial are plotted in the right graph.

```
>> xscheb = gridcheb([-1 1],m);
>> subplot(ax(2));
>> legend hide;
>> plotsample(xscheb,zeros(size(xscheb)));
>> plotfun(@(x) prod(x-xscheb),[-1 1]);
>> axis([-1 1 -1.5e-3 1.5e-3]);
>> xlabel('x');
```

For the uniform grid, the monic polynomial shows the large oscillation characteristic of the Runge phenomenon. For the Chebyshev grid, the monic polynomial stays small throughout the interval. ∎

The polynomial $l(x)$ equals zero at every interpolation node x_i, regardless of the nodes' placement. For a Chebyshev grid, the polynomial stays close to zero between the nodes as well, which in turn helps to keep the interpolation error small.

10.4 ▪ Proof of barycentric weights

Our proof of Theorem 10.1, on the barycentric weights for a Chebyshev grid, relies on the *Chebyshev polynomials of the second kind,*

$$U_m(x) = \frac{\sin((m + 1)\cos^{-1} x)}{\sqrt{1 - x^2}}, \tag{10.6}$$

which are defined for $m = 0, 1, 2, 3, \ldots$. L'Hôpital's rule shows that $\lim_{x \to -1} U_m(x) = (-1)^m(m + 1)$ and $\lim_{x \to 1} U_m(x) = m + 1$, so to enforce continuity, the values at the endpoints and defined to be

$$U_m(-1) = (-1)^m(m + 1), \tag{10.7}$$
$$U_m(1) = m + 1. \tag{10.8}$$

Proposition 10.7. *The function* $U_m(x)$ *is a polynomial of degree m on $[-1, 1]$ with leading coefficient 2^m.*

Proof. Let $\theta = \cos^{-1} x$, and note that $\sin\theta = \sqrt{1 - x^2}$. The proof is by induction. There are two base cases. First,

$$U_0(x) = U_0(\cos\theta) = \frac{\sin\theta}{\sin\theta} = 1.$$

Second, $U_1(x)$ can be simplified using the identity $\sin(2\theta) = 2\sin(\theta)\cos(\theta)$:

$$U_1(x) = U_1(\cos\theta) = \frac{\sin(2\theta)}{\sin\theta} = \frac{2\sin(\theta)\cos(\theta)}{\sin\theta} = 2\cos\theta = 2x.$$

Thus, the proposition is proved for $m = 0, 1$.

For the induction step, suppose that $U_i(x)$ is a polynomial of degree i with leading coefficient 2^i for $i = 0, \ldots, m - 1$. The next function in the sequence is simplified using the trigonometric identity

$$\sin((m + 1)\theta) = 2\cos(\theta)\sin(m\theta) - \sin((m - 1)\theta). \tag{10.9}$$

We find

$$U_m(x) = U_m(\cos\theta) = \frac{\sin((m + 1)\theta)}{\sin\theta} = \frac{2\cos(\theta)\sin(m\theta) - \sin((m - 1)\theta)}{\sin\theta}$$

$$= 2x\frac{\sin(m\cos^{-1}x)}{\sqrt{1 - x^2}} - \frac{\sin((m - 1)\cos^{-1}x)}{\sqrt{1 - x^2}} = 2xU_{m-1}(x) - U_{m-2}(x).$$

Because $U_{m-1}(x)$ and $U_{m-2}(x)$ are polynomials of degrees $m - 1$ and $m - 2$, respectively, the combination $2xU_{m-1}(x) - U_{m-2}(x)$ is a polynomial of degree m. The leading coefficient of $U_m(x)$ is twice that of $U_{m-1}(x)$, hence 2^m.   □

The Chebyshev polynomials of the second kind are related to the barycentric weights through the monic polynomial $l(x)$ of (10.5).

Lemma 10.8. *If $x_0, \ldots, x_m$ are the nodes of the Chebyshev grid of degree m on $[-1, 1]$, then*

$$l(x) = 2^{-m+1}(x^2 - 1)U_{m-1}(x). \tag{10.10}$$

Proof. Note that

$$\frac{l(x)}{x^2 - 1} = \prod_{i=1}^{m-1}(x - x_i)$$

and $2^{-m+1}U_{m-1}(x)$ are both monic polynomials of degree $m - 1$. Hence, to show that these functions are equal, it suffices to show that they have identical zeros. Of course, the zeros of $l(x)/(x^2 - 1)$ are at $x = x_i$, $i = 1, \ldots, m - 1$. Because

$$\sin(m\cos^{-1}x_i) = \sin\left(m\left(\pi - \frac{i\pi}{m}\right)\right) = \sin((m - i)\pi) = 0,$$

it is true that $2^{-m+1}U_{m-1}(x_i) = 0$ as well.   □

Our technique for computing the barycentric weight $w_i = C/l_i(x_i)$ is to express the Lagrange basis polynomial $l_i(x)$ as

$$l_i(x) = \prod_{k\neq i}(x - x_k) = \frac{l(x)}{x - x_i}$$

for $x \neq x_i$ and to take the limit $x \to x_i$ to recover $l_i(x_i)$. The derivation of the limit uses l'Hôpital's rule, and hence we need to understand the derivative of $U_m(x)$. The derivative can be expressed with the help of a *Chebyshev polynomial of the first kind*,

$$T_m(x) = \cos(m \cos^{-1} x).$$

Proposition 10.9. *We have*

$$U_m'(x) = \frac{m+1}{x^2-1} T_{m+1}(x) - \frac{x}{x^2-1} U_m(x).$$

Proof. Let $\theta = \cos^{-1} x$. Basic calculus shows

$$\frac{d}{d\theta}[U_m(\cos\theta)] = \frac{d}{d\theta} \frac{\sin((m+1)\theta)}{\sin\theta} = \frac{(m+1)\cos((m+1)\theta)}{\sin\theta} - \frac{\sin((m+1)\theta)\cos(\theta)}{\sin^2\theta}$$

$$= \frac{m+1}{\sin\theta} T_{m+1}(\cos\theta) - \frac{\cos\theta}{\sin\theta} U_m(\cos\theta).$$

Thus,

$$U_m'(x) = \frac{d}{d\theta}[U_m(\cos\theta)] \frac{1}{dx/d\theta} = \frac{m+1}{-\sin^2\theta} T_{m+1}(\cos\theta) - \frac{\cos\theta}{-\sin^2\theta} U_m(\cos\theta)$$

$$= \frac{m+1}{x^2-1} T_{m+1}(x) - \frac{x}{x^2-1} U_m(x). \qquad \square$$

Finally, we can derive the barycentric weights.

Proof of Theorem **10.1.** Assume that the grid is on the interval $[-1, 1]$. The proof for intervals other than $[-1, 1]$ is Exercise 12.

The ith barycentric weight is given by the expression $C/l_i(x_i)$. Because $l_i(x)$ is continuous and

$$l_i(x) = \frac{l(x)}{x - x_i}, \quad x \neq x_i,$$

we have

$$l_i(x_i) = \lim_{x \to x_i} \frac{l(x)}{x - x_i} = \lim_{x \to x_i} \frac{2^{-m+1}(x^2-1)U_{m-1}(x)}{x - x_i}.$$

By l'Hôpital's rule,

$$l_i(x_i) = 2^{-m+1}(2x_i U_{m-1}(x_i) + (x_i^2 - 1)U_{m-1}'(x_i)).$$

Applying Proposition 10.9, we find

$$l_i(x_i) = 2^{-m+1}(m T_m(x_i) + x_i U_{m-1}(x_i)). \tag{10.11}$$

For $i = 0, \dots, m$, we have

$$T_m(x_i) = \cos(m \cos^{-1}(-\cos(i\pi/m))) = \cos((m-i)\pi) = (-1)^{m-i}.$$

For $i = 1, \dots, m-1$, we have

$$U_{m-1}(x_i) = \frac{\sin(m \cos^{-1}(-\cos(i\pi/m)))}{\sqrt{1 - \cos(i\pi/m)^2}} = \frac{\sin((m-i)\pi)}{\sqrt{1 - \cos(i\pi/m)^2}} = 0.$$

At the extreme nodes $x_0 = -1$ and $x_m = 1$, the values are, according to (10.7)–(10.8), $U_{m-1}(x_0) = (-1)^{m-1}m$ and $U_{m-1}(x_m) = m$. Plugging the specific values for $T_m(x_i)$ and $U_{m-1}(x_i)$ into (10.11) gives

$$l_i(x_i) = \begin{cases} 2^{-m+1}m(-1)^m \times 2 & \text{if } i = 0, \\ 2^{-m+1}m(-1)^{m-i} & \text{if } i = 1, \ldots, m-1, \\ 2^{-m+1}m \times 2 & \text{if } i = m. \end{cases}$$

Let $C = (-1)^m 2^{-m+1}m$. Then

$$\frac{C}{l_i(x_i)} = \begin{cases} 1/2 & \text{if } i = 0, \\ (-1)^i & \text{if } i = 1, \ldots, m-1, \\ (-1)^m/2 & \text{if } i = m. \end{cases}$$

This is equivalent to the expression for w_i in (10.3). $\square$

10.5 ▪ Piecewise-Chebyshev interpolation

We will see in the next chapter that Chebyshev interpolation converges relatively slowly when a function fails to be smooth everywhere. In this situation, points where the function is not smooth can be isolated with a *piecewise-Chebyshev grid*.

Let $[a_j, b_j]$, $j = 1, \ldots, n$, be a partition of an interval $[a, b]$. The subintervals are not required to be of identical width. A piecewise-Chebyshev grid of degree m on the partition consists of a sequence of Chebyshev grids of degree m, one on each subinterval. A piecewise-polynomial function in which the jth piece is defined by its values on the jth Chebyshev grid is a *piecewise-Chebyshev interpolant*.

The provided function gridchebpw constructs a piecewise-Chebyshev grid, and functions samplechebpw and interpchebpw sample and interpolate, respectively, on such a grid. Each of these routines requires an argument asbs that specifies the partition $[a_j, b_j]$, $j = 1, \ldots, n$, as a 2-by-n matrix:

$$\begin{bmatrix} a_1 & a_2 & \cdots & a_n \\ b_1 & b_2 & \cdots & b_n \end{bmatrix}.$$

All three routines are demonstrated in the next example.

Example 10.10. Interpolate $f(x) = |\sin(\pi x^2)|$, $0 \le x \le \sqrt{3}$, using a piecewise-Chebyshev interpolant. Partition the interval into three subintervals $[0, 1]$, $[1, \sqrt{2}]$, and $[\sqrt{2}, \sqrt{3}]$, and use a polynomial of degree 4 on each subinterval. Plot the interpolant, showing the partition and sample points.

Solution. The function is sampled, and the interpolant is constructed.

```
>> f = @(x) abs(sin(pi*x^2));
>> a = 0;
>> b = sqrt(3);
>> asbs = [ 0 1; 1 sqrt(2); sqrt(2) sqrt(3) ]'
asbs =
                        0    1.000000000000000    1.414213562373095
        1.000000000000000    1.414213562373095    1.732050807568877
>> m = 4;
```

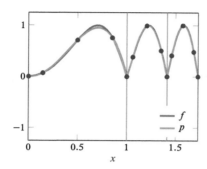

Figure 10.7. *A piecewise-Chebyshev interpolant.*

```
>> ps = samplechebpw(f,asbs,m)
ps =
                    0    0.000000000000000    0.000000000000002
   0.067325545123602    0.382683432365089    0.408130155815340
   0.707106781186547    0.990934541947620    0.996854127758270
   0.753108748852774    0.503333358969775    0.479202841429385
   0.000000000000000    0.000000000000002    0.000000000000002
>> p = interpchebpw(ps,asbs);
```

Next, it is plotted.

```
>> newfig;
>> plotfun(f,[a b],'displayname','f');
>> plotfun(p,[a b],'displayname','p');
>> plotpartition(asbs);
>> xs = gridchebpw(asbs,m);
>> plotsample(xs,ps);
>> ylim([-1.25 1.25]);
>> xlabel('x');
```

The plot is shown in Figure 10.7.   ∎

Locating and isolating singularities by hand is often impractical. Exercise 30 considers an *adaptive* scheme for automatically partitioning an interval.

Notes

The story of Chebyshev grids begins with a pair of papers from the 1850's [14, 15]. Chebyshev was particularly concerned with the problem of best polynomial approximation, in the sense of choosing a polynomial p among all degree-m polynomials to minimize $\|f - p\|_\infty$. Interpolation through a grid of Chebyshev nodes produces a polynomial whose fit is nearly optimal in this sense, as proved by Ehlich and Zeller in 1966 [27]. Interpolation is easier to implement and faster to execute than best approximation and so is often preferred. Trefethen compares best approximation and interpolation and discusses historical attitudes towards them [75].

Books by Boyd [8] and Trefethen [76] introduce Chebyshev interpolation and explore in greater depth various problems that can be solved with Chebyshev interpolants. Many properties of the Chebyshev polynomials are cataloged in the reference by Mason and Handscomb [51].

The barycentric weights for a Chebyshev grid are due to Salzer [65]. The weights are not only simple; they are well behaved in the sense of varying little in magnitude. Compare with Exercise 28 in Chapter 9.

Exercises

Exercises 1–10: Interpolate the given function using Chebyshev grids of increasing degree. Do the interpolants converge numerically? If so, at approximately what degree? Concede defeat if numerical convergence does not occur by degree 300.

1. $1/x$, $1 \leq x \leq 2$
2. $\sqrt{x}$, $1/4 \leq x \leq 1$
3. $\sqrt{x}$, $0 \leq x \leq 1$
4. $x^{3/2}$, $0 \leq x \leq 1$
5. $(x^2 - 3x - 1)/(3x + 10)$, $-3 \leq x \leq 15$
6. e^x, $-1 \leq x \leq 1$
7. $\log x$, $1/2 \leq x \leq 1$
8. $\tanh x$, $-4 \leq x \leq 4$
9. $\arccos x$, $-1 \leq x \leq 1$
10. $\arctan x$, $-3 \leq x \leq 3$

11. Repeat Example 9.5 using Chebyshev grids instead of uniform grids. Do Chebyshev interpolants perform any better than uniform interpolants on this example?

12. Prove that (10.3) is a sequence of barycentric weights for the Chebyshev grid on any interval $[a, b]$.

13. Let $l(x) = \prod_{i=0}^{m}(x - x_i)$ be the monic polynomial associated with the degree-m Chebyshev grid on $[-1, 1]$. Prove that $|l(x)| \leq 2^{-m+1}\sqrt{1 - x^2} \leq 2^{-m+1}$ for all $x \in [-1, 1]$.

14. Expand the Chebyshev polynomials $U_m(x)$, $m = 0, 1, 2, 3$, in the monomial basis, expressing each in the form $U_m(x) = c_0 + c_1 x + \cdots + c_m x^m$.

15. Use the identity $\sin(\alpha + \beta) = \sin \alpha \cos \beta + \cos \alpha \sin \beta$ to prove $\sin(\phi) + \sin(\psi) = 2 \sin((\phi + \psi)/2) \cos((\phi - \psi)/2)$.

16. Use the result of Exercise 15 to prove (10.9).

17. Use the identity $\cos(\alpha + \beta) = \cos \alpha \cos \beta - \sin \alpha \sin \beta$ to prove $\cos(\phi) + \cos(\psi) = 2 \cos((\phi + \psi)/2) \cos((\phi - \psi)/2)$.

18. Use the result of Exercise 17 to prove $\cos((m + 1)\theta) = 2 \cos(\theta) \cos(m\theta) - \cos((m - 1)\theta)$.

19. Prove that the Chebyshev polynomials of the first kind satisfy the recurrence relation $T_{m+1}(x) = 2x T_m(x) - T_{m-1}(x)$ and the base cases $T_0(x) = 1$ and $T_1(x) = x$.

20. Prove that the Chebyshev polynomial $T_m(x)$ is a degree-m polynomial and that its leading coefficient equals 1 if $m = 0$ or 2^{m-1} if $m \geq 1$.

21. Expand the Chebyshev polynomials $T_m(x)$, $m = 0, 1, 2, 3$, in the monomial basis, expressing each in the form $T_m(x) = c_0 + c_1 x + \cdots + c_m x^m$.

22. Prove $T_m'(x) = m U_{m-1}(x)$.

23. Prove that the local extrema of $T_m(x)$ on $[-1, 1]$ are located at $x_i = \cos(i\pi/m)$, $i = 1, \ldots, m - 1$. (For this reason, the Chebyshev nodes are also called the "Chebyshev extrema nodes.")

Exercises 24–28: These exercises explore an alternative grid

$$x_i = \cos\frac{(2i + 1)\pi}{2m + 2}, \quad i = 0, \ldots, m. \tag{10.12}$$

24. Prove that the zeros of $T_{m+1}(x)$ are precisely given by (10.12). *Hint.* Use the result of Exercise 20.

25. Prove that the monic polynomial $l(x) = \prod_{i=0}^{m}(x - x_i)$ for the nodes (10.12) equals $2^{-m}T_{m+1}(x)$.

26. Prove that $w_i = (-1)^i \sin((2i + 1)\pi/(2m + 2))$, $i = 0, \ldots, m$, is a sequence of barycentric weights for the grid (10.12).

27. Write MATLAB routines analogous to gridcheb and samplecheb but for the grid (10.12). Demonstrate your routines.

28. Write a MATLAB routine analogous to interpcheb to construct a polynomial interpolant on the grid (10.12). Demonstrate it.

29. Construct a piecewise-Chebyshev interpolant for $|(x^2 + x - 3)/(x^2 - 2x + 4)|$ on $[-5, 5]$ using interpchebpw. Adjust the degree m and the partition of $[-5, 5]$ to construct as accurate an interpolant as possible.

30. Implement the following adaptive interpolation procedure in code. The inputs should be a function f, an interval $[a, b]$, an error tolerance ε, and a maximum recursion depth. The output should be a piecewise-Chebyshev sample of f for which the interpolating piecewise-polynomial p achieves $\|f - p\|_\infty < \varepsilon$. The routine should first construct a degree-32 Chebyshev interpolant p for f. If $\|f - p\|_\infty$ is estimated to be below ε, then the process can terminate immediately. Otherwise, the domain should be split into two equal halves and the interpolation process repeated recursively on each half. The process should be aborted if the desired accuracy is not achieved within the specified maximum recursion depth. Demonstrate the routine on $f(x) = \sqrt{x}$, $0 \le x \le 1$, with an error tolerance of 10^{-8} and a maximum recursion depth of 50.

Chapter 11

Accuracy of Chebyshev interpolation

To analyze the accuracy of Chebyshev interpolation, functions are organized into four classes, based on their smoothness. Functions that are very smooth over large regions are easiest to fit by polynomials, and Chebyshev interpolation performs best on these functions.

Behavior in the complex plane is relevant here. See Appendix B for a brief introduction to complex functions.

11.1 ▪ Functions analytic on the entire complex plane

Our first class of functions consists of those that are *analytic* over the entire complex plane. A function is analytic at a point z if it is complex differentiable in a neighborhood of z. At a *singularity*, a complex function fails to be analytic. The graphs of the real and imaginary parts of a complex function often show puncturing or tearing or creasing at singularities.

The complex exponential function is an example of a function that is analytic on the entire complex plane, also known as an *entire function*. At any complex number z, the derivative of the exponential function exists and is $\frac{d}{dz}[e^z] = e^z$. Graphs of the real and imaginary parts of the exponential function (Figure 11.1) show no evidence of puncturing or tearing or creasing, unlike the graphs of Runge's function (Figure 9.2), which display singularities where division by zero occurs. Other examples of entire functions are polynomials and the sine and cosine functions.

Theorem 11.1. *On functions that are analytic over the entire complex plane, Chebyshev interpolation converges supergeometrically.*

The proof is in the next section.

Example 11.2. Measure and classify the rate of convergence for Chebyshev interpolation on $f(x) = e^x$, $-1 \le x \le 1$.

Solution. The function is sampled on Chebyshev grids of increasing degree, and the interpolation errors are measured.

```
>> f = @(x) exp(x);
>> a = -1;
```

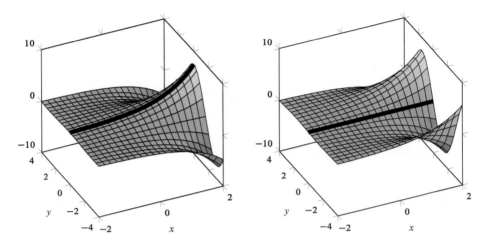

Figure 11.1. *Real (left) and imaginary (right) parts of the natural exponential function* e^z *in the complex plane.*

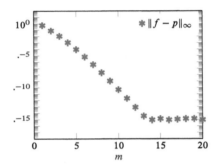

Figure 11.2. *Rate of convergence for Chebyshev interpolation on* e^x, $-1 \le x \le 1$.

```
>> b = 1;
>> m = 1:20;
>> err = nan(size(m));
>> for k = 1:length(m)
     ps = samplecheb(f,[a b],m(k));
     p = interpcheb(ps,[a b]);
     err(k) = infnorm(@(x) f(x)-p(x),[a b],100);
   end
```

The results are plotted in Figure 11.2.

```
>> newfig;
>> ylog; ylim([1e-18 1e2]);
>> plot(m,err,'*','displayname','||f-p||');
>> xlabel('m');
```

Because the errors converge to zero faster than a straight line on a log-linear plot, the convergence is supergeometric. This behavior is guaranteed by the previous theorem. ∎

The sampling cost for a degree-m Chebyshev grid is $m + 1$ function evaluations. Therefore, a graph of error-versus-degree such as Figure 11.2 provides a measure of efficiency with respect to sampling cost.

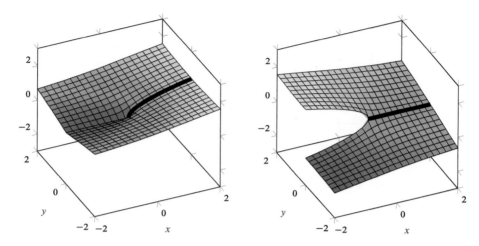

Figure 11.3. *Real (left) and imaginary (right) parts of the square root function in the complex plane.*

11.2 ▪ Functions analytic in a region

Our second class of functions consists of those that are analytic in a region of, but not necessarily the entire, complex plane. Specifically, a function is allowed to have singularities, but those singularities must be located away from the interval of the real line where the function will be interpolated.

There are various types of singularities. For example, Runge's function $f(z) = 1/(1 + 25z^2)$ has a pair of isolated singularities called poles at $z = \pm(1/5)i$, where the real and imaginary parts diverge to infinity. For another example, the square root function $\sqrt{z}$ has a *branch cut* along the negative real axis, where there is a "tear" in its imaginary part, but it is analytic on the rest of the complex plane; see Figure 11.3.

Our main theorem applies to a function that is analytic in a *Bernstein ellipse* of the complex plane. The simplest type of Bernstein ellipse is an ellipse with foci at $z = \pm 1$. Such an ellipse consists of the points $z = x + yi$ that satisfy

$$\frac{x^2}{((\rho + \rho^{-1})/2)^2} + \frac{y^2}{((\rho - \rho^{-1})/2)^2} = 1 \tag{11.1}$$

for some real number $\rho > 1$. The quantities $(\rho + \rho^{-1})/2$ and $(\rho - \rho^{-1})/2$ are the semimajor and semiminor axis lengths, respectively. More generally, for real numbers $a < b$ and $\rho > 1$, the Bernstein ellipse $E_\rho(a, b)$ is the ellipse with foci at $z = a$ and $z = b$ consisting of the points satisfying

$$\frac{u^2}{((\rho + \rho^{-1})/2)^2} + \frac{v^2}{((\rho - \rho^{-1})/2)^2} = 1, \tag{11.2}$$

in which $u = (2/(b - a))(x - (a + b)/2)$ and $v = (2/(b - a))y$. A few Bernstein ellipses are plotted in Figure 11.4. Larger values of ρ produce more rotund ellipses. To identify the foci in each plot, a line segment is drawn between them.

When a function is analytic in a Bernstein ellipse whose foci are located at the endpoints of the interpolation interval, Chebyshev interpolation converges geometrically. The following theorem makes this precise.

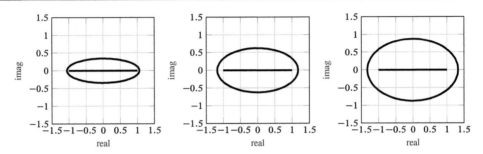

Figure 11.4. *Bernstein ellipses with foci at $a = -1$ and $b = 1$: $\rho = 1.4$ (left), $\rho = 1.8$ (middle), and $\rho = 2.2$ (right).*

Theorem 11.3. *Let p be the degree-m Chebyshev interpolant for a function f on $[a, b]$. If f is analytic in the interior of a Bernstein ellipse $E_\rho(a, b)$ and $|f(z)| \leq M$ for all z in the interior of the ellipse, then*

$$\|f - p\|_\infty \leq \frac{4M}{\rho - 1}\rho^{-m}. \tag{11.3}$$

A proof of a slightly weaker result is given in Appendix C. The proof requires more complex analysis than is expected for the rest of the book.

The previous theorem makes two requirements of f: that it is analytic and that it is bounded ($|f(z)| \leq M$). The following corollary applies when no bound is available.

Corollary 11.4. *Let p be the degree-m Chebyshev interpolant for a function f on $[a, b]$, and suppose that f is analytic (but not necessarily bounded) in the interior of a Bernstein ellipse $E_\rho(a, b)$. Then, for any $\rho_* < \rho$,*

$$\|f - p\|_\infty \leq C\rho_*^{-m} \tag{11.4}$$

for some coefficient C that is constant with respect to m.

Proof. Let $\bar{D}_*$ be the closed region consisting of the Bernstein ellipse $E_{\rho_*}(a, b)$ and its interior. Because this region is contained in the interior of the original ellipse, f is analytic on this compact (closed and bounded) region. An analytic function is continuous, and a continuous function on a compact space is bounded. Thus, $|f(z)|$ is bounded by some number M_* for all $z \in \bar{D}_*$. By the preceding theorem,

$$\|f - p\|_\infty \leq \frac{4M_*}{\rho_* - 1}\rho_*^{-m}. \qquad \square$$

To apply either theorem, one wants to find the largest ellipse in which a given function is analytic—that is, the ellipse with the largest value of ρ. The following test is useful for this purpose. For a given complex number z, let

$$t = \frac{2}{b - a}\left(z - \frac{a + b}{2}\right). \tag{11.5}$$

Then z is on $E_\rho(a, b)$ if and only if

$$|t \pm \sqrt{t^2 - 1}| = \{\rho^{-1}, \rho\} \tag{11.6}$$

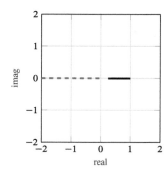 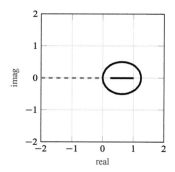

Figure 11.5. *An interpolation interval and a line of singularities along a branch cut (left) and a Bernstein ellipse (right).*

(Exercise 20).

Example 11.5. Use Theorem 11.3 to predict the rate of convergence of Chebyshev interpolation on $f(x) = \sqrt{x}$, $1/4 \leq x \leq 1$.

Solution. The interpolation interval $[1/4, 1]$ is plotted as a solid line segment in the left plot of Figure 11.5.

```
>> ax = newfig(1,2); legend hide;
>> axis([-2 2 -2 2]);
>> axis square; grid on;
>> plot([1/4 1],[0 0],'k-');
>> xlabel('real'); ylabel('imag');
```

Recall that the square root function has a branch cut along the negative real axis. The function is singular at every point on this ray, which is indicated by a dashed line in the same plot.

```
>> plot([-2 0],[0 0],'--');
```

Theorem 11.3 requires a Bernstein ellipse with foci at $a = 1/4$ and $b = 1$ in which the function is analytic, and the largest such ellipse provides the tightest bound on the rate of convergence. The nearest singularity is at $z = 0$. With an eye on (11.5)–(11.6), let

$$t = \frac{2}{1 - (1/4)} \left(0 - \frac{(1/4) + 1}{2} \right) = -\frac{5}{3},$$

and evaluate

$$\left| t \pm \sqrt{t^2 - 1} \right| = \left| -\frac{5}{3} \pm \sqrt{\left(-\frac{5}{3} \right)^2 - 1} \right| = \{1/3, 3\}.$$

We discard $1/3$ because it is less than 1 and take $\rho = 3$. The nearest singularity to $[1/4, 1]$ lies on the Bernstein ellipse $E_3(1/4, 1)$. This ellipse is plotted in the right graph of Figure 11.5 with the provided bernstein routine.

```
>> subplot(ax(2)); legend hide;
>> axis([-2 2 -2 2]);
>> axis square; grid on;
>> plot([-2 0],[0 0],'--');
>> bernstein([1/4 1],3);
>> xlabel('real'); ylabel('imag');
```

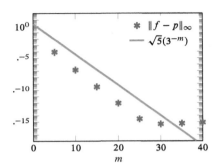

Figure 11.6. *Rate of convergence for Chebyshev interpolation on* $\sqrt{x}$, $1/4 \le x \le 1$.

To apply the theorem, a bound M on $|f(z)| = |\sqrt{z}|$ is required. The absolute value of the square root function increases with distance from the origin, and the point on the ellipse that is farthest from the origin is $z = 5/4$. Thus, $|f(z)| < \sqrt{5/4} = \sqrt{5}/2$ for every point z in the interior of the ellipse.

The theorem implies

$$\|f - p\|_\infty \le \frac{4(\sqrt{5}/2)}{3 - 1} 3^{-m} = \sqrt{5}(3^{-m}).$$ ∎

Example 11.6. Test the prediction of the previous exercise experimentally.

Solution. Several interpolants are computed, and their errors are measured.

```
>> f = @(x) sqrt(x);
>> a = 1/4;
>> b = 1;
>> m = 5:5:40;
>> err = nan(size(m));
>> for k = 1:length(m)
      ps = samplecheb(f,[a b],m(k));
      p = interpcheb(ps,[a b]);
      err(k) = infnorm(@(x) f(x)-p(x),[a b],100);
   end
```

The results are plotted in Figure 11.6 and compared against the bound from the previous example.

```
>> newfig;
>> ylog; ylim([1e-18 1e2]);
>> plot(m,err,'*','displayname','||f-p||');
>> plotfun(@(m) sqrt(5)*3^(-m),[0 m(end)] ...
         ,'displayname','sqrt(5)*3^{-m}');
>> xlabel('m');
```

The observed errors lie below the bound $\sqrt{5}(3^{-m})$ as expected, until numerical convergence is achieved. ∎

Example 11.7. Use Theorem 11.3 to predict the rate of convergence of Chebyshev interpolation on Runge's function over $-1 \le x \le 1$.

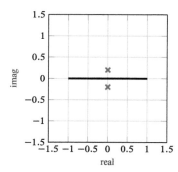 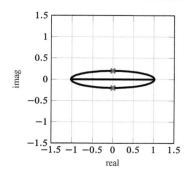

Figure 11.7. *Singularities of Runge's function (left) and a Bernstein ellipse (right).*

Solution. Runge's function $f(z) = 1/(1 + 25z^2)$ has singularities at $z = \pm(1/5)i$, which are indicated by ×'s in Figure 11.7.

```
>> ax = newfig(1,2); legend hide;
>> axis([-1.5 1.5 -1.5 1.5]);
>> axis square; grid on;
>> plot([-1 1],[0 0],'k');
>> plot([0 0],[-1/5 1/5],'x');
>> xlabel('real'); ylabel('imag');
```

We would like to find the largest Bernstein ellipse with foci at $a = -1$ and $b = 1$ whose interior contains neither singularity. Let $t = (1/5)i$ and consider

$$|t \pm \sqrt{t^2 - 1}| = \left| \frac{1}{5}i \pm \sqrt{\left(\frac{1}{5}i\right)^2 - 1} \right| = \left\{ \frac{-1 + \sqrt{26}}{5}, \frac{1 + \sqrt{26}}{5} \right\}.$$

Only the second value is greater than 1, so $\rho = (1 + \sqrt{26})/5 = 1.22\ldots$. The ellipse $E_\rho(-1, 1)$ is shown in the plot on the right.

```
>> subplot(ax(2)); legend hide;
>> axis([-1.5 1.5 -1.5 1.5]);
>> axis square; grid on;
>> plot([0 0],[-1/5 1/5],'x');
>> bernstein([-1 1],(1+sqrt(26))/5);
>> xlabel('real'); ylabel('imag');
```

Runge's function is analytic in the interior of the ellipse.

Corollary 11.4 shows that, for any $\rho_* < \rho = 1.22\ldots$,

$$\|f - p\|_\infty \le C\rho_*^{-m}$$

for a constant coefficient C. ■

Example 11.8. Compare the prediction of the previous exercise with experimental evidence.

Solution. Several interpolants are tested.

```
>> f = @(x) 1/(1+25*x^2);
>> a = -1;
>> b = 1;
>> m = 40:40:200;
>> err = nan(size(m));
```

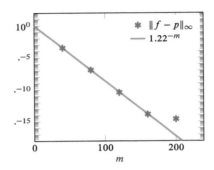

Figure 11.8. *Rate of convergence for Chebyshev interpolation on Runge's function.*

```
>> for k = 1:length(m)
     ps = samplecheb(f,[a b],m(k));
     p = interpcheb(ps,[a b]);
     err(k) = infnorm(@(x) f(x)-p(x),[a b],100);
   end
```

The results are plotted in Figure 11.8.

```
>> newfig;
>> ylog; ylim([1e-18 1e2]);
>> plot(m,err,'*','displayname','||f-p||');
```

The rate of convergence appears to be very close to 1.22^{-m}. This is consistent with the prediction of the previous exercise.

```
>> plotfun(@(m) 1.22^(-m),[0 240],'displayname','1.22^{-m}');
>> xlabel('m');
```
■

Note the significance of the previous two examples: Theorem 11.3 guarantees that Chebyshev interpolation converges geometrically on Runge's function and other analytic functions. The Runge phenomenon is soundly defeated.

We can now prove the earlier theorem on entire functions as well.

***Proof of Theorem* 11.1.** If f is analytic over the entire complex plane, then it is analytic and bounded in any Bernstein ellipse $E_\rho(a,b)$. By Theorem 11.3, the rate of convergence is as fast as ρ^{-m} for any ρ.   □

11.3 ▪ Functions differentiable on the real line

Our third class is the class of functions that are real differentiable but not complex differentiable.

The appropriate notion of differentiability is a subtle one. We say that a function f is *weakly differentiable* on $[a,b]$ if there exists a function g for which

$$f(x) = f(a) + \int_a^x g(\xi)\,d\xi \tag{11.7}$$

for all $x \in [a,b]$. In words, f is weakly differentiable if it is the integral of another function. The function g is called a *weak derivative* for f. A function is r-times weakly differentiable if it has an $(r-1)$th derivative that is weakly differentiable.

Weak differentiability is a more inclusive notion than differentiability in the classical sense.

Example 11.9. Show that the absolute value function $|x|$ is not differentiable in the classical sense at $x = 0$. Then show that the function is weakly differentiable on $[-1, 1]$.

Solution. The classical definition of the derivative is

$$f'(x) = \lim_{h \to 0} \frac{f(x+h) - f(x)}{h}. \tag{11.8}$$

For the absolute value function at $x = 0$, the one-sided derivatives are not equal:

$$\lim_{h \to 0^+} \frac{|0 + h| - |0|}{h} = \lim_{h \to 0^+} \frac{h}{h} = 1,$$

$$\lim_{h \to 0^-} \frac{|0 + h| - |0|}{h} = \lim_{h \to 0^-} \frac{-h}{h} = -1.$$

Hence, the two-sided limit does not exist and the function is not differentiable at $x = 0$ in the classical sense.

For a weak derivative, let

$$g(x) = \begin{cases} -1 & \text{if } x < 0, \\ 1 & \text{if } x \geq 0. \end{cases} \tag{11.9}$$

For $x < 0$, we have

$$1 + \int_{-1}^{x} g(\xi)\, d\xi = 1 + \int_{-1}^{x} -1\, d\xi = 1 - (x - (-1)) = -x,$$

and for $x \geq 0$, we have

$$1 + \int_{-1}^{x} g(\xi)\, d\xi = 1 + \int_{-1}^{0} -1\, d\xi + \int_{0}^{x} 1\, d\xi = 1 - 1 + (x - 0) = x.$$

Thus,

$$1 + \int_{-1}^{x} g(\xi)\, d\xi = |x|$$

for all $x \in [-1, 1]$. The absolute value function is weakly differentiable on $[-1, 1]$, and g is a weak derivative. ∎

The absolute value function and its weak derivative are plotted in Figure 11.9. Note that a weak derivative is not uniquely defined because its value can be changed at a few points without affecting its integral. In particular, in the weak derivative for $|x|$, we could have chosen $g(0) = -1$, or any other value for that matter.

Also note that a weakly differentiable function is necessarily continuous because the integral on the right-hand side of (11.7) is continuous with respect to x.

Theorem 11.10. *Suppose a function f is r-times weakly differentiable on $[a, b]$ for some $r \geq 1$, with an rth derivative that satisfies a technical condition known as bounded variation. Then, for any $m > r$, the degree-m Chebyshev interpolant p satisfies*

$$\|f - p\|_\infty \leq C m^{-r} \tag{11.10}$$

for a coefficient C that is constant with respect to m.

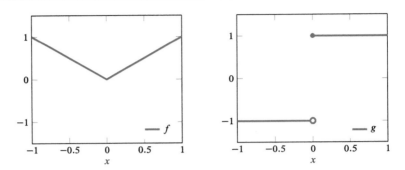

Figure 11.9. *The absolute value function and a derivative in the weak sense.*

The proof is omitted.

For a function that is real differentiable, Chebyshev interpolation converges algebraically, with a rate guarantee proportional to m^{-r}. The exponent r is the number of times that f can be differentiated in the weak sense.

The technical condition of bounded variation is elaborated in Exercises 22–24. It is useful to know that every function of bounded variation is bounded (has infinity norm $<\infty$) and that every monotonic and bounded function has bounded variation.

Example 11.11. Predict the rate of convergence for Chebyshev interpolation on $f(x) = |x|$, $-1 \le x \le 1$. Compare with experimental evidence.

Solution. From the previous example, we know that the absolute value function has a weak first derivative g with a jump discontinuity at $x = 0$. Because g has a discontinuity, it is not differentiable even in the weak sense. Hence, f is once weakly differentiable but not twice weakly differentiable. By the theorem,

$$\|f - p\|_\infty \le Cm^{-1}$$

for some constant coefficient C, for $m > 1$. (See also Exercise 24 on the bounded variation condition.)

To verify this, the interpolation errors are measured experimentally.

```
>> f = @(x) abs(x);
>> a = -1;
>> b = 1;
>> m = 5.^(1:4);
>> err = nan(size(m));
>> for k = 1:length(m)
      ps = samplecheb(f,[a b],m(k));
      p = interpcheb(ps,[a b]);
      err(k) = infnorm(@(x) f(x)-p(x),[a b],1000);
   end
```

They are compared with m^{-1} in Figure 11.10.

```
>> newfig;
>> xlog; ylog; ylim([1e-18 1e2]);
>> plot(m,err,'*','displayname','||f-p||');
>> plotfun(@(m) m^(-1),[2 2000],'displayname','m^{-1}');
>> xlabel('m');
```

The rates of convergence appear to match. ∎

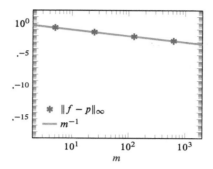

Figure 11.10. *Rate of convergence for Chebyshev interpolation on $f(x) = |x|$, $-1 \leq x \leq 1$.*

Example 11.12. Predict the rate of convergence for Chebyshev interpolation on $f(x) = \frac{1}{2}x\,|x|$, $-1 \leq x \leq 1$. Compare with experimental evidence.

Solution. The function f can be expressed piecewise to avoid absolute value notation:

$$f(x) = \begin{cases} \frac{1}{2}x^2 & \text{if } x \geq 0, \\ -\frac{1}{2}x^2 & \text{if } x < 0. \end{cases}$$

It has a first derivative in the classical sense:

$$f'(x) = \begin{cases} x & \text{if } x \geq 0, \\ -x & \text{if } x < 0. \end{cases}$$

And it has a second derivative in the weak sense:

$$f''(x) = \begin{cases} 1 & \text{if } x \geq 0, \\ -1 & \text{if } x < 0. \end{cases}$$

The second derivative equals the function g that is plotted in Figure 11.9. Theorem 11.10 implies

$$\|f - p\|_\infty \leq C m^{-2}$$

for some constant C, for $m > 2$.

To verify this, the interpolation errors are measured experimentally.

```
>> f = @(x) (1/2)*x*abs(x);
>> a = -1;
>> b = 1;
>> m = 5.^(1:4);
>> err = nan(size(m));
>> for k = 1:length(m)
       ps = samplecheb(f,[a b],m(k));
       p = interpcheb(ps,[a b]);
       err(k) = infnorm(@(x) f(x)-p(x),[a b],100);
   end
```

The data are compared with the power function m^{-2} in Figure 11.11.

```
>> newfig;
>> xlog; ylog; ylim([1e-18 1e2]);
>> plot(m,err,'*','displayname','||f-p||');
>> plotfun(@(m) m^(-2),[3 2000],'displayname','m^{-2}');
>> xlabel('m');
```

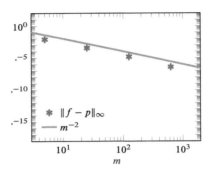

Figure 11.11. *Rate of convergence for Chebyshev interpolation on* $\frac{1}{2}x\,|x|$, $-1 \le x \le 1$.

The interpolation error appears to converge to zero as fast as the power function. ■

Note that the function in Example 11.12 is smoother than the function in Example 11.11 in the sense of possessing a higher-order derivative, and Chebyshev interpolation converges more rapidly on the smoother function.

11.4 ▪ Functions with discontinuities

When a function has a discontinuity, all bets are off.

Example 11.13. How does Chebyshev interpolation perform on the following function?

$$f(x) = \begin{cases} 0 & \text{if } x < 0, \\ 1 & \text{if } x \ge 0. \end{cases}$$

Solution. Because this function has a jump discontinuity, it cannot be differentiable in the classical or weak sense. Thus, Theorem 11.10 does not apply.

Let's see what happens, using Chebyshev interpolants of degrees $m = 21, 41$, and 61.

```
>> f = @(x) iif(x<0,@()0,@()1);
>> a = -1;
>> b = 1;
>> m = [21 41 61];
>> ax = newfig(1,3);
>> for k = 1:3
      ps = samplecheb(f,[a b],m(k));
      p = interpcheb(ps,[a b]);
      subplot(ax(k));
      plotfun(f,[a b],'displayname','f');
      plotfun(p,[a b],'displayname','p');
      ylim([-0.3 1.3]);
      xlabel('x');
   end
```

See Figure 11.12. The polynomials show oscillations around the discontinuity that do not diminish in height as m increases. It appears that $\|f - p\|_\infty$ does not converge to zero. ■

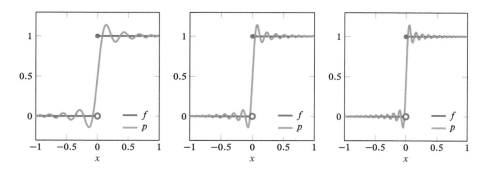

Figure 11.12. *A function with a jump discontinuity and a Chebyshev interpolant of degree m: m = 21 (left), m = 41 (middle), and m = 61 (right).*

The oscillations in Figure 11.12 exemplify the *Gibbs phenomenon*. This behavior commonly occurs when interpolating over a jump discontinuity. Unsurprisingly, it is unwise to fit a jump discontinuity by a polynomial.

11.5 ▪ Summary of convergence rates

In summary, we have considered Chebyshev interpolation on four classes of functions:

function class	rate of convergence
entire	supergeometric
analytic in an ellipse	geometric
real differentiable	algebraic
not differentiable	no guarantees

For a function analytic in an ellipse, the size of the ellipse, in terms of the parameter ρ, controls the rate of convergence. For a differentiable function, the number of times that the function can be differentiated controls the rate of convergence.

11.6 ▪ Lebesgue constant

Recall that the Lebesgue constant measures the sensitivity of a polynomial interpolant to sampling errors. For Chebyshev interpolation to be viable in finite precision, its Lebesgue constant must stay reasonably small.

Theorem 11.14. *The Lebesgue constant Λ_m for a degree-m Chebyshev grid satisfies*

$$\lim_{m \to \infty} \frac{\Lambda_m}{\log m} = \frac{2}{\pi}. \tag{11.11}$$

The proof is omitted.
In alternative notation,

$$\Lambda_m \sim \frac{2}{\pi} \log m \quad (m \to \infty). \tag{11.12}$$

Thus, the Lebesgue constant for a Chebyshev grid grows at the slow rate of $\log m$. In contrast, the Lebesgue constant for a uniform grid grows nearly as fast as 2^m; see Theorem 9.4.

Notes

The bound in Theorem 11.3 is inequality (8.3) in Trefethen's *Approximation Theory and Approximation Practice* [76]. Trefethen cites related results by Bernstein in 1912 [5] and Tadmor in 1986 [69]. A 1965 paper by Elliott relies more heavily on complex analysis and less heavily on the notion of a Chebyshev series, which we have not introduced in this chapter [28].

The bound in Theorem 11.10 is based on inequality (7.5) in Trefethen's book [76]. An expression for the coefficient C and additional references can be found there. Differentiability in the weak sense is traditionally called *absolute continuity*.

The asymptotic expression (11.12) for the Lebesgue constant is due to work of Bernstein in 1931 [4] and Ehlich and Zeller in 1966 [27].

Exercises

Exercises 1–4: Chebyshev interpolation converges geometrically on the given function. Find a numerical value for ρ in the rate ρ^{-m} and justify using the theory of this chapter. Check your prediction experimentally.

1. $1/(x-3)$, $-1 \le x \le 1$
2. $(3x-4)/(x^2+2)$, $-2 \le x \le 2$
3. $(5x^2+7)/(x^2-2x+2)$, $-3 \le x \le 3$
4. $1/(x-2)^2$, $0 \le x \le 1$

Exercises 5–16: Does Chebyshev interpolation converge supergeometrically, geometrically, algebraically, or not at all on the given problem? Predict using the theory from this chapter, and then verify experimentally.

5. $1/x$, $1 \le x \le 2$
6. $\sqrt{x}$, $1 \le x \le 2$
7. $x^{3/2}$, $0 \le x \le 1$
8. $x \vee 0$, $-1 \le x \le 1$
9. $\log x$, $1/2 \le x \le 1$
10. e^{-x^2}, $-6 \le x \le 6$
11. $\sinh x$, $-4 \le x \le 4$
12. $\cos(x^3)$, $-3 \le x \le 3$
13. $\exp(\sin x)$, $-1 \le x \le 1$
14. $\arctan x$, $-3 \le x \le 3$
15. $\arctan(1/x)$, $-3 \le x \le 3$
16. $\begin{cases} e^x & \text{if } x \ge 0, \\ 1 + x + \frac{1}{2}x^2 & \text{if } x < 0, \end{cases}$ $-1 \le x \le 1$

17. Prove that z lies on the Bernstein ellipse $E_\rho(-1, 1)$ if and only if

$$z = \left(\frac{\rho + \rho^{-1}}{2}\right)\cos\theta + \mathrm{i}\left(\frac{\rho - \rho^{-1}}{2}\right)\sin\theta$$

for some $\theta \in [0, 2\pi)$.

18. Let $z = (w + w^{-1})/2$. Prove that w lies on the circle of radius $\rho > 1$ centered at the origin of the complex plane if and only if z lies on the Bernstein ellipse $E_\rho(-1, 1)$.

19. Let $w_+ = z + \sqrt{z^2 - 1}$ and $w_- = z - \sqrt{z^2 - 1}$. Prove that z lies on the Bernstein ellipse $E_\rho(-1, 1)$ if and only if w_+ and w_- lie on circles of radii ρ and ρ^{-1} centered at the origin in the complex plane. *Hint.* Use Exercise 18.

20. Prove that z lies on the Bernstein ellipse $E_\rho(a, b)$ if and only if (11.5)–(11.6) hold. *Hint.* Use Exercise 19.

21. How do we know that a function that is differentiable in the classical sense of (11.8) is also weakly differentiable?

Exercises 22–24: These exercises consider the *total variation* of a function f on an interval $[a, b]$. This is defined to be

$$V(f) = \sup \sum_{i=1}^{n} |f(x_i) - f(x_{i-1})|,$$

in which the supremum is taken over all partitions $a = x_0 < x_1 < \cdots < x_{n-1} < x_n = b$. If the set of sums is unbounded, we write $V(f) = \infty$. Total variation measures the total change in a function over an interval, without regard to the direction of the change. In particular, when $f(t)$ represents the position of an object along a line at time t, the total variation of f equals the distance traveled by the object, as opposed to the displacement.

22. Prove that if f is monotonic on $[a, b]$, then its total variation equals $|f(b) - f(a)|$.

23. A function is bounded on an interval if its infinity norm is finite. A function has bounded variation on an interval if its total variation is finite.

    (a) Prove that every monotonic, bounded function has bounded variation.

    (b) Prove that a function of bounded variation must be bounded.

24. Find the total variation of the function g of (11.9) on $[-1, 1]$.

25. What do Theorems 11.3 and 11.10 imply, if anything, about the rate of convergence of Chebyshev interpolation on $f(x) = \sqrt{x}$, $0 \le x \le 1$? What happens in practice?

26. What do Theorems 11.3 and 11.10 imply, if anything, about the rate of convergence of Chebyshev interpolation on $f(x) = e^{-1/x}$, $0 \le x \le 1$? What happens in practice? *Hint.* $\lim_{x \to 0+} f^{(k)}(x) = 0$ for $k = 0, 1, 2, \ldots$.

27. Compute numerically the Lebesgue constant for a Chebyshev grid of degree $m = 1, 2, 3, 20$.

Chapter 12

Evaluation matrices

The barycentric form of a polynomial, first seen in Chapter 6, provides a simple, efficient, and accurate way to evaluate a polynomial at a point. In addition, as shown in this chapter, the barycentric form allows polynomial evaluation to be expressed as a matrix-vector product. This proves to be useful in the next part of the book, which considers differentiation and integration.

12.1 ▪ General grids

Let $\mathbf{x} = (x_0, \ldots, x_m)$ be a grid, let $p(x)$ be a polynomial of degree at most m, and let

$$\mathbf{p} = p|_{\mathbf{x}} = \begin{bmatrix} p(x_0) \\ \vdots \\ p(x_m) \end{bmatrix}.$$

The problem for this chapter is to compute

$$\begin{bmatrix} p(x_0') \\ \vdots \\ p(x_l') \end{bmatrix}$$

for any given points $x_0', \ldots, x_l'$. Although we already know how to evaluate a polynomial using the barycentric form, we want to express the answer here as a matrix-vector product. This will allow us in the future to apply tools from linear algebra to problems involving polynomial interpolation.

The *evaluation matrix* for the nodes above is the $(l + 1)$-by-$(m + 1)$ matrix

$$\mathbf{E}_{\mathbf{x}',\mathbf{x}} = \begin{bmatrix} e_{00} & e_{01} & e_{02} & \cdots & e_{0m} \\ e_{10} & e_{11} & e_{12} & \cdots & e_{1m} \\ \vdots & \vdots & \vdots & \ddots & \vdots \\ e_{l0} & e_{l1} & e_{l2} & \cdots & e_{lm} \end{bmatrix} \tag{12.1}$$

with entries

$$
e_{ij} = \begin{cases} \dfrac{w_j}{x_i' - x_j} \bigg/ \displaystyle\sum_{k=0}^{m} \dfrac{w_k}{x_i' - x_k} & \text{if } x_i' \notin \{x_0, \ldots, x_m\}, \\[2ex] \delta_{jk} & \text{if } x_i' = x_k, \end{cases} \tag{12.2}
$$

in which $\delta_{jk} = 1$ for $j = k$ and $\delta_{jk} = 0$ for $j \neq k$ and $w_0, \ldots, w_m$ are barycentric weights associated with the nodes $x_0, \ldots, x_m$.

Theorem 12.1. *If p is a polynomial of degree at most m, then*

$$
\begin{bmatrix} p(x_0') \\ \vdots \\ p(x_l') \end{bmatrix} = \mathbf{E}_{\mathbf{x}', \mathbf{x}} \begin{bmatrix} p(x_0) \\ \vdots \\ p(x_m) \end{bmatrix}. \tag{12.3}
$$

Proof. The barycentric form for $p(x)$ is

$$
p(x) = \frac{1}{\sum_{k=0}^{m} \frac{w_k}{x - x_k}} \sum_{j=0}^{m} \frac{w_j \, p(x_j)}{x - x_j} = \sum_{j=0}^{m} \left(\frac{w_j}{x - x_j} \bigg/ \sum_{k=0}^{m} \frac{w_k}{x - x_k} \right) p(x_j).
$$

If x_i' is not equal to any x_k, then

$$
p(x_i') = \sum_{j=0}^{m} \left(\frac{w_j}{x_i' - x_j} \bigg/ \sum_{k=0}^{m} \frac{w_k}{x_i' - x_k} \right) p(x_j) = \sum_{j=0}^{m} e_{ij} \, p(x_j).
$$

If, on the other hand, $x_i' = x_k$ for some k, then

$$
p(x_i') = \sum_{j=0}^{m} \delta_{jk} \, p(x_j) = \sum_{j=0}^{m} e_{ij} \, p(x_j).
$$

We have shown that each entry of the left-hand-side vector in (12.3) equals the corresponding entry of the right-hand-side vector. $\square$

The evaluation matrix is constructed by `evalmatrixgen` and its helper `evalmatrix_`. These routines are shown in Listing 12.1.

Example 12.2. Let $x_i = i/4$, $i = 0, \ldots, 4$, and let $x_i' = i/3$, $i = 0, \ldots, 3$. Construct the evaluation matrix using `evalmatrixgen`.

Solution. The evaluation matrix is as follows:

```
>> xs = (0:4)'/4;
>> xs_ = (0:3)'/3;
>> format short;
>> E = evalmatrixgen(xs_,xs)
E =
    1.0000         0         0         0         0
   -0.0412    0.6584    0.4938   -0.1317    0.0206
    0.0206   -0.1317    0.4938    0.6584   -0.0412
         0         0         0         0    1.0000
>> format long;
```

Listing 12.1. Construct an evaluation matrix.

```
function E = evalmatrixgen(xs_,xs)
ws = baryweights(xs);
E = evalmatrix_(xs_,xs,ws);
```

```
function E = evalmatrix_(xs_,xs,ws)
m = length(xs_)-1;
l = length(xs)-1;
E = nan(m+1,l+1);
for i = 1:m+1
  if any(xs==xs_(i))
    E(i,:) = xs_(i)==xs;
  else
    E(i,:) = ws./(xs_(i)-xs);
    E(i,:) = E(i,:)/sum(E(i,:));
  end
end
```

Note in the previous example that $x_0' = x_0$ and thus the evaluation matrix has a particularly simple first row. The 1 in the top-left entry extracts the first sample value $p(x_0)$ from any given vector **p**, and the 0's in the rest of the row ignore the remaining sample values. Similarly, since $x_3' = x_4$, the last row uses a single 1 in its last entry to extract the last sample value. The middle rows, in contrast, apply the barycentric form, making use of all sample values.

Example 12.3. Let $p(x)$ be the degree-4 polynomial passing through the following points:

x_i	0.00	0.25	0.50	0.75	1.00
y_i	0.50	0.70	0.60	0.30	0.10

Evaluate the polynomial at $x_i' = i/3$, $i = 0, \ldots, 3$, using an evaluation matrix. Verify the computation.

Solution. The original sample is constructed.

```
>> xs = (0:4)'/4;
>> ps = [0.5; 0.7; 0.6; 0.3; 0.1];
```

Then the polynomial is evaluated on the coarser grid.

```
>> xs_ = (0:3)'/3;
>> E = evalmatrixgen(xs_,xs);
>> ps_ = E*ps;
ps_ =
   0.500000000000000
   0.699176954732510
   0.407818930041152
   0.100000000000000
```

To verify that the above values are correct, we construct the polynomial using interpgen and then evaluate using the barycentric form.

```
>> p = interpgen(xs,ps);
>> arrayfun(p,xs_)
ans =
   0.500000000000000
```

```
0.699176954732510
0.407818930041152
0.100000000000000
```

These values are identical to the ones that are computed by the matrix-vector multiplication. ∎

An important special case is when a single polynomial value $p(x')$ is required, i.e., when $l = 0$. In this case, the evaluation matrix has just one row. Denoting the row vector $\mathbf{E}_{x',\mathbf{x}}$, we have

$$p(x') = \mathbf{E}_{x',\mathbf{x}}\mathbf{p},$$

in which $\mathbf{p} = p|_{\mathbf{x}}$.

12.2 ▪ Uniform grids

A common use of evaluation matrices is to *resample* from one grid to another on the same interval. When the new grid has fewer nodes than the original grid, the conversion is called *downsampling*, and when the new grid has more nodes than the original grid, the conversion is called *upsampling*.

Suppose $\mathbf{x}$ and $\mathbf{x}'$ are uniform grids of degrees m and l, respectively, on the same interval $[a, b]$. The evaluation matrix $\mathbf{E}_{x',\mathbf{x}}$ may also be called a *resampling matrix*. It turns out that this matrix depends only on m and l, not on the interval $[a, b]$.

Theorem 12.4. *Let $\mathbf{t}$ and $\mathbf{t}'$ be uniform grids of degrees m and l, respectively, on $[0, 1]$, and let $\mathbf{x}$ and $\mathbf{x}'$ be uniform grids of degrees m and l, respectively, on $[a, b]$. Then*

$$\mathbf{E}_{x',\mathbf{x}} = \mathbf{E}_{t',\mathbf{t}}. \tag{12.4}$$

Proof. The sequence $w_0, \ldots, w_m$ of (7.5) is a sequence of barycentric weights for both $\mathbf{x}$ and $\mathbf{t}$.

Note that $x_i = a + (b - a)t_i$ and $x_i' = a + (b - a)t_i'$. It follows that $x_i' = x_k$ if and only if $t_i' = t_k$. If $x_i' \notin \{x_0, \ldots, x_m\}$, then the entry e_{ij} of $\mathbf{E}_{x',\mathbf{x}}$ is

$$e_{ij} = \frac{w_j}{(a + (b - a)t_i') - (a + (b - a)t_j)} \bigg/ \sum_{k=0}^{m} \frac{w_k}{(a + (b - a)t_i') - (a + (b - a)t_k)}.$$

Canceling, factoring, and canceling again gives

$$e_{ij} = \frac{w_j}{t_i' - t_j} \bigg/ \sum_{k=0}^{m} \frac{w_k}{t_i' - t_k},$$

which equals the corresponding entry of $\mathbf{E}_{t',\mathbf{t}}$. If $x_i' = x_k$ for some k, then the e_{ij} entries of $\mathbf{E}_{x',\mathbf{x}}$ and $\mathbf{E}_{t',\mathbf{t}}$ both equal δ_{jk}. □

The theorem implies that once you've seen one resampling matrix from a degree-m uniform grid to a degree-l uniform grid, you've seen them all. Because the degrees are all that matter, we'll sometimes write $\mathbf{E}_{lm}$ in place of $\mathbf{E}_{x',\mathbf{x}}$. It must be clear from context that the grids are uniform.

Routine `resamplematrixuni` constructs a resampling matrix from one uniform grid to another, on the same interval. See Listing 12.2.

Listing 12.2. Construct a resampling matrix from one uniform grid to another.

```
function E = resamplematrixuni(l,m)

[xs,ws] = griduni([0 1],m,1);
xs_ = griduni([0 1],l,1);
E = evalmatrix_(xs_,xs,ws);
```

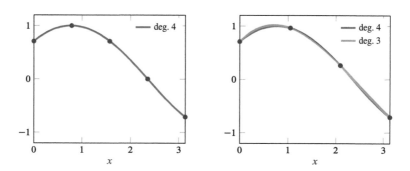

Figure 12.1. *A polynomial interpolant on a uniform grid of degree 4 (left) and a down-sampled polynomial of degree 3 (right).*

Example 12.5. The grids in Example 12.2 are uniform grids on the same interval. Verify that `resamplematrixuni` computes the same resampling matrix.

Solution. The resampling matrix is computed.

```
>> l = 3;
>> m = 4;
>> format short;
>> E = resamplematrixuni(l,m)
E =
    1.0000         0         0         0         0
   -0.0412    0.6584    0.4938   -0.1317    0.0206
    0.0206   -0.1317    0.4938    0.6584   -0.0412
         0         0         0         0    1.0000
>> format long;
```
Yes, the matrix is identical to the one found in Example 12.2. ∎

Example 12.6. Let $p(x)$ be the interpolating polynomial for $\sin(x+\pi/4)$ on a uniform grid of degree 4 over $[0, \pi]$. Plot this polynomial. Then downsample to a grid of degree 3 and plot the polynomial that interpolates the reduced sample.

Solution. The original polynomial $p(x)$ is constructed.

```
>> f = @(x) sin(x+pi/4);
>> a = 0;
>> b = pi;
>> m = 4;
>> ps = sampleuni(f,[a b],m,1);
>> p = interpuni(ps,[a b]);
```
It is plotted in the left plot of Figure 12.1, and the original degree-4 grid is shown.

```
>> ax = newfig(1,2);
>> xs = griduni([a b],m,1);
```

```
>> plotsample(xs,ps);
>> plotfun(p,[a b],'displayname','deg. 4');
>> ylim([-1.2 1.2]);
>> xlabel('x');
```

Next, the downsampled polynomial is constructed.

```
>> l = 3;
>> E = resamplematrixuni(l,m);
>> ps_ = E*ps;
>> p_ = interpuni(ps_,[a b]);
```

The original degree-4 polynomial and the downsampled degree-3 polynomial are plotted together in the right graph of Figure 12.1.

```
>> subplot(ax(2));
>> plotfun(p,[a b],'displayname','deg. 4');
>> xs_ = griduni([a b],l,1);
>> plotsample(xs_,ps_);
>> plotfun(p_,[a b],'displayname','deg. 3');
>> ylim([-1.2 1.2]);
>> xlabel('x');
```

Note that the interpolating polynomials for the two samples are different. Information is lost in downsampling. ∎

12.3 ▪ Chebyshev grids

For Chebyshev grids, the situation is very similar.

Theorem 12.7. *Let* **t** *and* **t'** *be Chebyshev grids of degrees m and l, respectively, on* $[-1, 1]$*, and let* **x** *and* **x'** *be Chebyshev grids of degrees m and l, respectively, on* $[a, b]$*. Then*

$$\mathbf{E}_{\mathbf{x'},\mathbf{x}} = \mathbf{E}_{\mathbf{t'},\mathbf{t}}. \tag{12.5}$$

Proof. The proof is very similar to the proof of Theorem 12.4, except that $x_i = (a + b)/2 + ((b - a)/2)t_i$ and $x'_i = (a + b)/2 + ((b - a)/2)t'_i$. □

As with uniform grids, when resampling from one Chebyshev grid to another on the same interval, the degrees matter, but the interval is irrelevant. Because of this, we'll sometimes write $\mathbf{E}_{lm}$ in place of $\mathbf{E}_{\mathbf{x'},\mathbf{x}}$. Because the notation $\mathbf{E}_{lm}$ may refer to a resampling matrix between uniform grids or between Chebyshev grids, we must be careful to make clear which is intended.

Routine resamplematrixcheb constructs a resampling matrix from one Chebyshev grid to another on the same interval. See Listing 12.3.

Listing 12.3. Construct a resampling matrix from one Chebyshev grid to another.

```
function E = resamplematrixcheb(l,m)

[xs,ws] = gridcheb([-1 1],m);
xs_ = gridcheb([-1 1],l);
E = evalmatrix_(xs_,xs,ws);
```

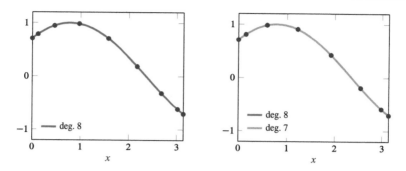

Figure 12.2. *A polynomial interpolant on a Chebyshev grid of degree 8 (left) and a downsampled polynomial of degree 7 (right).*

Example 12.8. Let $p(x)$ be the degree-8 Chebyshev interpolant for $\sin(x + \pi/4)$ on $[0, \pi]$. Resample the polynomial on a Chebyshev grid of degree 7. Plot the original and downsampled polynomials.

Solution. The polynomial is constructed.

```
>> f = @(x) sin(x+pi/4);
>> a = 0;
>> b = pi;
>> m = 8;
>> ps = samplecheb(f,[a b],m);
>> p = interpcheb(ps,[a b]);
```

It is plotted in the left plot of Figure 12.2, with its sample values pinpointed.

```
>> ax = newfig(1,2);
>> xs = gridcheb([a b],m);
>> plotsample(xs,ps);
>> plotfun(p,[a b],'displayname','deg. 8');
>> ylim([-1.2 1.2]);
>> xlabel('x');
```

The polynomial is downsampled.

```
>> l = 7;
>> E = resamplematrixcheb(l,m);
>> ps_ = E*ps;
>> p_ = interpcheb(ps_,[a b]);
```

The downsampled polynomial is plotted on top of the original polynomial in the graph on the right in Figure 12.2.

```
>> subplot(ax(2));
>> plotfun(p,[a b],'displayname','deg. 8');
>> xs_ = gridcheb([a b],l);
>> plotsample(xs_,ps_);
>> plotfun(p_,[a b],'displayname','deg. 7');
>> ylim([-1.2 1.2]);
>> xlabel('x');
```

Although the degree-7 and degree-8 polynomials are not identical, the difference is imperceptible to the eye. ∎

Although downsampling typically causes some loss of accuracy, the error can be small if the downsampled grid is fine enough.

Notes

Resampling matrices appear in a 2016 paper of Driscoll and Hale in the context of solving differential equations [25]. Our discussion follows a similar path starting in Chapter 23.

Exercises

Exercises 1–4: Construct by hand an evaluation matrix from the given source grid to the given destination points.

1. source: $0, 1$; destination: $1/2$
2. source: $0, 1/2, 1$; destination: $1/2$
3. source: $0, 1/2, 1$; destination: $1/4$
4. source: $0, 1/2, 1$; destination: $1/3, 2/3$

Exercises 5–8: Use `evalmatrixgen`, `resamplematrixuni`, or `resamplematrixcheb` to construct the given evaluation matrix. Then demonstrate that the evaluation matrix performs correctly on a polynomial of appropriate degree.

5. source: uniform grid of degree 5 on $[a, b]$; destination: $(a + b)/2$
6. source: uniform grid of degree 6 on $[a, b]$; destination: uniform grid of degree 4 on $[a, b]$
7. source: Chebyshev grid of degree 5 on $[a, b]$; destination: $(a + b)/2$
8. source: Chebyshev grid of degree 6 on $[a, b]$; destination: Chebyshev grid of degree 4 on $[a, b]$

9. What are the entries of $\mathbf{E}_{\mathbf{x},\mathbf{x}}$?

10. Let $p(x) = c_0 + c_1 x + c_2 x^2 + \cdots + c_m x^m$ be a polynomial represented in the monomial basis, and let $\mathbf{c}$ be the column vector $(c_0, c_1, c_2, \ldots, c_m)$. For any given real number a, find a column vector $\mathbf{e}_a$ such that $\mathbf{e}_a^T \mathbf{c} = p(a)$. Justify.

11. The ith row sum of a matrix is the sum of all entries in the ith row. Prove that every row sum of $\mathbf{E}_{\mathbf{x}',\mathbf{x}}$ equals 1.

12. Let $\mathbf{A}$ and $\mathbf{B}$ be m-by-n matrices. Prove that $\mathbf{A} = \mathbf{B}$ if and only if $\mathbf{A}\mathbf{x} = \mathbf{B}\mathbf{x}$ for all n-by-1 vectors $\mathbf{x}$.

13. Let $\mathbf{E}_{lm}$ denote a Chebyshev resampling matrix, as in the discussion after Theorem 12.7. Let k, l, and m be positive integers with $l \geq m$. Prove that $\mathbf{E}_{kl}\mathbf{E}_{lm} = \mathbf{E}_{km}$.

14. This exercise concerns a translation $p(x - \tau)$ of a polynomial p of degree m for some $\tau \geq 0$. Suppose $\mathbf{p}$ is a sample of $p(x)$ on a uniform grid $\mathbf{x}$ of degree m on $[a, b]$. The task is to create a matrix $\mathbf{E}$ for which

$$\mathbf{Ep} = \begin{bmatrix} p(x_i' - \tau) \\ \vdots \\ p(x_l' - \tau) \end{bmatrix},$$

in which $\mathbf{x}'$ is a uniform grid of degree l on the same interval. To avoid evaluating p outside of $[a, b]$, the index i is defined to be the minimum index for which $x_i' - \tau \geq a$. Write a MATLAB function to construct the matrix $\mathbf{E}$ and demonstrate it on an example.

15. Repeat Exercise 14 using Chebyshev grids for $\mathbf{x}$ and $\mathbf{x}'$.

16. Let $\mathbf{A}$ and $\mathbf{B}$ be matrices in which the number of columns in $\mathbf{A}$ equals the number of rows in $\mathbf{B}$. Prove that $(\mathbf{AB})^T = \mathbf{B}^T \mathbf{A}^T$.

17. The vec operator maps an m-by-n matrix to an mn-by-1 column vector by stacking its columns. Denoting the columns of the original m-by-n matrix $\mathbf{X}$ by $\mathbf{x}_1, \ldots, \mathbf{x}_n$, the vec operation is as follows:

$$\text{vec}(\mathbf{X}) = \text{vec}\left(\begin{bmatrix} \mathbf{x}_1 & \cdots & \mathbf{x}_n \end{bmatrix}\right) = \begin{bmatrix} \mathbf{x}_1 \\ \vdots \\ \mathbf{x}_n \end{bmatrix}.$$

For an l-by-p matrix $\mathbf{A} = [a_{ij}]$ and a matrix $\mathbf{B}$, the Kronecker product $\mathbf{A} \otimes \mathbf{B}$ is defined as follows:

$$\mathbf{A} \otimes \mathbf{B} = \begin{bmatrix} a_{11}B & \cdots & a_{1p}B \\ \vdots & \ddots & \vdots \\ a_{l1}B & \cdots & a_{lp}B \end{bmatrix}.$$

For an l-by-m matrix $\mathbf{C}$, prove that

$$\text{vec}(\mathbf{CX}) = (\mathbf{I} \otimes \mathbf{C}) \, \text{vec}(\mathbf{X}),$$

in which $\mathbf{I}$ denotes the identity matrix of appropriate size. (What size is this?)

18. The vec operator and the Kronecker product are defined in the previous exercise. Let $\mathbf{X}$ be m-by-n and $\mathbf{C}$ be l-by-n. Prove that $\text{vec}(\mathbf{XC}^T) = (\mathbf{C} \otimes \mathbf{I}) \, \text{vec}(\mathbf{X})$, in which $\mathbf{I}$ denotes the identity matrix of appropriate size. (What size is this?)

19. Let $(p_1(x), \ldots, p_r(x))$ be a vector-valued function in which each p_k is a polynomial of degree at most m. Let $\mathbf{x} = (x_0, \ldots, x_m)$ and $\mathbf{x}' = (x_0', \ldots, x_l')$ be grids and sample

$$p|_{\mathbf{x}} = \begin{bmatrix} p_1(x_0) & \cdots & p_1(x_m) \\ \vdots & \ddots & \vdots \\ p_r(x_0) & \cdots & p_r(x_m) \end{bmatrix}, \quad p|_{\mathbf{x}'} = \begin{bmatrix} p_1(x_0') & \cdots & p_1(x_l') \\ \vdots & \ddots & \vdots \\ p_r(x_0') & \cdots & p_r(x_l') \end{bmatrix}.$$

Prove

$$(\mathbf{E}_{\mathbf{x}',\mathbf{x}} \otimes \mathbf{I}) \, \text{vec}(p|_{\mathbf{x}}) = \text{vec}(p|_{\mathbf{x}'}).$$

(The vec operator and the Kronecker product are considered in Exercises 17–18.)

Part III

Integration

Chapter 13

Differentiation

The next few chapters are about differentiation and integration. Before considering the specifics, we expose a deep connection between calculus and linear algebra, one that pays dividends many times over.

13.1 ▪ Linear operators

One of the most important lessons of linear algebra is, "Matrices represent *linear transformations.*" This means that matrices are not merely inert grids of numbers, but rather they encode actions. Given a matrix $\mathbf{A}$, the matrix-vector product maps an input vector $\mathbf{x}$ to the output vector $\mathbf{Ax}$. The mapping

$$\mathbf{x} \mapsto \mathbf{Ax}$$

is the *transformation*. It is called *linear* because it obeys two equations:

1. $\mathbf{A}(c\mathbf{x}) = c\mathbf{Ax}$ for any scalar c and vector $\mathbf{x}$, and

2. $\mathbf{A}(\mathbf{x} + \mathbf{y}) = \mathbf{Ax} + \mathbf{Ay}$ for any pair of vectors $\mathbf{x}, \mathbf{y}$.

Certain notions from calculus are remarkably similar. Consider differentiation,

$$f \mapsto f'.$$

This transformation is also linear:

$$(cf)' = cf', \quad (f + g)' = f' + g'.$$

In this setting, functions replace column vectors, and the derivative operator replaces the matrix. (c is still a constant scalar.) The integral operator

$$g \mapsto \int g$$

satisfies the laws of linearity as well:

$$\int cg = c \int g, \quad \int (g + h) = \int g + \int h.$$

Because differentiation and integration are linear operators, many problems involving calculus can be translated into problems involving linear algebra. Polynomials p

and q can be replaced by samples $\mathbf{p}$ and $\mathbf{q}$, which are column vectors. Linear operators such as the derivative and the integral can be replaced by matrices. Over the next two chapters, we will develop the following correspondences, in which $\mathbf{D}$, $\mathbf{K}$, $\hat{\mathbf{K}}$, and $\mathbf{J}$ are matrices, and $\boldsymbol{\lambda}$ is a column vector:

calculus	matrix algebra
$p'(x) = q(x)$	$\mathbf{Dp} = \mathbf{q}$
$\int_a^b q(x)\,dx = I$	$\boldsymbol{\lambda}^T \mathbf{q} = I$
$\int_a^x q(\xi)\,d\xi = p(x)$	$\mathbf{Kq} = \mathbf{p}$
$\int q(x)\,dx = p(x) + C$	$\hat{\mathbf{K}}\mathbf{q} = \mathbf{Jp}$

The first row involves the derivative, the subject of the present chapter. The last three rows involve integrals of different types, which are defined and analyzed in the next chapter.

13.2 ▪ The differentiation matrix

Suppose a polynomial p is given and its derivative $q = p'$ is desired. If p has degree at most m, then it is uniquely identified by a sample $\mathbf{p} = p|_\mathbf{x}$ on a grid $\mathbf{x}$ of degree m. We want to compute a sample $\mathbf{q} = q|_{\mathbf{x}'}$ of its derivative on any desired grid $\mathbf{x}'$. As usual, each grid $\mathbf{x}$ and $\mathbf{x}'$ is required to be composed of distinct nodes.

The *differentation matrix* from grid $\mathbf{x} = (x_0, \ldots, x_m)$ to grid $\mathbf{x}' = (x_0', \ldots, x_l')$ is the $(l + 1)$-by-$(m + 1)$ matrix

$$\mathbf{D}_{\mathbf{x}',\mathbf{x}} = \begin{bmatrix} d_{00} & d_{01} & \cdots & d_{0,m-1} & d_{0m} \\ d_{10} & d_{11} & \cdots & d_{1,m-1} & d_{1m} \\ \vdots & \vdots & \ddots & \vdots & \vdots \\ d_{l0} & d_{l1} & \cdots & d_{l,m-1} & d_{lm} \end{bmatrix} \tag{13.1}$$

with entries

$$d_{ij} = \frac{l_j'(x_i')}{l_j(x_j)}, \tag{13.2}$$

in which

$$l_j(x) = \prod_{\substack{k=0,\ldots,m \\ k \neq j}} (x - x_k)$$

is the jth Lagrange basis polynomial for $\mathbf{x}$ and $l_j'(x)$ is its derivative.

Theorem 13.1. *Let p be a polynomial of degree at most m, and let q be a polynomial of degree at most $m - 1$. Sample $\mathbf{p} = p|_\mathbf{x}$ and $\mathbf{q} = q|_{\mathbf{x}'}$ on grids $\mathbf{x}$ and $\mathbf{x}'$ defined as above. If*

$$q = p', \tag{13.3}$$

then

$$\mathbf{q} = \mathbf{D}_{\mathbf{x}',\mathbf{x}}\mathbf{p}. \tag{13.4}$$

If $l \geq m - 1$, then the converse is true as well.

Proof. First suppose $q = p'$. The polynomial p in Lagrange form is

$$p(x) = \sum_{j=0}^{m} \frac{p(x_j)}{l_j(x_j)} l_j(x).$$

Its derivative at $x = x_i'$ equals

$$q(x_i') = p'(x_i') = \sum_{j=0}^{m} \frac{p(x_j)}{l_j(x_j)} l_j'(x_i') = \sum_{j=0}^{m} d_{ij}\, p(x_j);$$

i.e., the $(i+1)$th entry of $\mathbf{q}$ equals the $(i+1)$th entry of $\mathbf{D}_{x',x}\mathbf{p}$.

For the converse statement, suppose $\mathbf{q} = \mathbf{D}_{x',x}\mathbf{p}$ and $\mathbf{x}'$ has degree $l \geq m - 1$. Then $q(x_i') = p'(x_i')$ for $i = 0, \ldots, l$. Because q and p' are polynomials of degree at most $m - 1$ and they are equal at $l + 1 \geq m$ points, they must be identical.   □

Example 13.2. Let $\mathbf{x} = (-1, 0, 1)$ and $\mathbf{x}' = (-1, 1)$. Compute the differentiation matrix $\mathbf{D}_{x',x}$. Check its correctness on the polynomial $p(x) = x^2$.

Solution. For the grid $\mathbf{x}$, the Lagrange basis polynomials and their derivatives are

$$l_0(x) = (x - 0)(x - 1), \quad l_0'(x) = (x - 0) + (x - 1),$$
$$l_1(x) = (x + 1)(x - 1), \quad l_1'(x) = (x + 1) + (x - 1),$$
$$l_2(x) = (x + 1)(x - 0), \quad l_2'(x) = (x + 1) + (x - 0).$$

The differentiation matrix is

$$\mathbf{D}_{x',x} = \begin{bmatrix} \dfrac{l_0'(-1)}{l_0(-1)} & \dfrac{l_1'(-1)}{l_1(0)} & \dfrac{l_2'(-1)}{l_2(1)} \\[2mm] \dfrac{l_0'(1)}{l_0(-1)} & \dfrac{l_1'(1)}{l_1(0)} & \dfrac{l_2'(1)}{l_2(1)} \end{bmatrix} = \begin{bmatrix} \dfrac{-3}{2} & \dfrac{-2}{-1} & \dfrac{-1}{2} \\[2mm] \dfrac{1}{2} & \dfrac{2}{-1} & \dfrac{3}{2} \end{bmatrix}$$

$$= \begin{bmatrix} -1.5 & 2 & -0.5 \\ 0.5 & -2 & 1.5 \end{bmatrix}.$$

The polynomials are $p(x) = x^2$ and $p'(x) = 2x$, and their samples are

$$\mathbf{p} = \begin{bmatrix} p(-1) \\ p(0) \\ p(1) \end{bmatrix} = \begin{bmatrix} 1 \\ 0 \\ 1 \end{bmatrix}, \quad \mathbf{q} = \begin{bmatrix} p'(-1) \\ p'(1) \end{bmatrix} = \begin{bmatrix} -2 \\ 2 \end{bmatrix}.$$

The matrix-vector product $\mathbf{D}_{x',x}\mathbf{p}$ is

$$\mathbf{D}_{x',x}\mathbf{p} = \begin{bmatrix} -1.5 & 2 & -0.5 \\ 0.5 & -2 & 1.5 \end{bmatrix} \begin{bmatrix} 1 \\ 0 \\ 1 \end{bmatrix} = \begin{bmatrix} -2 \\ 2 \end{bmatrix}.$$

As expected, $\mathbf{D}_{x',x}\mathbf{p} = \mathbf{q}$.   ■

In an abstract sense, differentiation is accomplished. However, the definition of d_{ij} includes the derivative $l_j'(x_i')$, which can get messy. Over the remainder of the chapter, we develop an alternative approach to computing the differentiation matrix. The approach is in two parts. First, the special case $\mathbf{D}_{x,x}$, in which p and q are sampled on the same grid, is considered. Second, an evaluation matrix from Chapter 12 is used to compute the general case $\mathbf{D}_{x',x}$. Because the degree of $q = p'$ is one less than the degree of p, many of our computations in subsequent chapters use a grid $\mathbf{x}'$ of degree one less than $\mathbf{x}$ and therefore require the more general case.

13.3 ▪ Differentiation on a single grid

Let $\mathbf{x} = (x_0, \ldots, x_m)$ be a grid of distinct nodes, and let $w_0, \ldots, w_m$ be a sequence of barycentric weights for the grid. The following theorem applies to the special case of (13.1) in which $\mathbf{x}' = \mathbf{x}$.

Theorem 13.3. *In the differentiation matrix* $\mathbf{D}_{\mathbf{x},\mathbf{x}}$,

$$d_{ij} = \frac{w_j}{w_i(x_i - x_j)}, \quad i \neq j, \tag{13.5}$$

and

$$d_{ii} = - \sum_{\substack{j=0,\ldots,m \\ j \neq i}} d_{ij}. \tag{13.6}$$

Before beginning the proof, we need a lemma.

Lemma 13.4. *For* $x \neq x_j$,

$$l_j'(x) = \frac{1}{x - x_j} \sum_{k \neq j} l_k(x). \tag{13.7}$$

Proof. The first step is logarithmic differentiation. Taking logarithms of both sides of $l_j(x) = \prod_{k \neq j}(x - x_k)$ gives

$$\log|l_j(x)| = \sum_{k \neq j} \log|x - x_k|,$$

and differentiating gives

$$\frac{l_j'(x)}{l_j(x)} = \sum_{k \neq j} \frac{1}{x - x_k}.$$

This is valid for $x \notin \{x_0, \ldots, x_{j-1}, x_{j+1}, \ldots, x_m\}$. Next we solve for $l_j'(x)$, assuming $x \neq x_j$, and perform some algebraic manipulations. In the following, $l(x) = \prod_{k=0}^{m}(x - x_k)$:

$$l_j'(x) = l_j(x) \sum_{k \neq j} \frac{1}{x - x_k} = \frac{l(x)}{x - x_j} \sum_{k \neq j} \frac{1}{x - x_k}$$

$$= \sum_{k \neq j} \frac{l(x)}{(x - x_j)(x - x_k)} = \sum_{k \neq j} \frac{l_k(x)}{x - x_j} = \frac{1}{x - x_j} \sum_{k \neq j} l_k(x).$$

Now (13.7) has been justified for $x \notin \{x_0, \ldots, x_m\}$. The equation extends to all $x \neq x_j$ because both sides of the equation are continuous at all $x \neq x_j$. More explicitly, for $i \neq j$, we have

$$l_j'(x_i) = \lim_{x \to x_i} l_j'(x) = \lim_{x \to x_i} \frac{1}{x - x_j} \sum_{k \neq j} l_k(x) = \frac{1}{x_i - x_j} \sum_{k \neq j} l_k(x_i). \qquad \square$$

Proof of Theorem 13.3. In the notation of (13.1)–(13.2), $x_i' = x_i$, so

$$d_{ij} = \frac{l_j'(x_i)}{l_j(x_j)}.$$

First consider the case $i \neq j$. By the lemma,

$$l_j'(x_i) = \frac{1}{x_i - x_j} \sum_{k \neq j} l_k(x_i).$$

Because $l_k(x_i) = 0$ for $k \neq i$, all but one of the terms in the sum vanish:

$$l_j'(x_i) = \frac{1}{x_i - x_j} \left(l_i(x_i) + \sum_{k \neq i,j} l_k(x_i) \right) = \frac{l_i(x_i)}{x_i - x_j}.$$

From the definition (6.3) of barycentric weights, we find

$$l_j'(x_i) = \frac{C}{w_i(x_i - x_j)}.$$

With the numerator in hand, we can simplify d_{ij}:

$$d_{ij} = \frac{l_j'(x_i)}{l_j(x_j)} = \frac{C}{w_i(x_i - x_j)} \frac{w_j}{C} = \frac{w_j}{w_i(x_i - x_j)}.$$

To cover the case $i = j$, we differentiate the function $p(x) = 1$. Its derivative is $q(x) = 0$, and sampling produces $\mathbf{p} = p|_{\mathbf{x}} = \mathbf{1} = (1, 1, \ldots, 1)$, and $\mathbf{q} = q|_{\mathbf{x}} = \mathbf{0} = (0, 0, \ldots, 0)$. According to Theorem 13.1, $\mathbf{D}_{\mathbf{x},\mathbf{x}}\mathbf{1} = \mathbf{0}$. In other words, every row sum of the differentiation matrix equals zero:

$$\sum_{j=0}^{m} d_{ij} \times 1 = 0, \quad i = 0, \ldots, m.$$

Solving for d_{ii} gives $d_{ii} = -\sum_{j \neq i} d_{ij}$.   □

The matrix $\mathbf{D}_{\mathbf{x},\mathbf{x}}$ is constructed by `diffmatrixsquare_` in Listing 13.1. With it, any polynomial can be differentiated automatically.

Listing 13.1. Construct a differentiation matrix with identical source and destination grids.

```
function D = diffmatrixsquare_(xs,ws)

m = length(xs)-1;
D = nan(m+1);
for i = 1:m+1
  for j = [1:i-1 i+1:m+1]
    D(i,j) = ws(j)./(ws(i)*(xs(i)-xs(j)));
  end
  D(i,i) = -sum(D(i,[1:i-1 i+1:end]));
end
```

13.4 ▪ Differentiation on arbitrary grids

Now we have simple formulas for the entries of $\mathbf{D}_{\mathbf{x},\mathbf{x}}$, the special case of the differentiation matrix in which the source and target grids are identical. These can be leveraged to compute any differentiation matrix $\mathbf{D}_{\mathbf{x}',\mathbf{x}}$ in a simple and efficient way. In the following theorem, $\mathbf{E}_{\mathbf{x}',\mathbf{x}}$ refers to an evaluation matrix of Chapter 12.

Theorem 13.5. *The differentiation matrix* $\mathbf{D}_{\mathbf{x}',\mathbf{x}}$ *satisfies*

$$\mathbf{D}_{\mathbf{x}',\mathbf{x}} = \mathbf{E}_{\mathbf{x}',\mathbf{x}}\mathbf{D}_{\mathbf{x},\mathbf{x}}. \tag{13.8}$$

Proof. Let $\mathbf{p}$ be any vector of length $m + 1$. Let p be the polynomial of degree at most m for which $p|_{\mathbf{x}} = \mathbf{p}$. Let $q = p'$ and let $\mathbf{q} = q|_{\mathbf{x}'}$. On one hand, $\mathbf{q} = \mathbf{D}_{\mathbf{x}',\mathbf{x}}\mathbf{p}$ by Theorem 13.1. On the other hand, $\mathbf{q} = \mathbf{E}_{\mathbf{x}',\mathbf{x}}\mathbf{D}_{\mathbf{x},\mathbf{x}}\mathbf{p}$ by Theorems 12.1 and 13.1. Because $\mathbf{D}_{\mathbf{x}',\mathbf{x}}\mathbf{p} = \mathbf{E}_{\mathbf{x}',\mathbf{x}}\mathbf{D}_{\mathbf{x},\mathbf{x}}\mathbf{p}$ for any vector $\mathbf{p}$, the matrices $\mathbf{D}_{\mathbf{x}',\mathbf{x}}$ and $\mathbf{E}_{\mathbf{x}',\mathbf{x}}\mathbf{D}_{\mathbf{x},\mathbf{x}}$ have to be identical.   □

An arbitrary differentiation matrix $\mathbf{D}_{\mathbf{x}',\mathbf{x}}$ can be computed with the routine diffmatrixgen, shown in Listing 13.2.

Listing 13.2. Construct a differentiation matrix for given source and destination grids.

```
function D = diffmatrixgen(xs_,xs)

ws = baryweights(xs);
D = diffmatrix_(xs_,xs,ws);
```

```
function D = diffmatrix_(xs_,xs,ws)

Dxx = diffmatrixsquare_(xs,ws);
E = evalmatrix_(xs_,xs,ws);
D = E*Dxx;
```

Example 13.6. Sample $f(x) = \sin(x)$ on the uniform grid $\mathbf{x}$ of degree 6 over $[0, 2\pi]$. Compute the derivative of the interpolating polynomial on the uniform grid $\mathbf{x}'$ of degree 5 over the same interval. Plot the polynomials.

Solution. The original function is sampled on the finer grid.

```
>> f = @(x) sin(x); a = 0; b = 2*pi;
>> m = 6;
>> xs = griduni([a b],m);
>> ps = arrayfun(f,xs);
>> p = interpgen(xs,ps);
```

Then its derivative is computed on the coarser grid.

```
>> xs_ = griduni([a b],m-1);
>> D = diffmatrixgen(xs_,xs);
>> qs = D*ps;
>> q = interpgen(xs_,qs);
```

The polynomial interpolant p and its derivative $q = p'$ are plotted in Figure 13.1.

```
>> newfig;
>> plotsample(xs,ps);
>> plotfun(p,[a b],'displayname','p');
```

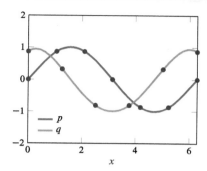

Figure 13.1. *A polynomial interpolant p for the sine function and its derivative q.*

```
>> plotsample(xs_,qs);
>> plotfun(q,[a b],'displayname','q');
>> ylim([-2 2]);
>> xlabel('x');
```
Note that q resembles the cosine function. ■

In the example above, p and q are sampled on grids $\mathbf{x}$ and $\mathbf{x}'$ of degrees 6 and 5, respectively. This is natural because taking a derivative decreases the degree of a polynomial by one. The general differentiation matrix $\mathbf{D}_{\mathbf{x}',\mathbf{x}}$ proves useful.

13.5 ▪ Accuracy analysis?

The differentiation matrix allows a polynomial to be differentiated exactly, up to round-off error. Given a function that is *not* a polynomial, we might consider differentiating a polynomial interpolant in its place. However, there is a fundamental difficulty illustrated by the following example.

Example 13.7. Let $f(x) = \sin(x) + 0.02 \sin(20x)$. Sample f on the uniform grid of degree 6 over $[0, 2\pi]$ and let p be the interpolating polynomial. Approximate f' by p' and comment on the accuracy.

Solution. The numerical approach to the derivative begins by sampling and interpolating f.

```
>> f = @(x) sin(x)+0.02*sin(20*x); a = 0; b = 2*pi;
>> m = 6;
>> xs = griduni([a b],m);
>> ps = arrayfun(f,xs);
>> p = interpgen(xs,ps);
```
The given function f and the interpolating polynomial p are plotted in the left panel of Figure 13.2. Notice the close agreement.

```
>> ax = newfig(1,2);
>> plotfun(f,[a b],'displayname','f');
>> plotsample(xs,ps);
>> plotfun(p,[a b],'displayname','p');
>> ylim([-3 3]);
>> xlabel('x');
```
The polynomial derivative p' is computed on a grid of degree $6 - 1 = 5$. We use a uniform grid again.

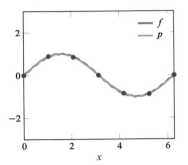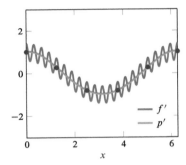

Figure 13.2. *The function $f(x) = \sin(x) + 0.02\sin(20x)$ and a polynomial interpolant $p(x)$ (left) and the derivatives $f'(x)$ and $p'(x)$ (right).*

```
>> xs_ = griduni([a b],m-1);
>> D = diffmatrixgen(xs_,xs);
>> qs = D*ps;
>> q = interpgen(xs_,qs);
```
The desired derivative is

$$f'(x) = \frac{d}{dx}[\sin(x) + 0.02\sin(20x)] = \cos(x) + 0.4\cos(20x).$$

The polynomial derivative p' and the desired derivative f' are compared in the right panel of Figure 13.2.

```
>> subplot(ax(2));
>> g = @(x) cos(x)+0.4*cos(20*x);
>> plotfun(g,[a b],'displayname','f''');
>> plotsample(xs_,qs);
>> plotfun(q,[a b],'displayname','p''');
>> ylim([-3 3]);
>> xlabel('x');
```
Compared with the wild oscillations of f', the polynomial p' varies slowly. Even though p fits f closely, p' does not provide an accurate approximation for f'. ∎

Differentiation can magnify a small blip or high-frequency oscillation into a sizable feature. In the example, the term $0.02\sin(20x)$ seems relatively insignificant in the graph of f, but its derivative $0.4\cos(20x)$ is quite significant in the graph of f'. A polynomial interpolant may miss such features, leading to an inaccurate derivative approximation.

By comparison, the accuracy analysis for integration is straightforward. The next chapter moves immediately to integration, and later in the book, we avoid numerical derivatives by restating problems in terms of integrals. Our work on differentiation is not wasted, however; it proves useful in developing integration methods in the next chapter.

Notes

Equation (13.5) for the off-diagonal elements of a differentiation matrix is a special case of a recurrence given by Welfert in 1997 [84]. The question of numerical stability when computing the diagonal elements by (13.6) is discussed briefly by Weideman and Reddy in an article from 2000 [83]. The Weideman–Reddy article also contains additional references on numerical stability.

Equations (13.5) and (13.6) are for arbitrary grids. Differentiation formulas for some specific grids predate these equations. The Weideman–Reddy article mentioned above provides some background and references [83].

Exercises

1. Prove the linearity equations $\mathbf{A}(c\mathbf{x}) = c\mathbf{A}\mathbf{x}$ and $\mathbf{A}(\mathbf{x} + \mathbf{y}) = \mathbf{A}\mathbf{x} + \mathbf{A}\mathbf{y}$ for a 2-by-2 matrix $\mathbf{A}$ and 2-by-1 vectors $\mathbf{x}$ and $\mathbf{y}$.

2. Prove that the derivative is linear using the definition

$$f'(x) = \lim_{h \to 0} \frac{f(x + h) - f(x)}{h}.$$

3. Consider the linear transformation that maps a column vector $\mathbf{x} = (x_1, x_2, \ldots, x_n)$ to the vector of differences $\mathbf{A}\mathbf{x} = (x_2 - x_1, x_3 - x_2, \ldots, x_n - x_{n-1})$. Describe the matrix $\mathbf{A}$ that represents this transformation. How many rows and columns does it have? What are its entries?

4. Consider the linear transformation that maps a column vector $\mathbf{x} = (x_1, x_2, \ldots, x_n)$ to the vector of averages $\mathbf{A}\mathbf{x} = ((x_1 + x_2)/2, (x_2 + x_3)/2, \ldots, (x_{n-1} + x_n)/2)$. Describe the matrix $\mathbf{A}$ that represents this transformation. How many rows and columns does it have? What are its entries?

Exercises 5–8: Compute the entries of $\mathbf{D}_{\mathbf{x}',\mathbf{x}}$ by hand for the given grids $\mathbf{x}$ and $\mathbf{x}'$.

5. $\mathbf{x} = (0, 1), \mathbf{x}' = 1/2$

6. $\mathbf{x} = (0, 1/2), \mathbf{x}' = 1/4$

7. $\mathbf{x} = \mathbf{x}' = (0, 1, 2)$

8. $\mathbf{x} = (0, 1, 2), \mathbf{x}' = (0, 1)$

9. Let $\mathbf{x} = (-1, -1/3, 1/3, 1)$ and $\mathbf{x}' = (-1, 0, 1)$. Construct $\mathbf{D}_{\mathbf{x}',\mathbf{x}}$ using `diffmatrixgen`. Verify that it correctly differentiates an example polynomial of degree 3.

10. Let $\mathbf{x} = (0, 1/4, 1/2, 3/4, 1)$ and $\mathbf{x}' = (0, 1/3, 2/3, 1)$. Construct $\mathbf{D}_{\mathbf{x}',\mathbf{x}}$ using `diffmatrixgen`. Verify that it correctly differentiates an example polynomial of degree 4.

11. Let $\mathbf{t}$ and $\mathbf{t}'$ be uniform grids of degrees m and l, respectively, on $[0, 1]$. Let $\mathbf{x}$ and $\mathbf{x}'$ be uniform grids of degrees m and l, respectively, on $[a, b]$. Prove that $\mathbf{D}_{\mathbf{x},\mathbf{x}'} = c\mathbf{D}_{\mathbf{t},\mathbf{t}'}$ for a scalar c. Provide a formula for c.

12. Let $p(x) = c_0 + c_1 x + c_2 x^2 + \cdots + c_m x^m$ be a polynomial represented in the monomial basis, and let $\mathbf{c}$ be the column vector $(c_0, c_1, c_2, \ldots, c_m)$. Find a matrix $\mathbf{D}$ such that the entries of $\mathbf{D}\mathbf{c}$ are the coefficients of $p'(x)$ in the monomial basis. Justify.

13. Let $\mathbf{D}_{\mathbf{x}',\mathbf{x}}$ be the differentiation matrix for grids $\mathbf{x}$ and $\mathbf{x}'$ of degrees $m + 1$ and m, respectively (and each having distinct nodes as usual). Prove that $\mathbf{D}_{\mathbf{x}',\mathbf{x}}$ has a one-dimensional null space. (In other words, the equation $\mathbf{D}_{\mathbf{x}',\mathbf{x}}\mathbf{p} = \mathbf{0}$ has a one-dimensional solution space.) What is special about a polynomial p whose sample $\mathbf{p} = p|_{\mathbf{x}}$ is in the null space?

14. Let $\mathbf{D}_{\mathbf{x'},\mathbf{x}}$ be the differentiation matrix for grids $\mathbf{x}$ and $\mathbf{x'}$ of degrees $m + 1$ and m, respectively. Let $\mathbf{q}$ be an $(m + 1)$-by-1 column vector. Suppose $\mathbf{p}_1$ and $\mathbf{p}_2$ are solutions to $\mathbf{Dp} = \mathbf{q}$. How are the entries of $\mathbf{p}_1$ and $\mathbf{p}_2$ related? Prove it.

15. Let $\mathbf{x}$ be a uniform grid of degree 2 on $[a, b]$, let p be a quadratic polynomial, and let $\mathbf{p} = p|_{\mathbf{x}}$. Construct a second-derivative matrix $\mathbf{D}$ such that $\mathbf{Dp}$ equals $p''((a + b)/2)$.

16. Let $\mathbf{x}$ be a uniform grid of degree 3 on $[a, b]$, let p be a cubic polynomial, and let $\mathbf{p} = p|_{\mathbf{x}}$. Let $\mathbf{x''}$ be the uniform grid of degree 1 on $[a, b]$. Construct a second-derivative matrix $\mathbf{D}$ such that $\mathbf{Dp}$ equals $p''|_{\mathbf{x''}}$.

17. Let $f(x) = e^{-cx^2}$. Sample f on the Chebyshev grid of degree 50 over $[-1, 1]$ and let p be the interpolating polynomial. Find a value for c for which $\|f - p\|_\infty < 10^{-3}$ but $\|f' - p'\|_\infty > 10^{10}$.

Chapter 14

Integration of a polynomial interpolant

Let q be a polynomial. We would like to compute three different types of integrals: the definite integral

$$\int_a^b q(x)\,dx,$$

the specific antiderivative

$$p_a + \int_a^x q(\xi)\,d\xi,$$

and the indefinite integral

$$\int q(x)\,dx.$$

The first is a number; the second is a function; and the third is an infinite family of functions. For example,

$$\int_0^1 x^2\,dx = \left[\frac{x^3}{3}\right]_0^1 = \frac{1}{3},$$

$$4 + \int_0^x \xi^2\,d\xi = 4 + \left[\frac{\xi^3}{3}\right]_0^x = 4 + \frac{x^3}{3},$$

$$\int x^2\,dx = \frac{x^3}{3} + C.$$

These problems are easy enough when q is expressed in the form $q(x) = c_0 + c_1 x + \cdots + c_m x^m$, but of course our polynomials are typically specified by samples on grids.

We establish relationships of the following forms between polynomials p and q of degrees $m+1$ and m, respectively, and samples $\mathbf{p}$ and $\mathbf{q}$ of these polynomials. The

157

boxes represent vectors and matrices:

calculus	matrix algebra

$$I = \int_a^b q(x)\,\mathrm{d}x$$

$$I = \boxed{\quad \boldsymbol{\lambda}^T \quad}\; \boxed{\mathbf{q}}$$

$$p(x) = p_a + \int_a^x q(\xi)\,\mathrm{d}\xi$$

$$\boxed{\mathbf{p}} = p_a \boxed{\mathbf{1}} + \boxed{\quad \mathbf{K} \quad}\;\boxed{\mathbf{q}}$$

$$p(x) + C = \int q(x)\,\mathrm{d}x$$

$$\boxed{\quad \mathbf{J} \quad}\;\boxed{\mathbf{p}} = \boxed{\quad \hat{\mathbf{K}} \quad}\;\boxed{\mathbf{q}}$$

14.1 ▪ The integration matrix

The key construction is an *integration matrix* that allows questions about polynomials to be translated into questions about finite sample vectors. Given grids $\mathbf{x} = (x_0, \ldots, x_{m+1})$ and $\mathbf{x}' = (x'_0, \ldots, x'_m)$ and a lower limit of integration a, the integration matrix $\mathbf{K}_{\mathbf{x},\mathbf{x}'}$ is defined to be the $(m+2)$-by-$(m+1)$ matrix

$$\mathbf{K}_{\mathbf{x},\mathbf{x}'} = \begin{bmatrix} k_{00} & k_{01} & \cdots & k_{0m} \\ k_{10} & k_{11} & \cdots & k_{1m} \\ \vdots & \vdots & \ddots & \vdots \\ k_{m0} & k_{m1} & \cdots & k_{mm} \\ k_{m+1,0} & k_{m+1,1} & \cdots & k_{m+1,m} \end{bmatrix} \tag{14.1}$$

with entries

$$k_{ij} = \frac{1}{l_j(x'_j)} \int_a^{x_i} l_j(x)\,\mathrm{d}x, \tag{14.2}$$

in which

$$l_j(x) = \prod_{\substack{k=0,\ldots,m \\ k \neq j}} (x - x'_k)$$

is the jth Lagrange basis polynomial for the grid $\mathbf{x}'$.

Theorem 14.1. *Let p and q be polynomials of degrees at most $m+1$ and m, respectively. Sample $\mathbf{p} = p|_{\mathbf{x}}$ and $\mathbf{q} = q|_{\mathbf{x}'}$. Then*

$$p(x) = \int_a^x q(\xi)\,\mathrm{d}\xi \tag{14.3}$$

if and only if

$$\mathbf{p} = \mathbf{K}_{x,x'}\mathbf{q}. \qquad (14.4)$$

Proof. The Lagrange form for q is

$$q(x) = \sum_{j=0}^{m} \frac{q(x'_j)}{l_j(x'_j)} l_j(x).$$

We find

$$\int_{a}^{x_i} q(x)\,dx = \int_{a}^{x_i} \sum_{j=0}^{m} \frac{q(x'_j)}{l_j(x'_j)} l_j(x)\,dx$$

$$= \sum_{j=0}^{m} \left(\frac{1}{l_j(x'_j)} \int_{a}^{x_i} l_j(x)\,dx \right) q(x'_j) = \sum_{j=0}^{m} k_{ij} q(x'_j).$$

Note that the final sum equals the $(i+1)$th entry of $\mathbf{K}_{x,x'}\mathbf{q}$. Hence, $\mathbf{p} = \mathbf{K}_{x,x'}\mathbf{q}$ if and only if $p(x_i) = \int_{a}^{x_i} q(x)\,dx$, $i = 0,\ldots,m+1$. Because p and the antiderivative of q are polynomials of degree $m+1$, this is true if and only if $p(x) = \int_{a}^{x} q(\xi)\,d\xi$ for all x. $\quad\square$

Example 14.2. Let $\mathbf{x} = (0, 0.25, 0.5)$ and $\mathbf{x}' = (0, 0.5)$. Evaluate the integration matrix using these grids and the lower limit of integration $a = 0$.

Solution. The integration matrix is

$$\mathbf{K}_{x,x'} = \begin{bmatrix} \frac{1}{l_0(x'_0)} \int_{a}^{x_0} l_0(x)\,dx & \frac{1}{l_1(x'_1)} \int_{a}^{x_0} l_1(x)\,dx \\ \frac{1}{l_0(x'_0)} \int_{a}^{x_1} l_0(x)\,dx & \frac{1}{l_1(x'_1)} \int_{a}^{x_1} l_1(x)\,dx \\ \frac{1}{l_0(x'_0)} \int_{a}^{x_2} l_0(x)\,dx & \frac{1}{l_1(x'_1)} \int_{a}^{x_2} l_1(x)\,dx \end{bmatrix}$$

$$= \begin{bmatrix} \frac{1}{0-0.5} \int_{0}^{0} (x-0.5)\,dx & \frac{1}{0.5-0} \int_{0}^{0} (x-0)\,dx \\ \frac{1}{0-0.5} \int_{0}^{0.25} (x-0.5)\,dx & \frac{1}{0.5-0} \int_{0}^{0.25} (x-0)\,dx \\ \frac{1}{0-0.5} \int_{0}^{0.5} (x-0.5)\,dx & \frac{1}{0.5-0} \int_{0}^{0.5} (x-0)\,dx \end{bmatrix}$$

$$= \begin{bmatrix} 0 & 0 \\ 3/16 & 1/16 \\ 1/4 & 1/4 \end{bmatrix}. \qquad \blacksquare$$

14.2 ▪ The definite integral

With the integration matrix $\mathbf{K}_{x,x'}$ defined, the three forms of the integral follow quickly. First in line is the definite integral. By the fundamental theorem of calculus, we have $\int_{a}^{b} q(x)\,dx = p(b) - p(a)$, in which p is an antiderivative of q. Let $\mathbf{E}_{b,x}$ be the matrix that evaluates a polynomial at $x = b$ given a sample on $\mathbf{x}$, as defined by (12.1)–(12.2). Then define $\boldsymbol{\lambda}$ to be the column vector whose transpose is

$$\boldsymbol{\lambda}^T = \mathbf{E}_{b,x}\mathbf{K}_{x,x'}.$$

Theorem 14.3. *Let* $\mathbf{q} = q|_{\mathbf{x}'}$ *be a sample of a polynomial* q *of degree at most* m. *Then*

$$\int_a^b q(x)\,dx = \boldsymbol{\lambda}^T \mathbf{q}. \tag{14.5}$$

Proof. Define $p(x) = \int_a^x q(\xi)\,d\xi$ and $\mathbf{p} = p|_{\mathbf{x}}$. By Theorems 12.1 and 14.1,

$$\boldsymbol{\lambda}^T \mathbf{q} = \mathbf{E}_{b,\mathbf{x}}\mathbf{K}_{\mathbf{x},\mathbf{x}'}\mathbf{q} = \mathbf{E}_{b,\mathbf{x}}\mathbf{p} = p(b) = \int_a^b q(\xi)\,d\xi = \int_a^b q(x)\,dx. \qquad \Box$$

Remark 14.4. Often, the last node in $\mathbf{x}$ is $x_{m+1} = b$. In this case,

$$\mathbf{E}_{b,\mathbf{x}} = \begin{bmatrix} 0 & 0 & \cdots & 0 & 1 \end{bmatrix},$$

and so $\boldsymbol{\lambda}^T$ is just the last row of the integration matrix $\mathbf{K}_{\mathbf{x},\mathbf{x}'}$.

Example 14.5. Compute $\int_0^{0.5}(2x+3)\,dx$ using Theorem 14.3 with the grids from Example 14.2. Check the result using basic calculus.

Solution. The integration matrix

$$\mathbf{K}_{\mathbf{x},\mathbf{x}'} = \begin{bmatrix} 0 & 0 \\ 3/16 & 1/16 \\ 1/4 & 1/4 \end{bmatrix}$$

was computed in Example 14.2. Because the last node in $\mathbf{x}$ coincides with the upper limit of integration, $\boldsymbol{\lambda}^T$ equals the last row of the integration matrix:

$$\boldsymbol{\lambda}^T = \begin{bmatrix} 1/4 & 1/4 \end{bmatrix}.$$

The sample of the integrand $q(x) = 2x + 3$ is

$$\mathbf{q} = \begin{bmatrix} q(0) \\ q(0.5) \end{bmatrix} = \begin{bmatrix} 3 \\ 4 \end{bmatrix}.$$

Therefore,

$$\int_0^{0.5} q(x)\,dx = \boldsymbol{\lambda}^T \mathbf{q} = \begin{bmatrix} 1/4 & 1/4 \end{bmatrix}\begin{bmatrix} 3 \\ 4 \end{bmatrix} = 1.75.$$

The result can be checked with basic calculus rules: $\int_0^{0.5}(2x+3)\,dx = [x^2 + 3x]_0^{0.5} = (0.5)^2 + 3(0.5) - 0^2 - 3(0) = 1.75.$ ∎

14.3 ▪ A specific antiderivative

The function $p(x) = p_a + \int_a^x q(\xi)\,d\xi$ is the specific antiderivative of $q(x)$ that satisfies $p(a) = p_a$. Its computation can be reduced to a matrix-vector product followed by an addition.

Theorem 14.6. *Let p and q be polynomials of degrees at most $m+1$ and m, respectively. Sample $\mathbf{p} = p|_\mathbf{x}$ and $\mathbf{q} = q|_{\mathbf{x}'}$. Then*

$$p(x) = p_a + \int_a^x q(\xi)\,d\xi \tag{14.6}$$

if and only if

$$\mathbf{p} = p_a\mathbf{1} + \mathbf{K}_{\mathbf{x},\mathbf{x}'}\mathbf{q}, \tag{14.7}$$

in which $\mathbf{1}$ is an $(m+2)$-by-1 vector of ones.

Proof. We have $p(x) = p_a + \int_a^x q(\xi)\,d\xi$ if and only if $p(x) - p_a = \int_a^x q(\xi)\,d\xi$. By Theorem 14.1, this is true if and only if $\mathbf{p} - p_a\mathbf{1} = \mathbf{K}_{\mathbf{x},\mathbf{x}'}\mathbf{q}$.   □

Example 14.7. Compute

$$10 + \int_0^x (2\xi + 3)\,d\xi$$

using Theorem 14.6 with the grids of Example 14.2. Check the result using basic calculus.

Solution. Let $q(x) = 2x + 3$. Using the integration matrix $\mathbf{K}_{\mathbf{x},\mathbf{x}'}$ of Example 14.2 and the sample $\mathbf{q}$ of the previous example, we compute

$$\mathbf{p} = 10 \times \mathbf{1} + \mathbf{K}_{\mathbf{x},\mathbf{x}'}\mathbf{q} = \begin{bmatrix} 10 \\ 10 \\ 10 \end{bmatrix} + \begin{bmatrix} 0 & 0 \\ 3/16 & 1/16 \\ 1/4 & 1/4 \end{bmatrix} \begin{bmatrix} 3 \\ 4 \end{bmatrix} = \begin{bmatrix} 10.0000 \\ 10.8125 \\ 11.7500 \end{bmatrix}.$$

Hence, $p(x) = 10 + \int_0^x q(\xi)\,d\xi$ is the unique quadratic polynomial that interpolates $(0, 10)$, $(0.25, 10.8125)$, $(0.5, 11.75)$.

To check this, note that the integral is $p(x) = 10 + \int_0^x (2\xi + 3)\,d\xi = 10 + [\xi^2 + 3\xi]_0^x = x^2 + 3x + 10$. This has values $p(0) = 10$, $p(0.25) = 10.8125$, $p(0.5) = 11.75$, and it is the unique quadratic polynomial with these values.   ∎

14.4 ▪ The indefinite integral

The third integral to consider is the indefinite integral $\int q(x)\,dx$. Rather than generate an infinite family of functions, we develop a test to determine whether a given degree-$(m+1)$ polynomial p is an antiderivative of a degree-m polynomial q, based on the samples $\mathbf{p} = p|_\mathbf{x}$ and $\mathbf{q} = q|_{\mathbf{x}'}$. There are a number of such tests; the formulation below admits a simple error analysis.

According to the fundamental theorem of calculus, p is an antiderivative of q if and only if $p(x) - p(a) = \int_a^x q(\xi)\,d\xi$. The next theorem shows that when p and q are polynomials of appropriate degrees, this relation is equivalent to a finite system of equations

$$\mathbf{J}_\mathbf{x}\mathbf{p} = \hat{\mathbf{K}}_{\mathbf{x},\mathbf{x}'}\mathbf{q}.$$

The matrices are constructed so that

$$\mathbf{J}_\mathbf{x}\mathbf{p} = \begin{bmatrix} p(x_1) - p(x_0) \\ p(x_2) - p(x_0) \\ \vdots \\ p(x_{m+1}) - p(x_0) \end{bmatrix}, \quad \hat{\mathbf{K}}_{\mathbf{x},\mathbf{x}'}\mathbf{q} = \begin{bmatrix} \int_{x_0}^{x_1} q(x)\,dx \\ \int_{x_0}^{x_2} q(x)\,dx \\ \vdots \\ \int_{x_0}^{x_{m+1}} q(x)\,dx \end{bmatrix}. \tag{14.8}$$

Matrix $\mathbf{J_x}$ is the $(m+1)$-by-$(m+2)$ matrix

$$\mathbf{J_x} = \begin{bmatrix} -1 & 1 & & & \\ & -1 & 1 & & \\ & \vdots & & \ddots & \\ & -1 & & & 1 \end{bmatrix}, \tag{14.9}$$

in which the blank entries represent zeros, and $\hat{\mathbf{K}}_{x,x'}$ is the square matrix obtained by deleting the first row of the $(m+2)$-by-$(m+1)$ matrix $\mathbf{K}_{x,x'}$.

Theorem 14.8. *Let p and q be polynomials of degrees at most $m+1$ and m, respectively. Then p is an antiderivative of q if and only if*

$$\mathbf{J_x}\mathbf{p} = \hat{\mathbf{K}}_{x,x'}\mathbf{q}. \tag{14.10}$$

Proof. Check that the definitions of $\mathbf{J_x}$ and $\hat{\mathbf{K}}_{x,x'}$ imply the matrix-vector products in (14.8).

If p is an antiderivative of q, then $p(x) - p(x_0) = \int_{x_0}^{x} q(\xi)\,d\xi$ for all x, so in particular, $p(x_i) - p(x_0) = \int_{x_0}^{x_i} q(x)\,dx$ for $i = 1, \ldots, m+1$. This implies $\mathbf{J_x}\mathbf{p} = \hat{\mathbf{K}}_{x,x'}\mathbf{q}$.

Conversely, if $\mathbf{J_x}\mathbf{p} = \hat{\mathbf{K}}_{x,x'}\mathbf{q}$, then the degree-$(m+1)$ polynomial $p(x) - p(x_0) - \int_{x_0}^{x} q(\xi)\,d\xi$ has zeros at $x = x_i$, $i = 0, \ldots, m+1$. As a degree-$(m+1)$ polynomial with at least $m+2$ zeros, $p(x) - p(x_0) - \int_{x_0}^{x} q(\xi)\,d\xi$ must equal zero everywhere, which shows that p is an antiderivative of q. $\square$

Example 14.9. Consider the indefinite integral $\int (2x+3)\,dx$. Let $q(x) = 2x+3$ be the integrand and $\mathbf{q}$ its sample on the grid $\mathbf{x'} = (0, 0.5)$. Let $p(x)$ be any of the infinitely many antiderivatives of $q(x)$, and let $\mathbf{p}$ be its sample on the grid $\mathbf{x} = (0, 0.25, 0.5)$. Find a linear system of equations that characterizes the possible $\mathbf{p}$ using Theorem 14.8.

Solution. The truncated integration matrix is obtained by deleting the first row of the integration matrix of Example 14.2, giving

$$\hat{\mathbf{K}}_{x,x'} = \begin{bmatrix} 3/16 & 1/16 \\ 1/4 & 1/4 \end{bmatrix}.$$

A quadratic polynomial p is an antiderivative of $q(x) = 2x+3$ if and only if $\mathbf{p} = p|_x$ satisfies

$$\mathbf{J_x}\mathbf{p} = \hat{\mathbf{K}}_{x,x'}\mathbf{q},$$

in which $\mathbf{q} = q|_{x'} = (3, 4)$. This is equivalent to

$$\begin{bmatrix} -1 & 1 & 0 \\ -1 & 0 & 1 \end{bmatrix} \mathbf{p} = \begin{bmatrix} 3/16 & 1/16 \\ 1/4 & 1/4 \end{bmatrix} \begin{bmatrix} 3 \\ 4 \end{bmatrix},$$

or more simply,

$$\begin{bmatrix} -1 & 1 & 0 \\ -1 & 0 & 1 \end{bmatrix} \mathbf{p} = \begin{bmatrix} 0.8125 \\ 1.7500 \end{bmatrix}. \tag{14.11}$$

This is the requested system of linear equations.

Check. The set of solutions to the linear system is the set of vectors $\mathbf{p}$ of the form

$$\mathbf{p} = \begin{bmatrix} C \\ C + 0.8125 \\ C + 1.7500 \end{bmatrix}.$$

The antiderivative found in the previous example has a sample of this form, as does every other antiderivative. ∎

14.5 ▪ Implementation

It is time for computer code. The first matter is the integration matrix $\mathbf{K}_{\mathbf{x},\mathbf{x}'}$, whose entries are $k_{ij} = \frac{1}{l_j(x'_j)} \int_a^{x_i} l_j(x)\, dx$. Fortunately, these laborious integrals do not have to be computed one by one. Instead, the entire matrix can be computed as the solution to a single matrix equation, as shown in the upcoming theorem.

The statement of the theorem uses block-matrix arithmetic. Note that block matrices can be added and multiplied in the natural way. For example,

$$\begin{bmatrix} \mathbf{A} & \mathbf{B} \\ \mathbf{C} & \mathbf{D} \end{bmatrix}\begin{bmatrix} \mathbf{E} & \mathbf{F} \\ \mathbf{G} & \mathbf{H} \end{bmatrix} = \begin{bmatrix} \mathbf{AE} + \mathbf{BG} & \mathbf{AF} + \mathbf{BH} \\ \mathbf{CE} + \mathbf{DG} & \mathbf{CF} + \mathbf{DH} \end{bmatrix},$$

as long as the blocks have compatible sizes that make the multiplications and additions legal.

In the upcoming theorem, $\mathbf{D}_{\mathbf{x}',\mathbf{x}}$ is the $(m + 1)$-by-$(m + 2)$ differentiation matrix defined by (13.1)–(13.2), $\mathbf{E}_{a,\mathbf{x}}$ is the 1-by-$(m + 2)$ evaluation matrix defined by (12.1)–(12.2), and $\mathbf{K}_{\mathbf{x},\mathbf{x}'}$ is the $(m + 2)$-by-$(m + 1)$ integration matrix introduced earlier in this chapter.

Theorem 14.10. *If $\mathbf{x}$ and $\mathbf{x}'$ are grids of degrees $m + 1$ and m, respectively, then the integration matrix $\mathbf{K}_{\mathbf{x},\mathbf{x}'}$ is the unique solution to*

$$\begin{bmatrix} \mathbf{D}_{\mathbf{x}',\mathbf{x}} \\ \mathbf{E}_{a,\mathbf{x}} \end{bmatrix} \mathbf{K}_{\mathbf{x},\mathbf{x}'} = \begin{bmatrix} \mathbf{I} \\ \mathbf{0} \end{bmatrix}. \tag{14.12}$$

Here, $\mathbf{I}$ is the $(m + 1)$-by-$(m + 1)$ identity matrix, and $\mathbf{0}$ is a 1-by-$(m + 1)$ row of zeros.

Proof. Let $\mathbf{q}$ be a vector of length $m + 1$ and p_a a real number. Define q to be the unique polynomial of degree at most m satisfying $q|_{\mathbf{x}'} = \mathbf{q}$, and define $p(x) = p_a + \int_a^x q(\xi)\, d\xi$ and $\mathbf{p} = p|_{\mathbf{x}}$. Also, let $\mathbf{1}$ be an $(m + 2)$-by-1 column of ones. By Theorem 14.6,

$$\begin{bmatrix} \mathbf{K}_{\mathbf{x},\mathbf{x}'} & \mathbf{1} \end{bmatrix}\begin{bmatrix} \mathbf{q} \\ p_a \end{bmatrix} = \mathbf{K}_{\mathbf{x},\mathbf{x}'}\mathbf{q} + p_a\mathbf{1} = \mathbf{p}.$$

Then, by Theorems 13.1 and 12.1,

$$\begin{bmatrix} \mathbf{D}_{\mathbf{x}',\mathbf{x}} \\ \mathbf{E}_{a,\mathbf{x}} \end{bmatrix}\begin{bmatrix} \mathbf{K}_{\mathbf{x},\mathbf{x}'} & \mathbf{1} \end{bmatrix}\begin{bmatrix} \mathbf{q} \\ p_a \end{bmatrix} = \begin{bmatrix} \mathbf{D}_{\mathbf{x}',\mathbf{x}} \\ \mathbf{E}_{a,\mathbf{x}} \end{bmatrix}\mathbf{p} = \begin{bmatrix} \mathbf{q} \\ p_a \end{bmatrix}.$$

Equivalently, for every vector $\mathbf{v}$,

$$\begin{bmatrix} \mathbf{D}_{\mathbf{x}',\mathbf{x}} \\ \mathbf{E}_{a,\mathbf{x}} \end{bmatrix}\begin{bmatrix} \mathbf{K}_{\mathbf{x},\mathbf{x}'} & \mathbf{1} \end{bmatrix}\mathbf{v} = \mathbf{v}.$$

Thus, it must be the case that

$$\begin{bmatrix} \mathbf{D}_{x',x} \\ \mathbf{E}_{a,x} \end{bmatrix} \begin{bmatrix} \mathbf{K}_{x,x'} & 1 \end{bmatrix} = \begin{bmatrix} \mathbf{I} & \mathbf{0} \\ \mathbf{0} & 1 \end{bmatrix}. \tag{14.13}$$

Equation (14.12) is obtained by deleting the last column on each side of (14.13).

The solution is unique because the square matrix

$$\begin{bmatrix} \mathbf{D}_{x',x} \\ \mathbf{E}_{a,x} \end{bmatrix}$$

is invertible according to (14.13).   □

Equation (14.12) is closely related to the fundamental theorem of calculus; see Exercises 11–12.

The integration matrix $\mathbf{K}_{x,x'}$ can be computed numerically by solving (14.12). Routines intmatrixgen and intmatrix_ in Listing 14.1 do just this. (In the MATLAB language, X = A\B computes the solution **X** to the equation **AX** = **B**. Also note that the MATLAB function eye constructs an identity matrix of specified size.)

Listing 14.1. Compute an integration matrix using provided grids.

```
function K = intmatrixgen(xs,xs_,a)

ws = baryweights(xs);
K = intmatrix_(xs,ws,xs_,a);
```

```
function K = intmatrix_(xs,ws,xs_,a)

m = length(xs_)-1;
D = diffmatrix_(xs_,xs,ws);
Ea = evalmatrix_(a,xs,ws);
I = eye(m+1);
K = [ D; Ea ]\[ I; zeros(1,m+1) ];
K(xs==a,:) = 0;
```

Example 14.11. Verify that intmatrixgen produces the correct integration matrix for the grids in Example 14.2.

Solution. The integration matrix is computed automatically.

```
>> xs = [0; 0.25; 0.5];
>> xs_ = [0; 0.5];
>> a = 0;
>> K = intmatrixgen(xs,xs_,a)
K =
                     0                     0
   0.187500000000000   0.062500000000000
   0.250000000000000   0.250000000000000
```

Check that this matrix equals the solution to Example 14.2.   ■

The three forms of integration—definite integration, finding a specific antiderivative, and indefinite integration—are implemented by definitegen, antiderivgen, and indefinitegen in Listings 14.2–14.4.

Listing 14.2. Compute a definite integral using provided grids.

```
function I = definitegen(qs,ab,xs,xs_)

a = ab(1);
b = ab(2);
K = intmatrixgen(xs,xs_,a);
Eb = evalmatrixgen(b,xs);
lambdas = (Eb*K)';
I = lambdas'*qs;
```

Listing 14.3. Compute an antiderivative using provided grids.

```
function ps = antiderivgen(pa,qs,xs,xs_,a)

m = length(qs)-1;
K = intmatrixgen(xs,xs_,a);
ps = pa*ones(m+2,1)+K*qs;
```

Example 14.12. Repeat Examples 14.5, 14.7, and 14.9 using as much automation as possible.

Solution. The grids are placed as in the earlier examples.

```
>> a = 0; b = 0.5;
>> xs = [0; 0.25; 0.5];
>> xs_ = [0; 0.5];
```

The integrand $q(x) = 2x + 3$ is sampled on $\mathbf{x}'$.

```
>> q = @(x) 2*x+3;
>> qs = arrayfun(q,xs_);
```

The definite integral $\int_0^{0.5} q(x)\, dx$ is computed.

```
>> I = definitegen(qs,[a b],xs,xs_)
I =
    1.750000000000000
```

This agrees with Example 14.5.

Now the antiderivative $p(x) = 10 + \int_0^x q(\xi)\, d\xi$ is computed.

```
>> ps = antiderivgen(10,qs,xs,xs_,a)
ps =
   10.000000000000000
   10.812500000000000
   11.750000000000000
```

This agrees with Example 14.7.

Finally, the indefinite integral is represented as the solution space of a linear system.

```
>> [J,K] = indefinitegen(xs,xs_);
```

The left-hand side of the linear system is $\mathbf{J_x}$:

```
>> J
J =
    -1     1     0
    -1     0     1
```

And the right-hand side is $\hat{\mathbf{K}}_{\mathbf{x,x'}}\mathbf{q}$:

```
>> K*qs
ans =
```

Listing 14.4. Construct matrices for indefinite integration on provided grids.

```
function [J,K] = indefinitegen(xs,xs_)

m = length(xs_)-1;
J = [ -ones(m+1,1) eye(m+1) ];
K = intmatrixgen(xs,xs_,xs(1));
K = K(2:end,:);
```

```
0.812500000000000
1.750000000000000
```
These agree with Example 14.9.   ■

14.6 ▪ Accuracy analysis for integration

We wish to show that the integral of a function g is accurately approximated by the integral of a polynomial interpolant q as long as the interpolation fit is good.

The following theorem, known as the triangle inequality for integrals, is the basis of our accuracy analysis.

Theorem 14.13. *If g is an integrable function on* $[a, b]$, *then*

$$\left| \int_a^b g(x)\, dx \right| \le \int_a^b |g(x)|\, dx. \tag{14.14}$$

The proof is Exercise 14.

The next theorem can be used to bound error from definite integration. Although the theorem applies quite broadly, it is most useful to us when q is a polynomial interpolant for g.

Theorem 14.14. *If g and q are integrable functions on* $[a, b]$, *then*

$$\left| \int_a^b g(x)\, dx - \int_a^b q(x)\, dx \right| \le (b-a)\, \|g - q\|_\infty. \tag{14.15}$$

Proof. By linearity and the triangle inequality for integrals,

$$\left| \int_a^b g(x)\, dx - \int_a^b q(x)\, dx \right| = \left| \int_a^b (g(x) - q(x))\, dx \right|$$

$$\le \int_a^b |g(x) - q(x)|\, dx \le \int_a^b \|g - q\|_\infty\, dx = (b-a)\|g - q\|_\infty. \qquad \Box$$

The next theorem bounds error from antidifferentiation. Again, it is useful to consider q as a polynomial interpolant for the integrand g, although the theorem remains true regardless of q's origin.

Theorem 14.15. *If $f(x) = f_a + \int_a^x g(\xi)\,d\xi$ and $p(x) = f_a + \int_a^x q(\xi)\,d\xi$ on $[a,b]$, then*

$$\|f - p\|_\infty \le (b-a)\|g - q\|_\infty. \tag{14.16}$$

Proof. By the previous theorem, for any $x \in [a,b]$,

$$|f(x) - p(x)| = \left| \int_a^x g(x)\,dx - \int_a^x q(x)\,dx \right|$$

$$\le (x-a)\|g-q\|_\infty \le (b-a)\|g-q\|_\infty. \qquad \square$$

The final bound concerns indefinite integration. We have already established that if p and q are polynomials with $p' = q$, then $\mathbf{J_x p} = \hat{\mathbf{K}}_{\mathbf{x},\mathbf{x}'}\mathbf{q}$. For functions f and g that are not polynomials, we would like $\mathbf{J_x f} - \hat{\mathbf{K}}_{\mathbf{x},\mathbf{x}'}\mathbf{g}$ to be small when $f' = g$.

The bound in the theorem involves the *vector infinity norm*. If $\mathbf{v} = (v_1, \dots, v_n)$ is a column vector, then its infinity norm is $\|\mathbf{v}\|_\infty = \max_i |v_i|$. Vector norms are studied in more depth in Chapter 21.

Theorem 14.16. *Suppose f is an antiderivative of g. Let p be the polynomial interpolant for $\mathbf{f} = f|_\mathbf{x}$, and let q be the polynomial interpolant for $\mathbf{g} = g|_{\mathbf{x}'}$. Then*

$$\|\mathbf{J_x f} - \hat{\mathbf{K}}_{\mathbf{x},\mathbf{x}'}\mathbf{g}\|_\infty \le 2\|f - p\|_\infty + (b-a)\|g-q\|_\infty. \tag{14.17}$$

Proof. We have

$$\mathbf{J_x f} - \hat{\mathbf{K}}_{\mathbf{x},\mathbf{x}'}\mathbf{g} = \begin{bmatrix} p(x_1) - p(x_0) - \int_{x_0}^{x_1} q(x)\,dx \\ \vdots \\ p(x_{m+1}) - p(x_0) - \int_{x_0}^{x_{m+1}} q(x)\,dx \end{bmatrix}.$$

Let $u_i = f(x_i) - f(x_0) - \int_{x_0}^{x_i} q(x)\,dx$ and $v_i = -f(x_i) + p(x_i)$ and $w_i = f(x_0) - p(x_0)$. Then the ith entry of $\mathbf{J_x f} - \hat{\mathbf{K}}_{\mathbf{x},\mathbf{x}'}\mathbf{g}$ equals $u_i + v_i + w_i$. By Theorem 14.15, $|u_i| \le (b-a)\|g-q\|_\infty$. Also, $|v_i|$ and $|w_i|$ are bounded above by $\|f - p\|_\infty$. By the triangle inequality,

$$\left| p(x_i) - p(x_0) - \int_{x_0}^{x_i} q(x)\,dx \right| \le |u_i| + |v_i| + |w_i|$$

$$\le (b-a)\|g-q\|_\infty + \|f-p\|_\infty + \|f-p\|_\infty. \qquad \square$$

The three preceding theorems bound integration error in terms of interpolation error. By incorporating earlier results on interpolation accuracy, we can soon establish rates of convergence for numerical integration on common grids.

Notes

The integration matrix $\mathbf{K}_{\mathbf{x},\mathbf{x}'}$ is computed by solving the linear system (14.12). Is this numerically stable? Chapter 21 introduces the *condition number* of a matrix, a key concept in analyzing the accuracy of solutions to linear systems. Exercises 16 and 17 in that chapter consider the specific case of (14.12).

Exercises

1. Let $\mathbf{x}' = (0, 1)$, $\mathbf{x} = (0, 1/2, 1)$, and $a = 0$. Compute the entries of $\mathbf{K}_{\mathbf{x},\mathbf{x}'}$ using (14.2).

2. Let $\mathbf{x}' = (-1, 0, 1)$, $\mathbf{x} = (-1, -1/3, 1/3, 1)$, and $a = -1$. Compute the bottom row of $\mathbf{K}_{\mathbf{x},\mathbf{x}'}$ using (14.2).

3. Let $\mathbf{x}' = (0, 1/2, 1)$ and $\mathbf{x} = (0, 1/3, 2/3, 1)$. Compute $\int_0^1 (x^2 + 1)\,dx$ by constructing $\boldsymbol{\lambda}$, sampling $\mathbf{q}$, and evaluating $\boldsymbol{\lambda}^T\mathbf{q}$.

4. Let $\mathbf{x}' = (-1, 1)$ and $\mathbf{x} = (-1, 0, 1)$. Let $q(x) = 4x - 1$ and $p(x) = 2 + \int_{-1}^x q(\xi)\,d\xi$. Sample $\mathbf{q} = q|_{\mathbf{x}'}$ and then compute $\mathbf{p} = p|_{\mathbf{x}}$ using (14.7).

5. Let $\mathbf{x}' = (-1, 1)$, $\mathbf{x} = (-1, 0, 1)$, and $q(x) = 4x - 1$. Find a linear system of the same form as (14.11) whose solution set consists of samples $\mathbf{p} = p|_{\mathbf{x}}$ of all possible antiderivatives p of q. Use $a = -1$ as the lower limit of integration.

6. The definite integral of a polynomial can be computed using $\boldsymbol{\lambda}^T\mathbf{q}$, in which $\boldsymbol{\lambda}$ is defined by $\boldsymbol{\lambda}^T = \mathbf{E}_{b,\mathbf{x}}\mathbf{K}_{\mathbf{x},\mathbf{x}'}$. Prove that $\boldsymbol{\lambda}^T = (\mathbf{E}_{b,\mathbf{x}} - \mathbf{E}_{a,\mathbf{x}})\mathbf{K}_{\mathbf{x},\mathbf{x}'}$ as well. (Note that this expression more directly reflects the equation $\int_a^b q(x)\,dx = p(b) - p(a)$ from the fundamental theorem of calculus.)

7. Let $\mathbf{x}$, $\mathbf{x}'$, and $\hat{\mathbf{x}}$ be grids of degrees $m + 1$, m, and m, respectively, and let a be a number that is distinct from the nodes in $\hat{\mathbf{x}}$. Write a MATLAB function to create matrices $\mathbf{J}$ and $\hat{\mathbf{K}}$ for which

$$\mathbf{Jp} = \begin{bmatrix} p(\hat{x}_0) - p(a) \\ \vdots \\ p(\hat{x}_m) - p(a) \end{bmatrix}, \quad \hat{\mathbf{K}}\mathbf{q} = \begin{bmatrix} \int_a^{\hat{x}_0} q(x)\,dx \\ \vdots \\ \int_a^{\hat{x}_m} q(x)\,dx \end{bmatrix},$$

assuming $\mathbf{p} = p|_{\mathbf{x}}$ is a sample of a degree-$(m + 1)$ polynomial p and $\mathbf{q} = q|_{\mathbf{x}'}$ is a sample of a degree-m polynomial q. (The matrices in (14.8) are for the special case in which $a = x_0$ and $\hat{x}_i = x_{i+1}$.) Demonstrate on an example.

8. Define $\mathbf{J}$, $\hat{\mathbf{K}}$, p, and q as in the previous exercise. Prove that p is an antiderivative of q if and only if $\mathbf{Jp} = \hat{\mathbf{K}}\mathbf{q}$.

9. Prove the following fact, which is used in the proof of Theorem 14.10: If $\mathbf{A}$ is an n-by-n matrix and $\mathbf{Ax} = \mathbf{x}$ for every n-by-1 vector $\mathbf{x}$, then $\mathbf{A}$ equals the identity matrix—the square matrix with 1's on its main diagonal and 0's elsewhere.

10. Let $\mathbf{x} = (x_0, \ldots, x_{m+1})$ and $\mathbf{x}' = (x_0', \ldots, x_m')$ be grids of degrees $m + 1$ and m, respectively, let $a = x_0$, and let $\mathbf{p}$ be an $(m + 2)$-by-1 vector. Under what circumstances does $\mathbf{K}_{\mathbf{x},\mathbf{x}'}\mathbf{q} = \mathbf{p}$ have a solution $\mathbf{q}$? Is the solution unique?

11. Equation (14.12) implies $\mathbf{D}_{\mathbf{x}',\mathbf{x}}\mathbf{K}_{\mathbf{x},\mathbf{x}'} = \mathbf{I}$. Explain how this equation is related to the fundamental theorem of calculus. How is it similar, and in what way is it more restricted?

12. Prove $\hat{\mathbf{K}}_{\mathbf{x},\mathbf{x}'}\mathbf{D}_{\mathbf{x}',\mathbf{x}} = \mathbf{J}_{\mathbf{x}}$. Then explain how this equation is related to the fundamental theorem of calculus.

13. Prove $\mathbf{K}_{\mathbf{x},\mathbf{x}'}\mathbf{D}_{\mathbf{x}',\mathbf{x}} + \mathbf{1}\mathbf{E}_{a,\mathbf{x}} = \mathbf{I}$. (See Theorem 14.10 and its proof for the notation.) Interpret this equation in the language of polynomials.

14. Prove Theorem 14.13. *Hint.* Use the fact that if $u(x) \le v(x)$ for all $x \in [a, b]$, then $\int_a^b u(x)\, dx \le \int_a^b v(x)\, dx$.

15. Give an example for which $\left| \int_a^b g(x)\, dx \right| < \int_a^b |g(x)|\, dx$ and another example for which the quantities are equal.

16. Suppose $\int_a^b g(x)\, dx$ is approximated by $\int_a^b q(x)\, dx$, in which q is a polynomial interpolant for g. However, the sample is inexact, and $\mathbf{q}$ in (14.5) is replaced by an approximation $\hat{\mathbf{q}}$ that satisfies $\|\mathbf{q} - \hat{\mathbf{q}}\|_\infty \le \delta$. Prove that

$$\left| \int_a^b g(x)\, dx - \boldsymbol{\lambda}^T \hat{\mathbf{q}} \right| \le (b-a)\|g - q\|_\infty + (b-a)\Lambda\delta,$$

in which Λ is the Lebesgue constant for the grid $\mathbf{x}'$ on which q is sampled.

17. Let $p(x) = (p_1(x), \ldots, p_r(x))$ be a vector-valued function in which each $p_k(x)$ is a polynomial of degree at most $m + 1$. Let $q(x) = (q_1(x), \ldots, q_r(x))$ be a vector-valued function in which each $q_k(x)$ is a polynomial of degree at most m. Let $\mathbf{x} = (x_0, \ldots, x_{m+1})$ and $\mathbf{x}' = (x_0', \ldots, x_m')$ be grids and sample

$$\mathbf{p} = \begin{bmatrix} p_0(x_0) & \cdots & p_0(x_{m+1}) \\ \vdots & \ddots & \vdots \\ p_r(x_0) & \cdots & p_r(x_{m+1}) \end{bmatrix}, \quad \mathbf{q} = \begin{bmatrix} q_0(x_0') & \cdots & q_0(x_m') \\ \vdots & \ddots & \vdots \\ q_r(x_0') & \cdots & q_r(x_m') \end{bmatrix}.$$

Prove that each p_k is an antiderivative of the corresponding q_k if and only if

$$(\mathbf{J_x} \otimes \mathbf{I})\, \text{vec}(\mathbf{p}) = (\hat{\mathbf{K}}_{\mathbf{x},\mathbf{x}'} \otimes \mathbf{I})\, \text{vec}(\mathbf{q}).$$

(The vec operator and the Kronecker product $\otimes$ are introduced in Exercises 17–18 in Chapter 12.) Also, how many rows and columns are in each matrix and vector in the above equation?

Chapter 15

Integration on a piecewise-uniform grid: Newton–Cotes quadrature

Developing a practical integration method from the theory of the previous chapter is the aim of *Newton–Cotes quadrature*. The method consists of two steps: (1) Replace an integrand by a piecewise-polynomial interpolant on a piecewise-uniform grid, and (2) integrate the interpolant in place of the original function. This process is illustrated for the definite integral in Figure 15.1. In each graph, the area under the blue curve is approximated by the shaded area under the orange interpolant. The case $m = 0$, illustrated in the left graph of Figure 15.1, is known as the *midpoint rule*; the case $m = 1$, illustrated in the middle graph, is known as the *trapezoid rule*; and the case $m = 2$, illustrated in the right graph, is known as *Simpson's rule*.

By the way, the term *quadrature* refers to the calculation of area, originally by transforming shapes into squares with equal area but in modern times through any numerical scheme.

The present chapter focuses on the mechanics of integrating a piecewise-polynomial function, and the next chapter analyzes the accuracy of the larger interpolation-and-integration scheme.

15.1 ▪ Integration matrix for a uniform grid

In the previous chapter, we saw how to compute an integration matrix $\mathbf{K}_{\mathbf{x},\mathbf{x}'}$ by solving

$$\begin{bmatrix} \mathbf{D}_{\mathbf{x}',\mathbf{x}} \\ \mathbf{E}_{a,\mathbf{x}} \end{bmatrix} \mathbf{K}_{\mathbf{x},\mathbf{x}'} = \begin{bmatrix} \mathbf{I} \\ \mathbf{0} \end{bmatrix}.$$

This can be expensive. Fortunately, uniform grids have a common structure that enables reuse. Specifically, an integration matrix for $[a, b]$ is easily constructed from the matrix for $[0, 1]$ of the same degree, as described in the next theorem.

Let $\mathbf{t}$ and $\mathbf{t}'$ be uniform grids of degrees $m + 1$ and m, respectively, on $[0, 1]$, and let

$$\mathbf{K}_m = \mathbf{K}_{\mathbf{t},\mathbf{t}'}, \tag{15.1}$$

the integration matrix for this pair of grids with a lower limit of integration at $t = 0$.

Theorem 15.1. *Let $\mathbf{x}$ and $\mathbf{x}'$ be uniform grids of degrees $m + 1$ and m, respectively, on an interval $[a, b]$, and let $\mathbf{K}_{\mathbf{x},\mathbf{x}'}$ be the integration matrix for these grids with a lower*

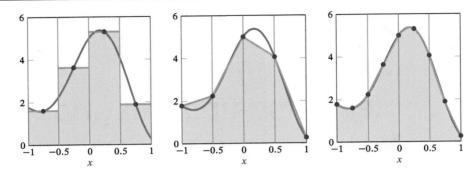

Figure 15.1. *Newton–Cotes quadrature of degree m: $m = 0$ (left), $m = 1$ (middle), and $m = 2$ (right).*

limit of integration at $x = a$. Then

$$\mathbf{K_{x,x'}} = (b - a)\mathbf{K}_m.$$

Proof. The $(i + 1), (j + 1)$-entry of $\mathbf{K_{x,x'}}$ is

$$k_{ij} = \frac{1}{\prod_{k \neq j}(x'_j - x'_k)} \int_a^{x_i} \prod_{k \neq j}(x - x'_k)\, dx.$$

Introduce the variable $t = (x - a)/(b - a)$, and check the following equations:

$$\begin{aligned}
x &= a + (b - a)t,\\
x'_k &= a + (b - a)t'_k,\\
x - x'_k &= (b - a)(t - t'_k),\\
x'_j - x'_k &= (b - a)(t'_j - t'_k).
\end{aligned}$$

Applying these substitutions and changing variables in the integral gives

$$\begin{aligned}
k_{ij} &= \frac{1}{\prod_{k \neq j}((b - a)(t'_j - t'_k))} \int_0^{t_i} \left(\prod_{k \neq j}((b - a)(t - t'_k)) \right)(b - a)\, dt\\
&= (b - a)\frac{1}{\prod_{k \neq j}(t'_j - t'_k)} \int_0^{t_i} \prod_{k \neq j}(t - t'_k)\, dt.
\end{aligned}$$

This differs from the $(i + 1), (j + 1)$-entry of $\mathbf{K}_m = \mathbf{K_{t,t'}}$ by a factor of $b - a$. $\qquad\square$

    The integration matrix $\mathbf{K}_m$ is computed by `intmatrixuni` in Listing 15.1. In the implementation, the real work is done by the call to `intmatrix_`. Several other lines are devoted to storing and retrieving integration matrices to avoid duplicate computation.

    The integration matrices turn out to have rational entries. Matrices $\mathbf{K}_m$, $m = 0, 1, 2, 3$, are shown here:

```
>> format rat
>> intmatrixuni(0)
ans =
        0
```

Listing 15.1. Construct an integration matrix for a uniform grid on [0, 1].

```
function K = intmatrixuni(m)

persistent Kstore
if isempty(Kstore), Kstore = cell(1,9); end
if m+1<=length(Kstore)&&~isempty(Kstore{m+1})
  % retrieve matrix if previously computed
  K = Kstore{m+1};
else
  % compute integration matrix
  [xs,ws] = griduni([0 1],m+1,1);
  xs_ = griduni([0 1],m,1);
  K = intmatrix_(xs,ws,xs_,0);
  % store for future use
  if m+1<=length(Kstore)
    Kstore{m+1} = K;
  end
end
```

```
           1
>> intmatrixuni(1)
ans =
         0                  0
       3/8                1/8
       1/2                1/2
>> intmatrixuni(2)
ans =
         0                  0                  0
     31/162             14/81              -5/162
     16/81              40/81              -2/81
       1/6                2/3                1/6
>> intmatrixuni(3)
ans =
         0                  0                  0                  0
    247/2048            363/2048          -123/2048           25/2048
     15/128             51/128             -3/128             1/128
    231/2048            891/2048           405/2048            9/2048
       1/8                3/8                3/8                1/8
>> format long
```

15.2 ▪ The definite integral

The setup for working on a piecewise-uniform grid is familiar. An interval $[a, b]$ is partitioned into n subintervals $[a_j, b_j]$, $j = 1, \ldots, n$, of equal width, and a uniform grid $\mathbf{x}'_j$ of degree m is placed on each subinterval. In this chapter, we consider integrands belonging to $P_{mn}[a, b]$, the class of piecewise-polynomial functions that have degree at most m on each subinterval. The next chapter extends the discussion to integrands that are not piecewise polynomial.

Our first goal is to compute the definite integral of $q \in P_{mn}[a, b]$, given samples $\mathbf{q}_j = q|_{\mathbf{x}'_j}$. For the definite integral, only the last row of the integration matrix $\mathbf{K}_m$ is needed. We denote this row $\boldsymbol{\lambda}_m^T$.

Theorem 15.2. *If $q \in P_{mn}[a, b]$, then*

$$\int_a^b q(x)\,dx = \frac{b-a}{n}\sum_{j=1}^n \lambda_m^T \mathbf{q}_j. \tag{15.2}$$

Proof. Let $p(x) = \int_a^x q(\xi)\,d\xi$. The desired integral is

$$\int_a^b q(x)\,dx = \sum_{j=1}^n \int_{a_j}^{b_j} q(x)\,dx = \sum_{j=1}^n (p(b_j) - p(a_j)).$$

Because b_j is the last node in $\mathbf{x}_j$, the quantity $p(b_j) - p(a_j)$ is the last entry of $\mathbf{p}_j = \frac{b-a}{n}\mathbf{K}_m\mathbf{q}_j$ by Theorems 14.1 and 15.1. With λ_m^T defined to be the last row of $\mathbf{K}_m$, the product $\frac{b-a}{n}\lambda_m^T\mathbf{q}_j$ equals the last entry of $\mathbf{p}_j$. Thus,

$$\int_a^b q(x)\,dx = \sum_{j=1}^n (p(b_j) - p(a_j)) = \sum_{j=1}^n \frac{b-a}{n}\lambda_m^T\mathbf{q}_j. \qquad \square$$

The entries of λ_m are called *Newton–Cotes weights*. We have already seen Newton–Cotes weights in the bottom rows of the matrices $\mathbf{K}_m$. Here they are again for $m = 0, 1, 2, 3$:

$$\lambda_0^T = \begin{bmatrix} 1 \end{bmatrix},$$
$$\lambda_1^T = \begin{bmatrix} 1/2 & 1/2 \end{bmatrix},$$
$$\lambda_2^T = \begin{bmatrix} 1/6 & 2/3 & 1/6 \end{bmatrix},$$
$$\lambda_3^T = \begin{bmatrix} 1/8 & 3/8 & 3/8 & 1/8 \end{bmatrix}.$$

Example 15.3. Approximate

$$\log 2 = \int_1^2 \frac{1}{x}\,dx$$

by Newton–Cotes quadrature with a 2-by-2 piecewise-uniform grid.

Solution. The interval $[a, b] = [1, 2]$ is partitioned into $n = 2$ subintervals, each of width $(b - a)/n = (2 - 1)/2 = 1/2$, and a uniform grid of degree $m = 2$ is placed on each subinterval:

$$\begin{bmatrix} \mathbf{x}_1' & \mathbf{x}_2' \end{bmatrix} = \begin{bmatrix} 1.00 & 1.50 \\ 1.25 & 1.75 \\ 1.50 & 2.00 \end{bmatrix}.$$

The integrand $g(x) = 1/x$ is sampled on this piecewise-uniform grid:

$$\begin{bmatrix} \mathbf{q}_1 & \mathbf{q}_2 \end{bmatrix} = \begin{bmatrix} g(1.00) & g(1.50) \\ g(1.25) & g(1.75) \\ g(1.50) & g(2.00) \end{bmatrix} = \begin{bmatrix} 1 & 2/3 \\ 4/5 & 4/7 \\ 2/3 & 1/2 \end{bmatrix}.$$

The Newton–Cotes weights are seen above to be

$$\lambda_2^T = \begin{bmatrix} 1/6 & 2/3 & 1/6 \end{bmatrix}.$$

The definite integral of the piecewise-quadratic interpolant is

$$\frac{1}{2}\left(\lambda_2^T \mathbf{q}_1 + \lambda_2^T \mathbf{q}_2\right)$$

$$= \frac{1}{2}\left(\begin{bmatrix} 1/6 & 2/3 & 1/6 \end{bmatrix}\begin{bmatrix} 1 \\ 4/5 \\ 2/3 \end{bmatrix} + \begin{bmatrix} 1/6 & 2/3 & 1/6 \end{bmatrix}\begin{bmatrix} 2/3 \\ 4/7 \\ 1/2 \end{bmatrix}\right) = \frac{1747}{2520},$$

or about 0.6933. This is a reasonably good estimate for $\log 2 = 0.6931\ldots$ considering the few arithmetic operations performed. ∎

Newton–Cotes quadrature is automated by `definiteuni`, shown in Listing 15.2.

Listing 15.2. Compute a definite integral on a piecewise-uniform grid.

```
function I = definiteuni(qs,ab)

a = ab(1);
b = ab(2);
m = size(qs,1)-1;
n = size(qs,2);
K = intmatrixuni(m);
lambdas = K(end,:)';
I = (b-a)/n*sum(lambdas'*qs);
```

Example 15.4. Compute

$$\log 2 = \int_1^2 \frac{1}{x}\,dx,$$

again by Newton–Cotes quadrature, this time with a 2-by-100 piecewise-uniform grid by using `definiteuni`.

Solution. The integrand is sampled.

```
>> g = @(x) 1/x; a = 1; b = 2;
>> m = 2;
>> n = 100;
>> qs = sampleuni(g,[a b],m,n);
```

The integral is computed.

```
>> definiteuni(qs,[a b])
ans =
    0.693147180579475
```

The computed value can be compared with the built-in `log` function.

```
>> log(2)
ans =
    0.693147180559945
```

The values agree to about 10 digits. ∎

15.3 ▪ A specific antiderivative

Next, we consider the antiderivative $p_a + \int_a^x q(\xi)\,d\xi$.

Let $q \in P_{mn}[a, b]$ as in the previous section, and introduce a second piecewise-polynomial $p \in P_{m+1,n}[a, b]$. Define $\mathbf{x}'_j$ and $\mathbf{q}_j$ as in the previous section, and let $\mathbf{x}_j$ be a uniform grid of degree $m + 1$ on $[a_j, b_j]$ and $\mathbf{p}_j = p|_{\mathbf{x}_j}$.

Computing the antiderivative reduces to one matrix-vector product plus one vector addition per subinterval.

Theorem 15.5. *If $p \in P_{m+1,n}[a, b]$ and $q \in P_{mn}[a, b]$, then*

$$p(x) = p_a + \int_a^x q(\xi) \, d\xi, \quad a \le x \le b, \tag{15.3}$$

if and only if the following hold:

$$\mathbf{p}_1 = p_a \mathbf{1} + \frac{b - a}{n} \mathbf{K}_m \mathbf{q}_1, \tag{15.4}$$

$$\mathbf{p}_j = p_{m+1,j-1} \mathbf{1} + \frac{b - a}{n} \mathbf{K}_m \mathbf{q}_j, \quad j = 2, \dots, n, \tag{15.5}$$

in which $\mathbf{1}$ is the $(m + 2)$-by-1 vector of ones and $p_{m+1,j-1}$ is the last entry of $\mathbf{p}_{j-1}$.

Proof. We prove by induction that $p(x) = p_a + \int_a^x q(\xi) \, d\xi$ on $[a, b_k]$ if and only if $\mathbf{p}_1, \dots, \mathbf{p}_k$ satisfy the given recurrence. The induction concludes at $k = n$, which gives the theorem.

The base case, that $p(x) = p_a + \int_a^x q(\xi) \, d\xi$ on $[a_1, b_1]$ if and only if (15.4) is true, follows immediately from Theorems 14.6 and 15.1.

For the induction step, suppose that the result has already been proved for some value of k. We have

$$p(x) = p_a + \int_a^x q(\xi) \, d\xi, \quad a \le x \le b_{k+1}, \tag{15.6}$$

if and only if

$$p(x) = p_a + \int_a^x q(\xi) \, d\xi, \quad a \le x \le b_k, \tag{15.7}$$

$$p(x) = p(a_{k+1}) + \int_{a_{k+1}}^x q(\xi) \, d\xi, \quad a_{k+1} \le x \le b_{k+1}. \tag{15.8}$$

Equation (15.7) is equivalent by the induction hypothesis to

$$\mathbf{p}_1 = p_a \mathbf{1} + \frac{b - a}{n} \mathbf{K}_m \mathbf{q}_1,$$

$$\mathbf{p}_j = p_{m+1,j-1} \mathbf{1} + \frac{b - a}{n} \mathbf{K}_m \mathbf{q}_j, \quad j = 2, \dots, k,$$

and (15.8) is equivalent to

$$\mathbf{p}_{k+1} = p_{m+1,k} \mathbf{1} + \frac{b - a}{n} \mathbf{K}_m \mathbf{q}_{k+1}$$

by Theorems 14.6 and 15.1. Hence, (15.6) is equivalent to

$$\mathbf{p}_1 = p_a \mathbf{1} + \frac{b - a}{n} \mathbf{K}_m \mathbf{q}_1,$$

$$\mathbf{p}_j = p_{m+1,j-1} \mathbf{1} + \frac{b - a}{n} \mathbf{K}_m \mathbf{q}_j, \quad j = 2, \dots, k + 1.$$

The induction step is complete.   □

Example 15.6. Approximate

$$\log x = \int_1^x \frac{1}{\xi} \, d\xi$$

on $1 \leq x \leq 2$ by interpolating the integrand on a 2-by-2 piecewise-uniform grid. Then compare with the true antiderivative.

Solution. The samples $\mathbf{q}_1$ and $\mathbf{q}_2$ are just as in Example 15.3, and the integration matrix is computed earlier in the chapter with `intmatrixuni(2)`:

$$\mathbf{K}_2 = \begin{bmatrix} 0 & 0 & 0 \\ 31/162 & 14/81 & -5/162 \\ 16/81 & 40/81 & -2/81 \\ 1/6 & 2/3 & 1/6 \end{bmatrix}.$$

The antiderivative computation proceeds one subinterval at a time:

$$\mathbf{p}_1 = p_a \mathbf{1} + \frac{b-a}{n} \mathbf{K}_m \mathbf{q}_1 = 0 \begin{bmatrix} 1 \\ 1 \\ 1 \\ 1 \end{bmatrix} + \frac{1}{2} \mathbf{K}_2 \begin{bmatrix} 1 \\ 4/5 \\ 2/3 \end{bmatrix} = \begin{bmatrix} 0 \\ 751/4860 \\ 70/243 \\ 73/180 \end{bmatrix} \approx \begin{bmatrix} 0 \\ 0.1545 \\ 0.2881 \\ 0.4056 \end{bmatrix}$$

and

$$\mathbf{p}_2 = p_{31} \mathbf{1} + \frac{b-a}{n} \mathbf{K}_m \mathbf{q}_2 = \frac{73}{180} \begin{bmatrix} 1 \\ 1 \\ 1 \\ 1 \end{bmatrix} + \frac{1}{2} \mathbf{K}_2 \begin{bmatrix} 2/3 \\ 4/7 \\ 1/2 \end{bmatrix}$$

$$= \begin{bmatrix} 73/180 \\ 4967/9720 \\ 20627/34020 \\ 1747/2520 \end{bmatrix} \approx \begin{bmatrix} 0.4056 \\ 0.5110 \\ 0.6063 \\ 0.6933 \end{bmatrix}.$$

The approximate antiderivative is the piecewise-cubic polynomial on $[1, 2]$ determined by the samples $\mathbf{p}_1$, $\mathbf{p}_2$ above.

The true antiderivative sampled on the same grid produces comparable values:

```
>> sampleuni(@(x) log(x),[1 2],3,2)
ans =
                  0    0.405465108108164
  0.154150679827258    0.510825623765991
  0.287682072451781    0.606135803570316
  0.405465108108164    0.693147180559945
```
■

Antidifferentiation on a piecewise-uniform grid is executed by `antiderivuni` in Listing 15.3.

Example 15.7. Compute the inverse tangent function

$$\arctan x = \int_0^x \frac{1}{1 + \xi^2} \, d\xi$$

Listing 15.3. Compute an antiderivative on a piecewise-uniform grid.

```
function ps = antiderivuni(pa,qs,ab)

a = ab(1);
b = ab(2);
m = size(qs,1)-1;
n = size(qs,2);
K = (b-a)/n*intmatrixuni(m);
ps = nan(m+2,n);
pinit = pa;
for j = 1:n
  ps(:,j) = pinit*ones(m+2,1)+K*qs(:,j);
  pinit = ps(end,j);
end
```

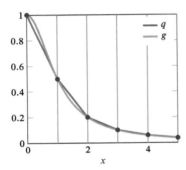 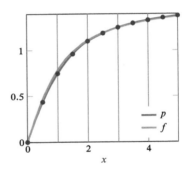

Figure 15.2. *The function $g(x) = 1/(1 + x^2)$ and a piecewise-polynomial interpolant $q(x)$ (left), and antiderivatives $f(x) = \int_0^x g(\xi)\,d\xi$ and $p(x) = \int_0^x q(\xi)\,d\xi$ (right).*

on $0 \le x \le 5$ by integrating. Use a 1-by-5 piecewise-uniform grid for the integrand. Plot the interpolated integrand and the resulting antiderivative.

Solution. The integrand and initial value are recorded.

```
>> g = @(x) 1/(1+x^2); a = 0; b = 5;
>> pa = 0;
```

The grid parameters are set.

```
>> m = 1;
>> n = 5;
```

The integrand g is replaced by a piecewise-polynomial interpolant q, which is then integrated.

```
>> qs = sampleuni(g,[a b],m,n);
>> q = interpuni(qs,[a b]);
>> ps = antiderivuni(pa,qs,[a b]);
>> p = interpuni(ps,[a b]);
```

The results are plotted in Figure 15.2.

```
>> ax = newfig(1,2);
>> plotfun(q,[a b],'displayname','q');
>> plotpartition([a b],n);
>> plotsample(griduni([a b],m,n),qs);
>> subplot(ax(2));
>> plotfun(p,[a b],'displayname','p');
```

```
>> plotpartition([a b],n);
>> plotsample(griduni([a b],m+1,n),ps);
```
Notice that because the integrand has pieces of degree $m = 1$, the antiderivative has pieces of degree $m + 1 = 2$.

The true integrand $g(x)$ and its antiderivative $f(x) = \arctan x$, as computed by the built-in atan function, are added to the plots.

```
>> f = @(x) atan(x);
>> subplot(ax(1));
>> plotfun(g,[a b],'displayname','g');
>> xlabel('x');
>> subplot(ax(2));
>> plotfun(f,[a b],'displayname','f');
>> xlabel('x');
```
∎

15.4 ▪ The indefinite integral

The last form of the integral is the indefinite integral $\int q(x)\,dx$. We give two tests of whether a piecewise-polynomial p is an antiderivative of another piecewise-polynomial q. The first test consists of a sequence of small matrix-vector equations, one per subinterval, while the second consists of a single large matrix-vector equation. The first makes for efficient computer code, while the second permits a simpler error analysis.

Let $p \in P_{m+1,n}[a,b]$ and $q \in P_{mn}[a,b]$, and define x_j, x'_j, $\mathbf{p}_j$, and $\mathbf{q}_j$ as in the previous section. The difference here is that there is no initial value p_a. The goal is to test whether p is *any* antiderivative of q. Note that we use the term *antiderivative* in the weak sense of Chapter 11. (If $m = 0$, then p is piecewise linear and likely not smooth.)

For the first test, let $\mathbf{J}_m$ be the $(m + 1)$-by-$(m + 2)$ matrix

$$
\mathbf{J}_m = \begin{bmatrix}
-1 & 1 & & & \\
-1 & & 1 & & \\
\vdots & & & \ddots & \\
-1 & & & & 1
\end{bmatrix},
$$

in which the blank entries represent zeros, and let $\hat{\mathbf{K}}_m$ be the $(m+1)$-by-$(m+1)$ matrix formed by deleting the first row of $\mathbf{K}_m$.

Theorem 15.8. *Let $p \in P_{m+1,n}[a,b]$ and $q \in P_{mn}[a,b]$. Then p is an antiderivative of q on $[a,b]$ if and only if both of the following hold:*

$$
\mathbf{J}_m \mathbf{p}_j = \frac{b-a}{n} \hat{\mathbf{K}}_m \mathbf{q}_j, \quad j = 1,\ldots,n, \tag{15.9}
$$

and

$$
p_{0j} = p_{m+1,j-1}, \quad j = 2,\ldots,n. \tag{15.10}
$$

Here, p_{0j} denotes the first entry of $\mathbf{p}_j$ and $p_{m+1,j-1}$ denotes the last entry of $\mathbf{p}_{j-1}$.

Proof. If p is an antiderivative of q on all of $[a,b]$, then p must be an antiderivative of q on each subinterval $[a_j, b_j]$, implying (15.9), and must also be continuous on $[a,b]$, implying (15.10).

To prove the converse, suppose that (15.9)–(15.10) are true. Then the system in (15.9) with $j = 1$ implies

$$p(x) - p(a) = \int_a^x q(\xi)\,d\xi, \quad a \le x \le b_1,$$

by Theorems 14.8 and 15.1, and induction shows that for $a \le x \le b_j$,

$$p(x) - p(a) = (p(b_{j-1}) - p(a)) + (p(x) - p(a_j))$$
$$= \int_a^{b_{j-1}} q(\xi)\,d\xi + \int_{a_j}^x q(\xi)\,d\xi = \int_a^x q(\xi)\,d\xi,$$

again using Theorems 14.8 and 15.1. Ultimately, $p(x) - p(a) = \int_a^x q(\xi)\,d\xi$ for all $x \in [a, b]$, which implies that p is an antiderivative of q. $\quad\square$

Example 15.9. Let

$$q(x) = \begin{cases} x & \text{if } 0 \le x < 0.5, \\ 1 - x & \text{if } 0.5 \le x \le 1. \end{cases}$$

Find a system of equations of the form (15.9)–(15.10) that characterizes the set of all antiderivatives.

Solution. Because $q \in P_{12}[0, 1]$, we take $a = 0$, $b = 1$, $m = 1$, and $n = 2$. The sample of q on this grid is

$$\mathbf{q}_1 = \begin{bmatrix} 0 \\ 0.5 \end{bmatrix}, \quad \mathbf{q}_2 = \begin{bmatrix} 0.5 \\ 0 \end{bmatrix}.$$

The integration matrices are

$$\mathbf{J}_1 = \begin{bmatrix} -1 & 1 & 0 \\ -1 & 0 & 1 \end{bmatrix}, \quad \hat{\mathbf{K}}_1 = \begin{bmatrix} 3/8 & 1/8 \\ 1/2 & 1/2 \end{bmatrix}.$$

(The full matrix $\mathbf{K}_1$ is computed below Theorem 15.1, and $\hat{\mathbf{K}}_1$ is obtained by deleting the leading row of zeros.) The right-hand sides in (15.9) become

$$\frac{b-a}{n}\hat{\mathbf{K}}_m\mathbf{q}_1 = \frac{1}{2}\begin{bmatrix} 3/8 & 1/8 \\ 1/2 & 1/2 \end{bmatrix}\begin{bmatrix} 0 \\ 0.5 \end{bmatrix} = \begin{bmatrix} 0.03125 \\ 0.125 \end{bmatrix},$$
$$\frac{b-a}{n}\hat{\mathbf{K}}_m\mathbf{q}_2 = \frac{1}{2}\begin{bmatrix} 3/8 & 1/8 \\ 1/2 & 1/2 \end{bmatrix}\begin{bmatrix} 0.5 \\ 0 \end{bmatrix} = \begin{bmatrix} 0.09375 \\ 0.125 \end{bmatrix}.$$

Let $p \in P_{22}[0, 1]$, a piecewise-quadratic function. From the theorem, we find that p is an antiderivative for q if and only if

$$\begin{bmatrix} -1 & 1 & 0 \\ -1 & 0 & 1 \end{bmatrix}\begin{bmatrix} p_{01} \\ p_{11} \\ p_{21} \end{bmatrix} = \begin{bmatrix} 0.03125 \\ 0.125 \end{bmatrix}$$

and

$$\begin{bmatrix} -1 & 1 & 0 \\ -1 & 0 & 1 \end{bmatrix}\begin{bmatrix} p_{02} \\ p_{12} \\ p_{22} \end{bmatrix} = \begin{bmatrix} 0.09375 \\ 0.125 \end{bmatrix}$$

and

$$p_{02} = p_{21}.$$

For an example, let $p(x) = (1/2)x^2$ for $0 \le x < 0.5$ and $p(x) = -(1/2)x^2 + x - (1/4)$ for $0.5 \le x \le 1$, a particular antiderivative of q. A sample of this function on a 2-by-2 piecewise-uniform grid satisfies the equations above. ■

The matrices $\mathbf{J}_m$ and $\hat{\mathbf{K}}_m$ are constructed by routine indefiniteuni, shown in Listing 15.4. This code is used in Chapter 25.

Listing 15.4. Construct indefinite integration matrices for a uniform grid.

```
function [J,K] = indefiniteuni(m)

J = [ -ones(m+1,1) eye(m+1) ];
K = intmatrixuni(m);
K = K(2:end,:);
```

For the second test, we want to work over the entire interval $[a, b]$ at once, rather than one subinterval at a time. Each vector $\mathbf{p}_j$ above is $(m + 2)$-by-1, and each vector $\mathbf{q}_j$ is $(m + 1)$-by-1. Stack $\mathbf{p}_1, \dots, \mathbf{p}_n$ into a column vector

$$\mathbf{p} = \begin{bmatrix} \mathbf{p}_1 \\ \vdots \\ \mathbf{p}_n \end{bmatrix} \tag{15.11}$$

with $(m + 2)n$ entries, and stack $\mathbf{q}_1, \dots, \mathbf{q}_n$ into a column vector

$$\mathbf{q} = \begin{bmatrix} \mathbf{q}_1 \\ \vdots \\ \mathbf{q}_n \end{bmatrix} \tag{15.12}$$

with $(m + 1)n$ entries. Let

$$\mathbf{J}_{mn} = \begin{bmatrix} \mathbf{J}_m & & & \\ -\mathbf{1}\mathbf{e}_1^T & \mathbf{I} & & \\ -\mathbf{1}\mathbf{e}_1^T & & \mathbf{I} & \\ \vdots & & & \ddots \\ -\mathbf{1}\mathbf{e}_1^T & & & & \mathbf{I} \end{bmatrix}, \tag{15.13}$$

in which $\mathbf{1}$ is an $(m + 2)$-by-1 vector of ones, $\mathbf{I}$ is the $(m + 2)$-by-$(m + 2)$ identity matrix, and $\mathbf{e}_1$ is the first column of $\mathbf{I}$, and let

$$\hat{\mathbf{K}}_{mn} = \begin{bmatrix} \hat{\mathbf{K}}_m & & & & \\ \mathbf{1}\lambda_m^T & \mathbf{K}_m & & & \\ \mathbf{1}\lambda_m^T & \mathbf{1}\lambda_m^T & \mathbf{K}_m & & \\ \vdots & \vdots & & \ddots & \\ \mathbf{1}\lambda_m^T & \mathbf{1}\lambda_m^T & \cdots & \mathbf{1}\lambda_m^T & \mathbf{K}_m \end{bmatrix}. \tag{15.14}$$

Both $\hat{\mathbf{K}}_{mn}$ and $\mathbf{J}_{mn}$ contain n blocks along their main diagonals.

Theorem 15.10. *Let $p \in P_{m+1,n}[a,b]$ and $q \in P_{mn}[a,b]$. Then p is an antiderivative of q if and only if*

$$\mathbf{J}_{mn}\mathbf{p} = \frac{b-a}{n}\hat{\mathbf{K}}_{mn}\mathbf{q}. \tag{15.15}$$

Proof. Denote the entries of $\mathbf{p}_j$ by p_{ij}, $i = 0,\ldots,m+1$.

The matrix-vector equation (15.15) is equivalent to

$$\mathbf{J}_m\mathbf{p}_1 = \frac{b-a}{n}\hat{\mathbf{K}}_m\mathbf{q}_1,$$

$$\mathbf{p}_j - p_{01}\mathbf{1} = \left(\sum_{k=0}^{j-1}\frac{b-a}{n}\lambda_m^T\mathbf{q}_k\right)\mathbf{1} + \frac{b-a}{n}\mathbf{K}_m\mathbf{q}_j, \quad j = 2,\ldots,n.$$

By Theorems 14.1, 15.1, 15.2, and 15.8, these equations are equivalent to

$$p_{i1} - p_{01} = \int_a^{x_{i1}} q(\xi)\,d\xi \tag{15.16}$$

for $i = 1,\ldots,m+1$ and

$$p_{ij} - p_{01} = \left(\sum_{k=0}^{j-1}\int_{a_k}^{b_k} q(\xi)\,d\xi\right) + \left(\int_{a_j}^{x_{ij}} q(\xi)\,d\xi\right) = \int_a^{x_{ij}} q(\xi)\,d\xi \tag{15.17}$$

for $i = 0,\ldots,m+1$, $j = 2,\ldots,n$.

If p is an antiderivative of q, then (15.16) and (15.17) are true.

For the converse, suppose that (15.16) and (15.17) are true. It's also true that $p_{01} - p_{01} = 0 = \int_a^{x_{01}} q(\xi)\,d\xi$, and so

$$p(x_{ij}) - p(a) = \int_a^{x_{ij}} q(\xi)\,d\xi, \quad i = 0,\ldots,m+1, \quad j = 1,\ldots,n.$$

These equations force $p(x) - p(a) = \int_a^x q(\xi)\,d\xi$ for all $x \in [a,b]$ since both functions belong to $P_{m+1,n}[a,b]$ (Exercise 15 in Chapter 7). Therefore, p is an antiderivative of q. $\square$

Example 15.11. Find a matrix-vector equation of the form (15.15) that characterizes the set of antiderivatives for the function q of Example 15.9.

Solution. In addition to all of the definitions in the solution to Example 15.9, we have

$$-\mathbf{1}\mathbf{e}_1^T = -\begin{bmatrix} 1 \\ 1 \\ 1 \end{bmatrix}\begin{bmatrix} 1 & 0 \end{bmatrix} = \begin{bmatrix} -1 & 0 \\ -1 & 0 \\ -1 & 0 \end{bmatrix}$$

and

$$\mathbf{1}\lambda_1^T = \begin{bmatrix} 1 \\ 1 \\ 1 \end{bmatrix}\begin{bmatrix} 1/2 & 1/2 \end{bmatrix} = \begin{bmatrix} 1/2 & 1/2 \\ 1/2 & 1/2 \\ 1/2 & 1/2 \end{bmatrix}$$

and

$$\mathbf{K}_1 = \begin{bmatrix} 0 & 0 \\ 3/8 & 1/8 \\ 1/2 & 1/2 \end{bmatrix}, \quad \mathbf{I} = \begin{bmatrix} 1 & 0 & 0 \\ 0 & 1 & 0 \\ 0 & 0 & 1 \end{bmatrix}.$$

The right-hand side of (15.15) is

$$
\frac{b-a}{n}\hat{\mathbf{K}}_{mn}\mathbf{q} = \frac{1}{2}
\left[\begin{array}{cc|cc}
3/8 & 1/8 & 0 & 0 \\
1/2 & 1/2 & 0 & 0 \\
1/2 & 1/2 & 0 & 0 \\
1/2 & 1/2 & 3/8 & 1/8 \\
1/2 & 1/2 & 1/2 & 1/2
\end{array}\right]
\left[\begin{array}{c}
0 \\
0.5 \\
0.5 \\
0
\end{array}\right]
=
\left[\begin{array}{c}
0.03125 \\
0.125 \\
0.125 \\
0.21875 \\
0.25
\end{array}\right].
$$

The matrix-vector equation is

$$
\left[\begin{array}{ccc|ccc}
-1 & 1 & 0 & 0 & 0 & 0 \\
-1 & 0 & 1 & 0 & 0 & 0 \\
-1 & 0 & 0 & 1 & 0 & 0 \\
-1 & 0 & 0 & 0 & 1 & 0 \\
-1 & 0 & 0 & 0 & 0 & 1
\end{array}\right]
\left[\begin{array}{c}
p_{01} \\
p_{11} \\
p_{21} \\
p_{02} \\
p_{12} \\
p_{22}
\end{array}\right]
=
\left[\begin{array}{c}
0.03125 \\
0.125 \\
0.125 \\
0.21875 \\
0.25
\end{array}\right].
$$
∎

Notes

Goldstine [34] provides an account of the early history of numerical integration, including contributions by Cavalieri, Gregory, Newton, and Cotes.

Exercises

Exercises 1–4: Compute the integral by applying Newton–Cotes quadrature by hand with the given values of m and n. Report the answer to four decimal places.

1. $\int_0^1 ((x-2)/(x+4))\,dx$; $m = 0, n = 4$
2. $\int_0^2 \sqrt{x^2 + 1}\,dx$; $m = 1, n = 4$
3. $\int_2^6 1/\sqrt{x}\,dx$; $m = 2, n = 2$
4. $\int_{-1}^1 ((x+3)/(x^2 - 2x - 4))\,dx$; $m = 3, n = 2$

Exercises 5–8: Compute the integral using definiteuni with the given value of m. Adjust n to achieve an absolute error below 10^{-6}. Measure the absolute error by comparing with the given known value.

5. $\int_0^1 (1/(1+x^2))\,dx$, $m = 0$; $\pi/4$
6. $\int_0^{1/\sqrt{2}} (1/\sqrt{1-x^2})\,dx$, $m = 1$; $\pi/4$
7. $\int_{\sqrt{2}}^2 (1/(x\sqrt{x^2-1}))\,dx$, $m = 2$; $\pi/12$
8. $\int_0^{1/\sqrt{2}} (\sqrt{1-x^2} - x)\,dx$, $m = 3$; $\pi/8$

Exercises 9–12: Compute the antiderivative on the given interval using antiderivuni with the specified value of m. Adjust n to achieve absolute error below 10^{-6}. Measure the absolute error by comparing with the built-in MATLAB function.

9. $\mathrm{erf}(x) = \frac{2}{\sqrt{\pi}} \int_0^x e^{-\xi^2}\,d\xi$, $[0, 5]$, $m = 0$; erf

10. $\arcsin x = \int_0^x (1/\sqrt{1-\xi^2})\,d\xi$, $[0, 0.9]$, $m = 1$; asin

11. $\log x = \int_0^x (1/\xi)\,d\xi$, $[1, 10]$, $m = 2$; log

12. $\mathrm{Si}(x) = \int_0^x ((\sin\xi)/\xi)\,d\xi$, $[0, 20]$, $m = 3$;
    1i/2*(expint(-1i*x)-expint(1i*x))+pi/2. *Hint.* Be careful at $\xi = 0$.

13. An ellipse with semimajor axis length 2 and semiminor axis length 1 is traced by the parametric equations $x(t) = 2\cos t$, $y(t) = \sin t$, $0 \le t \le 2\pi$. A sector of the ellipse is taken by restricting t to $[0, \pi/3]$. The arc length of this sector equals $\int_0^{\pi/3} \sqrt{x'(t)^2 + y'(t)^2}\,dt$. Compute this integral on a piecewise-uniform grid with $m = 2$. Increase n to achieve numerical convergence.

14. The trapezoid rule approximation for $\int_a^b g(x)\,dx$ is commonly written

    $$\frac{h}{2}\left(g(x_0) + 2g(x_1) + 2g(x_2) + \cdots + 2g(x_{n-2}) + 2g(x_{n-1}) + g(x_n)\right),$$

    in which $h = (b-a)/n$ and $x_j = a + jh$. Prove that the same value is produced by interpolating g on a 1-by-n piecewise-uniform grid over $[a, b]$ and then computing the definite integral of the piecewise-polynomial interpolant.

15. Simpson's rule for approximating $\int_a^b g(x)\,dx$ is commonly written

    $$\frac{h}{3}\left(g(x_0) + 4g(x_1) + 2g(x_2) + 4g(x_3) + 2g(x_4) + \cdots + 4g(x_{N-1}) + g(x_N)\right),$$

    in which $N = 2n$ is even, $h = (b-a)/N$, and $x_j = a + jh$. Prove that the same value is produced by interpolating g on a 2-by-n piecewise-uniform grid over $[a, b]$ and then computing the definite integral of the piecewise-polynomial interpolant.

16. Fix a positive integer n, and let I_m be the Newton–Cotes quadrature approximation for $\int_a^b g(x)\,dx$ on an m-by-n piecewise-uniform grid. Prove that $I_2 = \frac{2}{3}I_0 + \frac{1}{3}I_1$.

17. Prove that the Newton–Cotes weights form a palindrome for any degree m. That is, if the vector of Newton–Cotes weights is $(\lambda_0, \ldots, \lambda_m)$, then $\lambda_i = \lambda_{m-i}$ for all i.

Exercises 18–20: In these exercises, consider a simple uniform grid ($n = 1$) rather than partitioning the domain into subintervals. Recall that when the integrand q is a polynomial of degree m, the Newton–Cotes formula $(b - a)\lambda_m^T\mathbf{q}$ gives the integral $\int_a^b q(x)\,dx$ exactly.

18. Prove that Newton–Cotes quadrature on a uniform grid of degree 0 (just a single node at the midpoint) is not only exact for degree-0 polynomials but also for degree-1 polynomials. Also, draw a picture to illustrate.

19. Prove that if m is even, then the degree-m Newton–Cotes quadrature computes $\int_{-1}^1 q(x)\,dx$ exactly when q is a polynomial of degree at most $m+1$. *Hint.* Express q as the sum of even and odd functions and use Exercise 17.

20. Let m be a nonnegative even integer. Using the result of the previous exercise, prove that degree-m Newton–Cotes quadrature computes $\int_a^b q(x)\,dx$ exactly when q is a polynomial of degree $m + 1$, regardless of the integration limits a and b.

21. Let $p \in P_{m+1,n}[a, b]$ and $q \in P_{mn}[a, b]$. Sample $\mathbf{p}_j$ and $\mathbf{q}_j$ as usual, and stack the vectors as in (15.11)–(15.12). We would like to say that $p(x) = p_a + \int_a^x q(\xi)\,d\xi$ if and only if $\mathbf{p} = p_a\mathbf{1} + ((b-a)/n)\mathbf{K}_{mn}\mathbf{q}$. Specify a matrix $\mathbf{K}_{mn}$ that makes this statement true and justify.

Chapter 16

Accuracy analysis for Newton–Cotes quadrature

How effective is integration on a piecewise-uniform grid, also known as Newton–Cotes quadrature? As expected, based on our experience with piecewise-uniform interpolation, the convergence rate is algebraic. However, there is a pleasant surprise concerning the precise rate.

16.1 ▪ Accuracy analysis for the definite integral

In Chapter 14, we saw the triangle inequality for integrals, which gives

$$\left| \int_a^b g(x)\,dx - \int_a^b q(x)\,dx \right| \le \int_a^b |g(x) - q(x)|\,dx \le (b-a)\|g-q\|_\infty.$$

When integrating on a piecewise-uniform grid, the interpolation error bound

$$\|g-q\|_\infty \le \frac{1}{m+1}\|g^{(m+1)}\|_\infty (b-a)^{m+1} N^{-(m+1)}$$

of Theorem 8.1, for functions $g \in C^{m+1}[a,b]$, applies as well. Together, these produce the bound

$$\left| \int_a^b g(x)\,dx - \int_a^b q(x)\,dx \right| \le C N^{-(m+1)} \tag{16.1}$$

with $C = \|g^{(m+1)}\|_\infty (b-a)^{m+2}/(m+1)$.

Surprisingly, a faster rate of convergence can be proved when m is an even integer.

Theorem 16.1. *Suppose $q \in P_{mn}[a,b]$ interpolates g on an m-by-n piecewise-uniform grid over $[a,b]$. If $g \in C^{m+1}[a,b]$, then*

$$\left| \int_a^b g(x)\,dx - \int_a^b q(x)\,dx \right| \le C N^{-(m+1)} \tag{16.2}$$

with $C = \|g^{(m+1)}\|_\infty (b-a)^{m+2}/(m+1)$. If m is even and $g \in C^{m+2}[a,b]$, then the bound can be improved to

$$\left| \int_a^b g(x)\,dx - \int_a^b q(x)\,dx \right| \le C N^{-(m+2)}, \tag{16.3}$$

with $C = \|g^{(m+2)}\|_\infty (b-a)^{m+3}/(2(m+2))$. As always, $N = (m \vee 1)n$.

The first half of the theorem is a restatement of (16.1). The second half is proved in Appendix D.

An example illustrates the bounds of the theorem.

Example 16.2. Compute

$$\log 2 = \int_1^2 \frac{1}{x}\, dx$$

using piecewise-uniform grids. For $m = 0, 1, 2, 3$, measure the rate of convergence as $N \to \infty$.

Solution. The integrand and limits of integration are defined.

```
>> g = @(x) 1/x; a = 1; b = 2;
```

The grid parameters are set.

```
>> m = [0 1 2 3];
>> N = 6*4.^(0:5);
```

The computations are executed, and the absolute errors are measured.

```
>> I = log(2);
>> err = nan(length(m),length(N));
>> for k = 1:length(m)
      for l = 1:length(N)
         qs = sampleuni(g,[a b],m(k),N(l)/max(m(k),1));
         Iapprox = definiteuni(qs,[a b]);
         err(k,l) = abs(I-Iapprox);
      end
   end
```

Figure 16.1 shows the absolute errors along with lines that appear to converge to zero at the same rate.

```
>> s = [-2 -2 -4 -4];
>> ax = newfig(2,2); ax = ax';
>> for k = 1:length(m)
      subplot(ax(k));
      xlog; ylog; ylim([1e-18 1e2]);
      plot(N,err(k,:),'*','displayname','abs. error');
      plotfun(@(N) N^s(k),[1 1e4] ...
            ,'displayname',sprintf('N^{%d}',s(k)));
      xlim([1 1e4]);
      xlabel('N');
   end
```

For $m = 0, 1, 2$, and 3, the errors appear to be proportional to N^{-2}, N^{-2}, N^{-4}, and N^{-4}, respectively, as predicted by the theorem. ∎

In summary, Newton–Cotes quadrature converges algebraically, and at a higher-than-expected rate on grids of even degree.

Before moving on, let's take another look at the previous example.

Example 16.3. In Example 16.2, the computed value of $\int_1^2 (1/x)\, dx$ is compared against the value log(2) provided by the MATLAB library. In practical settings, of course, the desired answer is not already known, and error measurement is more difficult.

Repeat the computation of $\int_1^2 (1/x)\, dx$, but estimate the error without consulting the known value. Use piecewise-uniform grids with $m = 0$.

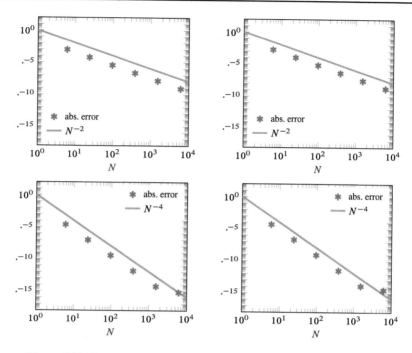

Figure 16.1. *Rate of convergence for definite integration on a piecewise-uniform grid of degree m: m = 0 (top-left), m = 1 (top-right), m = 2 (bottom-left), and m = 3 (bottom-right).*

Solution. The integral is defined as before.

```
>> g = @(x) 1/x; a = 1; b = 2;
```

The grids are the same as those used for the top-left plot in Figure 16.1.

```
>> m = 0;
>> N = 6*4.^(0:5);
```

The integral is approximated.

```
>> Iapprox = nan(size(N));
>> for k = 1:length(N)
     n = N(k);
     qs = sampleuni(g,[a b],m,n);
     Iapprox(k) = definiteuni(qs,[a b]);
   end
```

The results are as follows:

```
>> disptable('N',N,'%4d','approx. integral',Iapprox,'%17.15f');
   N      approx. integral
   6      0.692284320874979
  24      0.693092947672033
  96      0.693143789798411
 384      0.693146968632633
1536      0.693147167314470
6144      0.693147179732103
```

Let $I = \int_1^2 (1/x)\,dx = \log 2$ be the exact value, and let $I_0(N)$ be the value from Newton–Cotes quadrature with $m = 0$ and the given value of N. In place of the true error $I - I_0(N)$, we may start by studying the *Cauchy error* $I_0(4N) - I_0(N)$. (Generally, Cauchy error is the difference between consecutive approximations. In this particular example, the value of N increases by a factor of 4 with each data point.) The sequence

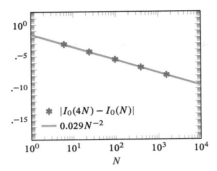

Figure 16.2. *Absolute Cauchy error for a definite integral computation.*

of absolute Cauchy errors is plotted in Figure 16.2.

```
>> cauchyerr = Iapprox(2:end)-Iapprox(1:end-1);
>> newfig;
>> xlog; ylog; ylim([1e-18 1e2]);
>> plot(N(1:end-1),abs(cauchyerr),'*' ...
      ,'displayname','|I_0(4N)-I_0(N)|');
>> xlabel('N');
```

We find that the Cauchy error is close to $0.029N^{-2}$.

```
>> cauchyerr(end)*N(end-1)^2
ans =
    0.029296871158294
>> plotfun(@(N) 0.029*N^(-2),[1 1e4],'displayname','0.029 N^{-2}');
```

What does the Cauchy error $I_0(4N) - I_0(N)$ imply about the true error $I - I_0(N)$? Intuitively, because $I_0(4N)$ is expected to be a considerably more accurate approximation than $I_0(N)$, the Cauchy error should be a reasonable estimate for the true error. We can be more precise. Considering Theorem 16.1, we hypothesize that $I - I_0(N) \approx CN^{-2}$ for some coefficient C. If this were true, then we would have

$$I_0(4N) - I_0(N) = (I - I_0(N)) - (I - I_0(4N)) \approx CN^{-2} - C(4N)^{-2} = \frac{15}{16}CN^{-2}.$$

Hence, our data are consistent with the hypothesis with $C \approx (16/15) \times 0.029 \approx 0.031$.

The evidence suggests that $|I - I_0(N)| \approx 0.031N^{-2}$. In particular, we estimate that the final absolute error is $|I - I_0(6144)| \approx 0.031 \times 6144^{-2} \approx 8 \times 10^{-10}$. Note that we have derived this estimate without peeking at the known value for I.   ∎

The error estimate in the previous example is on the money. The actual error, as best we can measure, is

```
>> abs(log(2)-0.693147179732103)
ans =
    8.278422392038465e-10
```

16.2 ▪ Accuracy analysis for a specific antiderivative

On a piecewise-uniform grid, antidifferentiation converges at the same algebraic rate as definite integration.

Theorem 16.4. *Suppose $f(x) = f_a + \int_a^x g(\xi)\,d\xi$. Let $q \in P_{mn}[a,b]$ be a piecewise-polynomial interpolant for g on an m-by-n piecewise-uniform grid over $[a,b]$, and let*

$p(x) = f_a + \int_a^x q(\xi)\, d\xi$. *If* $g \in C^{m+1}[a, b]$, *then*

$$\|f - p\|_\infty \leq C N^{-(m+1)} \tag{16.4}$$

for a coefficient C that is constant with respect to n. If m is even and $g \in C^{m+2}[a, b]$, *then the bound can be improved to*

$$\|f - p\|_\infty \leq C N^{-(m+2)} \tag{16.5}$$

for another coefficient C that is constant with respect to n. (As usual, $N = (m \vee 1)n$.)

The significance is that antidifferentiation converges at the rate $N^{-(m+1)}$ when m is odd and at the rate $N^{-(m+2)}$ when m is even.

Proof. The grid partitions $[a, b]$ into n contiguous subintervals $[a_j, b_j]$, $j = 1, \ldots, n$, of equal width. Let $x \in [a, b]$. Then x must belong to some $[a_j, b_j]$. By the triangle inequality,

$$\left| \int_a^x g(\xi)\, d\xi - \int_a^x q(\xi)\, d\xi \right|$$

$$\leq \left| \int_a^{b_{j-1}} g(\xi)\, d\xi - \int_a^{b_{j-1}} q(\xi)\, d\xi \right| + \left| \int_{a_j}^x g(\xi)\, d\xi - \int_{a_j}^x q(\xi)\, d\xi \right|. \tag{16.6}$$

The first term on the right-hand side of (16.6) is bounded by Theorem 16.1. Let $l = (b-a)/n$ and $h = (b-a)/N$. Note that $(b_{j-1}-a)/(j-1) = (j-1)l/(j-1) = l$ and $(b_{j-1} - a)/((m \vee 1)(j - 1)) = l/(m \vee 1) = h$. If m is odd, then

$$\left| \int_a^{b_{j-1}} g(\xi)\, d\xi - \int_a^{b_{j-1}} q(\xi)\, d\xi \right| \leq \frac{\|g^{(m+1)}\|_\infty (b_{j-1} - a)^{m+2}}{m + 1} (m(j - 1))^{-(m+1)}$$

$$= \frac{\|g^{(m+1)}\|_\infty (b_{j-1} - a)h^{m+1}}{m + 1} \leq C_1 N^{-(m+1)},$$

in which $C_1 = \|g^{(m+1)}\|_\infty (b - a)^{m+2}/(m + 1)$. If m is even, then

$$\left| \int_a^{b_{j-1}} g(\xi)\, d\xi - \int_a^{b_{j-1}} q(\xi)\, d\xi \right| \leq \frac{\|g^{(m+2)}\|_\infty (b_{j-1} - a)^{m+3}}{2(m + 2)} ((m \vee 1)(j-1))^{-(m+2)}$$

$$= \frac{\|g^{(m+2)}\|_\infty (b_{j-1} - a)h^{m+2}}{2(m + 2)} \leq C_2 N^{-(m+2)},$$

in which $C_2 = \|g^{(m+2)}\|_\infty (b - a)^{m+3}/(2(m + 2))$.

The second term on the right-hand side of (16.6) is bounded by applying Theorem 8.1 to $g(x) - q(x)$ on $[a_j, b_j]$. First note that

$$|g(x) - q(x)| \leq \frac{\|g^{(m+1)}\|_\infty (b_j - a_j)^{m+1}}{m + 1} (m \vee 1)^{-(m+1)} = \frac{\|g^{(m+1)}\|_\infty}{m + 1} h^{m+1}.$$

Thus,

$$\left| \int_{a_j}^x g(\xi)\, d\xi - \int_{a_j}^x q(\xi)\, d\xi \right| \leq \int_{a_j}^{b_j} |g(\xi) - q(\xi)|\, d\xi$$

$$\leq \frac{b - a}{n} \frac{\|g^{(m+1)}\|_\infty}{m + 1} h^{m+1} \leq C_3 N^{-(m+2)},$$

in which $C_3 = \|g^{(m+1)}\|_\infty (b - a)^{m+2}$.

The bounds can be combined to give the final result. When m is odd, the total error is bounded by $C_1 N^{-(m+1)} + C_3 N^{-(m+2)} \leq (C_1 + C_3) N^{-(m+1)}$. When m is even, the total error is bounded by $C_2 N^{-(m+2)} + C_3 N^{-(m+2)} = (C_2 + C_3) N^{-(m+2)}$.   □

Example 16.5. Measure the rate of convergence for the computation of

$$\arctan x = \int_0^x \frac{1}{1+\xi^2} \, d\xi$$

on an m-by-n piecewise-uniform grid over $0 \leq x \leq 5$. Use $m = 0, 1, 2, 3$ and measure the rate of convergence against $N = (m \vee 1)n$.

Solution. The problem is defined.

```
>> g = @(x) 1/(1+x^2); a = 0; b = 5;
```

The grid parameters are set.

```
>> m = [0 1 2 3];
>> N = 6*4.^(0:5);
```

The integrals are computed, and their absolute errors are measured in the infinity norm.

```
>> err = nan(length(m),length(N));
>> for k = 1:length(m)
      for l = 1:length(N)
        qs = sampleuni(g,[a b],m(k),N(l)/max(m(k),1));
        ps = antiderivuni(0,qs,[a b]);
        p = interpuni(ps,[a b]);
        err(k,l) = infnorm(@(x) atan(x)-p(x),[a b],100);
      end
    end
```

Errors are plotted against sampling cost in Figure 16.3 and compared against $N^{-(m+1)}$ for odd m and against $N^{-(m+2)}$ for even m.

```
>> s = [-2 -2 -4 -4];
>> ax = newfig(2,2); ax = ax';
>> for k = 1:length(m)
      subplot(ax(k));
      xlog; ylog; ylim([1e-18 1e2]);
      plot(N,err(k,:),'*','displayname','abs. error');
      plotfun(@(N) N^s(k),[1 1e4] ...
            ,'displayname',sprintf('N^{%d}',s(k)));
      xlim([1 1e4]);
      xlabel('N');
    end
```

The errors converge to zero at the expected rates.   ∎

16.3 ▪ Accuracy analysis for the indefinite integral

Now suppose that f is one of the infinitely many antiderivatives of a function g on $[a, b]$. The matrices $\mathbf{J}_{mn}$ and $\hat{\mathbf{K}}_{mn}$ of Theorem 15.10 should provide numerical confirmation that f is an antiderivative of g.

Partition the interval $[a, b]$ into contiguous subintervals $[a_j, b_j]$, $j = 1, \ldots, n$, of equal width, and sample $\mathbf{f}_j$ and $\mathbf{g}_j$ from f and g on uniform grids of degrees $m + 1$ and m, respectively, on $[a_j, b_j]$. Construct $\mathbf{f}$ and $\mathbf{g}$ by stacking the samples from the subintervals vertically, as $\mathbf{p}$ and $\mathbf{q}$ are constructed in (15.11)–(15.12).

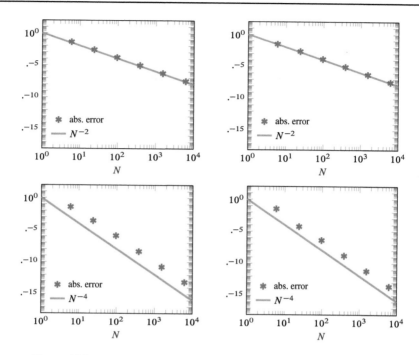

Figure 16.3. *Rate of convergence for antidifferentiation on a piecewise-uniform grid of degree m: m = 0 (top-left), m = 1 (top-right), m = 2 (bottom-left), and m = 3 (bottom-right).*

Theorem 16.6. *Suppose f is an antiderivative of g and $\mathbf{f}$ and $\mathbf{g}$ are sampled as above. If $g \in C^{m+1}[a, b]$, then*

$$\left\| \mathbf{J}_{mn}\mathbf{f} - \frac{b-a}{n}\hat{\mathbf{K}}_{mn}\mathbf{g} \right\|_\infty \leq C N^{-(m+1)} \tag{16.7}$$

for a coefficient C that is constant with respect to n. If m is even and $g \in C^{m+2}[a, b]$, then the bound can be improved to

$$\left\| \mathbf{J}_{mn}\mathbf{f} - \frac{b-a}{n}\hat{\mathbf{K}}_{mn}\mathbf{g} \right\|_\infty \leq C N^{-(m+2)} \tag{16.8}$$

for another coefficient C that is constant with respect to n. As usual, $N = (m \vee 1)n$.

Proof. Considering (15.13), every entry of $\mathbf{J}_{mn}\mathbf{f}$ has the form $f(x_{ij}) - f(a)$. By the proof of Theorem 15.10, the corresponding entry of $\frac{b-a}{n}\hat{\mathbf{K}}_{mn}\mathbf{g}$ equals $\int_a^{x_{ij}} q(x)\,dx$, in which q is the piecewise-polynomial interpolant for g. By Theorem 16.4, the difference is bounded as in the statement of the theorem. □

Chapter 26 relies heavily on Theorem 16.6.

Notes

Buck provides a nice illustration of how the midpoint rule ($m = 0$) can outperform the trapezoid rule ($m = 1$) [11].

Exercise 15 considers periodic integrands. Weideman investigates the phenomenon more deeply and provides several interesting examples [82].

Exercises

Exercises 1–4: Consider Newton–Cotes quadrature on the given integral with the given degree m. Can the absolute error be bounded by a function of the form CN^{-r} using the theory of this chapter? If so, provide a numerical value for r. Regardless, measure the rate of convergence experimentally.

1. $\int_0^1 (1/(1+x^2))\,dx; m = 0$

2. $\int_0^{\pi/4} \tan x \, dx; m = 1$

3. $\int_0^{1/\sqrt{2}} \sqrt{1-x^2}\,dx; m = 2$

4. $\int_0^4 \sqrt{x}\,dx; m = 3$

Exercises 5–8: When the specified antiderivative is computed on a piecewise-uniform grid of degree m, can the absolute error be bounded by a function of the form CN^{-r} using the theory of this chapter? If so, provide a numerical value for r. Regardless, measure the rate of convergence experimentally.

5. $\int_1^x (1/\xi)\,d\xi, 1 \le x \le 10; m = 0$

6. $\int_0^x \cos \xi \, d\xi, 0 \le x \le 2\pi; m = 1$

7. $\int_{-1}^x |\xi - e^{-1}|\,d\xi, -1 \le x \le 1; m = 2$

8. $\int_{-1/2}^x (1/(1-\xi^2))\,d\xi, -1/2 \le x \le 1/2; m = 3$

9. Let $g(x) = 0$ for $x < 1/\sqrt{2}$ and $g(x) = 1$ for $x \ge 1/\sqrt{2}$. Compute $\int_0^1 g(x)\,dx$ using Newton–Cotes quadrature with $m = 2$. Does the result converge to the correct answer as $n \to \infty$? If so, match the rate of convergence with N^{-r} for some value of r and provide numerical evidence.

10. Let $g \in C^2[a, b]$. Is the error bound of Theorem 16.1 smaller for $m = 0$, a.k.a. the midpoint rule, or $m = 1$, a.k.a., the trapezoid rule? Explain.

11. Let $I = \int_a^b g(x)\,dx$, and let $I_1(N)$ be the Newton–Cotes quadrature approximation using a 1-by-N grid. Consider the hypothesis that $I - I_1(N) \approx CN^{-2}$ for some constant C. (Note the $\approx$ in contrast to the $\le$ of Theorem 16.1.) Show that the hypothesis implies $I_1(2N) - I_1(N) \approx (3/4)CN^{-2}$. Measure the sequence of differences $I_1(2N) - I_1(N)$ on the example $\int_0^1 e^{-x^2}\,dx$. Are the data consistent with the hypothesis? If so, estimate the value of C from the differences $I_1(2N) - I_1(N)$.

12. Define I and $I_1(N)$ as in the previous exercise. Suppose $I - I_1(N) \approx CN^{-2}$. Find a linear combination $cI_1(N) + dI_1(2N)$ for which the absolute error

$$|I - (cI_1(N) + dI_1(2N))|$$

converges to zero asymptotically faster than N^{-2}. (The absolute error should fall toward zero faster than N^{-2} on a log-log graph.) Explain and test your combination on the example $\int_0^1 e^{-x^2}\,dx$. You may compare with the known value `sqrt(pi)/2*erf(1)`. (The above technique is known as *Richardson extrapolation*.)

13. Suppose $\int_{-1}^1 \sqrt{1-x^2}\,dx$ is computed by Newton–Cotes quadrature. What does Theorem 16.1 have to say about the rate of convergence for $m = 0, 1, 2, 3$? Measure the rate of convergence experimentally for each choice of m.

14. The integral

$$\int_0^1 \frac{1}{\sqrt{e^x - 1}} \, dx$$

is improper. Consider three substitutions: $x = u^{3/2}$, $x = u^2$, and $x = u^3$. Which of these substitutions remove the vertical asymptote at $x = 0$? Which substitution produces the best performance from Newton–Cotes quadrature? Demonstrate experimentally and explain using the theory of this chapter.

15. Newton–Cotes quadrature can converge especially quickly on periodic integrands. To investigate, consider the ellipse that is parameterized by $x(t) = 2 \cos t$, $y(t) = \sin t$. Compute first the arc length over $0 \le t \le \pi/3$ and second over $0 \le t \le 2\pi$. In both cases, use a 2-by-n piecewise-uniform grid and seek numerical convergence. Discuss the rates of convergence.

Chapter 17

Integration on a Chebyshev grid: Clenshaw–Curtis quadrature

When a Chebyshev grid is used in the integration framework of Chapter 14, highly accurate integrals can be computed with relatively little work. This approach is often called *Clenshaw–Curtis quadrature* in recognition of the 1960 work of Clenshaw and Curtis. We adopt the name as well, although the method developed here differs in a notable way from the original method; see the notes at the end of this chapter.

Clenshaw–Curtis quadrature is illustrated in Figure 17.1. The integrand is replaced by a Chebyshev interpolant, and the polynomial interpolant is integrated in place of the original function.

17.1 • Chebyshev integration matrix

The integration matrix $\mathbf{K}_{\mathbf{x},\mathbf{x}'}$ from Chapter 14 is again the key tool for integration. Because the computation of this matrix can be expensive, it is important to avoid duplicate work. The following theorem is crucial in this regard. In preparation, let $\mathbf{t}$ and $\mathbf{t}'$ be Chebyshev grids of degrees $m + 1$ and m, respectively, on $[-1, 1]$, with a lower limit of integration at $t = -1$, and define

$$\mathbf{K}_m = \mathbf{K}_{\mathbf{t},\mathbf{t}'}.$$

The notation $\mathbf{K}_m$ has also been used to refer to an integration matrix on a uniform grid, so in the future we'll have to be careful to clarify which sort of grid is in use.

Theorem 17.1. *Let* $\mathbf{x}$ *and* $\mathbf{x}'$ *be Chebyshev grids of degrees* $m + 1$ *and* m, *respectively, on an interval* $[a, b]$ *with a lower limit of integration at* $x = a$. *Then*

$$\mathbf{K}_{\mathbf{x},\mathbf{x}'} = \frac{b - a}{2} \mathbf{K}_m. \tag{17.1}$$

Proof. The proof is nearly identical to the proof of Theorem 15.1. Apply the substitution $t = 2(x - (a + b)/2)/(b - a)$ to the entries of $\mathbf{K}_{\mathbf{x},\mathbf{x}'}$ and compare with the entries of $\mathbf{K}_{\mathbf{t},\mathbf{t}'}$. ☐

Once the Chebyshev integration matrix $\mathbf{K}_m$ of degree m is computed, it can be reused later, even if the later integration is on a different interval.

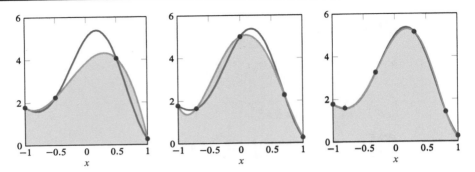

Figure 17.1. *Clenshaw–Curtis quadrature of degree m: m = 3 (left), m = 4 (middle), and m = 5 (right).*

Listing 17.1. Construct an integration matrix for a Chebyshev grid on $[-1, 1]$.

```
function K = intmatrixcheb(m)

persistent Kstore
if isempty(Kstore), Kstore = cell(1,400); end
if m+1<=length(Kstore)&&~isempty(Kstore{m+1})
  % retrieve matrix if previously computed
  K = Kstore{m+1};
else
  % compute integration matrix
  [xs,ws] = gridcheb([-1 1],m+1);
  xs_ = gridcheb([-1 1],m);
  K = intmatrix_(xs,ws,xs_,-1);
  % store for future use
  if m+1<=length(Kstore)
    Kstore{m+1} = K;
  end
end
end
```

The integration matrix is computed by `intmatrixcheb` in Listing 17.1.

For the definite integral, only the last row of $\mathbf{K}_m$ is needed. We define λ_m to be the vector whose transpose is the last row of $\mathbf{K}_m$.

17.2 ▪ Computation

Let p and q be polynomials of degrees at most $m + 1$ and m, respectively, and let $\mathbf{p} = p|_{\mathbf{x}}$ and $\mathbf{q} = q|_{\mathbf{x}'}$ be their samples on Chebyshev grids $\mathbf{x}$ and $\mathbf{x}'$ of the same degrees. From Theorems 14.3, 14.6, 14.8, 17.1 and Remark 14.4, we immediately find the following:

- The definite integral of q is

$$\int_a^b q(x)\, dx = \frac{b-a}{2}\lambda_m^T \mathbf{q}. \tag{17.2}$$

- p is the specific antiderivative $p(x) = p_a + \int_a^x q(\xi)\, d\xi$ if and only if

$$\mathbf{p} = p_a \mathbf{1} + \frac{b-a}{2}\mathbf{K}_m \mathbf{q}. \tag{17.3}$$

- p is any of the infinitely many antiderivatives of q if and only if

$$\mathbf{J}_m \mathbf{p} = \frac{b-a}{2} \hat{\mathbf{K}}_m \mathbf{q}, \tag{17.4}$$

in which $\mathbf{J}_m$ is another name for the matrix $\mathbf{J}_x$ of (14.9) and $\hat{\mathbf{K}}_m$ is formed by deleting the first row of $\mathbf{K}_m$.

The three forms of the integral are implemented in definitecheb, antiderivcheb, and indefinitecheb, shown in Listing 17.2.

Listing 17.2. Compute three different types of integrals on a Chebyshev grid.

```
function I = definitecheb(qs,ab)

a = ab(1);
b = ab(2);
m = size(qs,1)-1;
K = intmatrixcheb(m);
lambdas = K(end,:)';
I = (b-a)/2*lambdas'*qs;
```

```
function ps = antiderivcheb(pa,qs,ab)

a = ab(1);
b = ab(2);
m = length(qs)-1;
K = (b-a)/2*intmatrixcheb(m);
ps = pa*ones(m+2,1)+K*qs;
```

```
function [J,K] = indefinitecheb(m)

J = [ -ones(m+1,1) eye(m+1) ];
K = intmatrixcheb(m);
K = K(2:end,:);
```

Example 17.2. Approximate the definite integral

$$\log 2 = \int_1^2 \frac{1}{x}\, dx$$

using Chebyshev grids of increasing degree. Seek numerical convergence.

Solution. The integrand is $g(x) = 1/x$, and the limits of integration are $a = 1$ and $b = 2$.

```
>> g = @(x) 1/x;  a = 1;  b = 2;
```

We try degrees $m = 2, 4, 6, \ldots, 18$.

```
>> m = 2:2:18;
```

The definite integral is computed with definitecheb, and the results are displayed.

```
>> I = nan(size(m));
>> for k = 1:length(m)
     qs = samplecheb(g,[a b],m(k));
     I(k) = definitecheb(qs,[a b]);
   end
>> disptable('m',m,'%2d','I',I,'%17.15f');
```

m	I
2	0.694444444444445
4	0.693137254901961
6	0.693147117432832
8	0.693147179919533
10	0.693147180551082
12	0.693147180559802
14	0.693147180559943
16	0.693147180559945
18	0.693147180559945

The final value appears to be an accurate approximation for $\log 2$. ∎

Example 17.3. The integral

$$\int_0^1 x^{-1} \sin(\sqrt{x})\,dx,$$

like many interesting integrals, is improper. Ameliorate the singularity at $x = 0$ with a substitution of the form $x = t^\alpha$. Then compute the definite integral using Clenshaw–Curtis quadrature, seeking numerical convergence.

Solution. If $\alpha > 0$, then

$$\int_0^1 x^{-1} \sin(\sqrt{x})\,dx = \int_0^1 t^{-\alpha} \sin(t^{\alpha/2})(\alpha t^{\alpha-1})\,dt = \alpha \int_0^1 t^{-1} \sin(t^{\alpha/2})\,dt.$$

One solution is to take $\alpha = 2$. Then the integral becomes

$$2 \int_0^1 \frac{\sin t}{t}\,dt.$$

The integrand is well behaved; in particular,

$$\lim_{t \to 0} \frac{\sin t}{t} = 1$$

by l'Hôpital's rule.

There is a small hiccup in the computer implementation. Because the integrand $g(t) = (\sin t)/t$ is of the indeterminate form $0/0$ at $t = 0$, the function must be patched at $t = 0$ to ensure continuity.

```
>> g = @(t) iif(t==0,@()1,@()sin(t)/t); a = 0; b = 1;
>> m = 10;
>> qs = samplecheb(g,[a b],m);
>> 2*definitecheb(qs,[a b])
ans =
    1.892166140734368
```

Further experiments (not shown here) reveal that increasing m beyond 10 does not significantly change the computed value. ∎

Example 17.4. The incomplete elliptic integral of the second kind is

$$E(\phi, m) = \int_0^\phi \sqrt{1 - m \sin^2 \theta}\,d\theta.$$

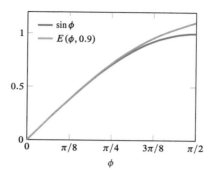

Figure 17.2. *The incomplete elliptic integrals* $E(\phi, 1) = \sin \phi$ *and* $E(\phi, 0.9)$.

Compute $E(\phi, 0.9)$ for $0 \le \phi \le \pi/2$ using Clenshaw–Curtis quadrature, seeking full accuracy. Plot next to $E(\phi, 1) = \int_0^\phi \sqrt{1 - \sin^2 \theta}\, d\theta = \sin \phi$ for comparison.

Solution. The integrand for $E(\phi, 0.9)$ is $g(\theta) = \sqrt{1 - 0.9 \sin^2 \theta}$, and the initial value is $E(0, 0.9) = 0$.

```
>> g = @(th) sqrt(1-0.9*sin(th)^2); a = 0; b = pi/2;
>> pa = 0;
```

Through trial and error, we find that a Chebyshev interpolant of degree 50 fits the integrand nearly as well as possible in finite precision.

```
>> m = 50;
>> qs = samplecheb(g,[a b],m);
>> q = interpcheb(qs,[a b]);
>> infnorm(@(th) g(th)-q(th),[a b],100)
ans =
     1.554312234475219e-15
```

The antiderivative is computed.

```
>> ps = antiderivcheb(pa,qs,[a b]);
>> p = interpcheb(ps,[a b]);
```

The sine function and the incomplete elliptic integral are plotted in Figure 17.2.

```
>> newfig;
>> plotfun(@(ph) sin(ph),[a b],'displayname','sin \phi');
>> plotfun(p,[a b],'displayname','E(\phi,0.9)');
>> xlabel('\phi');
```                                                                                                  ∎

Notes

A 1933 article by Fejér, which predates the Clenshaw–Curtis paper by 27 years, contains many of the key ideas in this chapter [30]. However, the name Clenshaw–Curtis quadrature is standard.

The method in the Clenshaw–Curtis paper [16] is superior to the method presented in this chapter when applied to antidifferentiation. The Clenshaw–Curtis method first expands the interpolating polynomial $q(x)$ as a linear combination $\sum_i c_i T_i(x)$ of Chebyshev polynomials. Integrating the series then reduces for the most part to integrating $T_i(x)$. This can all be done quickly and easily with the fast Fourier transform and an identity relating $\int T_i(x)\, dx$ to $T_{i-1}(x)$ and $T_{i+1}(x)$.

Exercises

Exercises 1–4: Compute the definite integral using Clenshaw–Curtis quadrature. Seek numerical convergence.

1. $\int_{-4}^{4} (2\pi)^{-1/2} e^{-x^2/2} \, dx$

2. $\int_{0}^{10} x^4 e^{-x^2/2} \, dx$

3. $\int_{0}^{\pi/4} \sqrt{\sin^2 x + 2\cos^2 x} \, dx$

4. $\int_{0}^{\pi/4} 2 \log(2 \cos x) \, dx$ (Catalan's constant)

Exercises 5–8: Compute the antiderivative on a Chebyshev grid over the given interval. Seek numerical convergence.

5. $-1 + \int_{-0.5}^{x} (1/(\xi^3 + \xi + 1)) \, d\xi; \; -0.5 \leq x \leq 1.5$

6. $F(x, 1/\sqrt{2}) = \int_{0}^{x} (1 - (1/2) \sin^2 \theta)^{-1/2} \, d\theta; \; 0 \leq x \leq \pi/2$ (elliptic integral)

7. $\gamma(3, x) = \int_{0}^{x} \xi^2 e^{-\xi} \, d\xi; \; 0 \leq x \leq 20$ (incomplete gamma)

8. $T_3(x) = (1/2) + \int_{0}^{x} (6\sqrt{3}/\pi)(3 + \xi^2)^{-2} \, d\xi; \; 0 \leq x \leq 10$ (CDF of Student's t)

9. Fill in the details for the proof of Theorem 17.1.

10. Construct a table of integrals $\int_{-z}^{z} (2\pi)^{-1/2} e^{-x^2/2} \, dx$ for $z = 0.05, 0.10, 0.15, \ldots,$ 4.00, using Clenshaw–Curtis quadrature. Seek numerical convergence.

11. Certain values of the gamma function can be expressed as especially simple integrals, e.g., $\Gamma(4/3) = \int_{0}^{\infty} e^{-t^3} \, dt$.

   (a) Show that for any $L \geq 1$, $\Gamma(4/3) = \int_{0}^{L} e^{-t^3} \, dt + \varepsilon$ with $|\varepsilon| \leq e^{-L^3}$.

   (b) Find an L for which $|\varepsilon| < 10^{-16}$.

   (c) Evaluate $\Gamma(4/3)$ by computing the truncated integral in part (a) with the value of L found in part (b). Use Clenshaw–Curtis quadrature and seek numerical convergence.

12. The Gompertz constant is $G = \int_{0}^{\infty} (e^{-x}/(1 + x)) \, dx$.

   (a) Prove that for any $L \geq 0$, $G = \int_{0}^{L} (e^{-x}/(1+x)) \, dx + \varepsilon$ with $|\varepsilon| \leq e^{-L}/(1 + L)$.

   (b) Find an L for which $|\varepsilon| < 10^{-16}$.

   (c) Calculate G by computing the truncated integral in part (a) with the value of L in part (b). Use Clenshaw–Curtis quadrature and seek numerical convergence.

13. The integral

$$\int_{0}^{1} \frac{1}{\sqrt{\sin x}} \, dx$$

is improper. Find a substitution of the form $x = t^\alpha$ that makes the integral proper and that enables you to achieve numerical convergence when computing the integral on a Chebyshev grid. *Hint.* You may need to patch the integrand at one of the limits of integration as in Example 17.3.

14. The period of a certain pendulum released at an angle θ_0 from its rest position is

$$\int_0^{\theta_0} \frac{1}{\sqrt{\cos\theta - \cos\theta_0}}\, d\theta.$$

Below, take $\theta_0 = \pi/3$.

   (a) Try computing the integral using `definitecheb`. What happens? Why?
   (b) Find a substitution of the form $\theta = c + dx^{\alpha}$ for which the resulting integrand is analytic over the interval of integration. (Technically, since the integrand above is undefined at $\theta = \theta_0$, the transformed integrand will likely also be undefined at one of the limits of integration, but it should be possible to plug the hole as in Example 17.3 to arrive at an analytic integrand.)
   (c) Compute the proper integral using `definitecheb` and seek numerical convergence.

15. The integral $\arcsin x = \int_0^x (1-\xi^2)^{-1/2}\, d\xi$ is improper when $x = 1$. Compute the arcsin function on $[0, 1]$ as follows:

   (a) Apply the substitution $\xi = 1 - t^2$ to the integral above.
   (b) Compute the transformed integral on a Chebyshev grid, achieving numerical convergence.
   (c) Reverse the change of variables to recover $\arcsin x$. Compare with the built-in `asin` function to measure the absolute error in the infinity norm.

16. Develop a MATLAB function for computing the definite integral of a piecewise-Chebyshev interpolant. Demonstrate on the interpolant of Example 10.10.

17. Develop a MATLAB function for computing an antiderivative of a piecewise-Chebyshev interpolant. Demonstrate it by computing $4 + \int_0^x q(\xi)\, d\xi$ for the interpolant q of Example 10.10.

Chapter 18

Accuracy analysis for Clenshaw–Curtis quadrature

Integration on a Chebyshev grid, i.e., Clenshaw–Curtis quadrature, converges every bit as fast as we should hope.

18.1 • Accuracy analysis for the definite integral

Let g be analytic and bounded in absolute value by some number M in the interior of a Bernstein ellipse $E_\rho(a, b)$. Let q be a degree-m Chebyshev interpolant for g on $[a, b]$.

Theorem 18.1. *If g is analytic and bounded as specified above, then the definite integral of its Chebyshev interpolant q satisfies*

$$\left| \int_a^b g(x)\, dx - \int_a^b q(x)\, dx \right| \leq \frac{4M(b-a)}{\rho - 1} \rho^{-m}. \tag{18.1}$$

Proof. By Theorems 11.3 and 14.14,

$$\left| \int_a^b g(x)\, dx - \int_a^b q(x)\, dx \right| \leq (b-a)\|g - q\|_\infty \leq (b-a)\frac{4M}{\rho - 1} \rho^{-m}. \qquad \square$$

It follows that Clenshaw–Curtis quadrature converges supergeometrically when the integrand is analytic on the entire complex plane.

Example 18.2. Approximate

$$\log 2 = \int_1^2 \frac{1}{x}\, dx$$

using Chebyshev grids of increasing degree. Measure the rate of convergence and compare with Theorem 18.1.

Solution. The integral is computed numerically on a sequence of more and more refined grids.

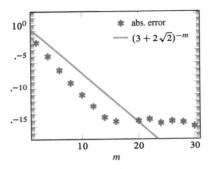

Figure 18.1. *Rate of convergence for definite integration with an analytic integrand and a Chebyshev grid.*

```
>> g = @(x) 1/x; a = 1; b = 2;
>> m = 2:2:30;
>> Iapprox = nan(size(m));
>> for k = 1:length(m)
       qs = samplecheb(g,[a b],m(k));
       Iapprox(k) = definitecheb(qs,[a b]);
   end
```

Because the value of the integral is known exactly as $\log 2$, the errors can be measured directly. The results are plotted in Figure 18.1.

```
>> I = log(2);
>> newfig;
>> ylog; ylim([1e-18 1e2]);
>> plot(m,abs(I-Iapprox),'*','displayname','abs. error');
>> xlabel('m');
```

What does Theorem 18.1 predict? Note that the integrand's only singularity is at $z = 0$. We must find a Bernstein ellipse $E_\rho(a, b)$ that intersects this singularity. Let

$$t = \frac{2}{b-a}\left(0 - \frac{a+b}{2}\right) = \frac{2}{2-1}\left(0 - \frac{1+2}{2}\right) = -3.$$

Because

$$\left|t \pm \sqrt{t^2 - 1}\right| = \left|-3 \pm \sqrt{(-3)^2 - 1}\right| = \{3 - 2\sqrt{2}, 3 + 2\sqrt{2}\},$$

the singularity lies on the Bernstein ellipse $E_\rho(a, b)$ with $a = 1$, $b = 2$, and $\rho = 3 + 2\sqrt{2}$. Although the integrand is unbounded in this ellipse, it is bounded in any strictly smaller ellipse. Hence, the rate of convergence is guaranteed to be as fast as ρ_*^{-m} for any $\rho_* < \rho$. The experimental data are compared with the curve $(3 + 2\sqrt{2})^{-m}$ in the figure.

```
>> plotfun(@(m) (3+2*sqrt(2))^(-m),[1 31] ...
          ,'displayname','(3+2*sqrt(2))^{-m}');
```

The errors appear to converge to zero geometrically and even faster than predicted. ∎

In the previous example, Clenshaw–Curtis quadrature performs even better than Theorem 18.1 would suggest. This is a common phenomenon, but the explanation is outside the scope of this book.

As we've seen, Clenshaw–Curtis quadrature converges geometrically on analytic integrands. For integrands that are real differentiable but not analytic, the rate of convergence may be algebraic (Exercise 15).

18.2 ▪ Accuracy analysis for a specific antiderivative

Theorem 18.3. *Suppose g is analytic and bounded as in the previous section and that q is a degree-m Chebyshev interpolant for g on* $[a, b]$. *Let* $f(x) = f_a + \int_a^x g(\xi) \, d\xi$ *and* $p(x) = f_a + \int_a^x q(\xi) \, d\xi$. *Then*

$$\|f - p\|_\infty \le \frac{4M(b-a)}{\rho - 1} \rho^{-m}. \tag{18.2}$$

Proof. By Theorems 11.3 and 14.15,

$$\|f - p\|_\infty \le (b-a)\|g - q\|_\infty \le (b-a)\frac{4M}{\rho - 1}\rho^{-m}. \qquad \square$$

Example 18.4. Approximate

$$\arctan x = \int_0^x \frac{1}{1 + \xi^2} \, d\xi$$

on $0 \le x \le 5$ using Chebyshev grids of increasing degree. Measure the rate of convergence and compare with Theorem 18.3.

Solution. The integrand and limits of integration are defined.
```
>> g = @(x) 1/(1+x^2); a = 0; b = 5;
```
The antiderivative is computed numerically and compared with the known solution.
```
>> f = @(x) atan(x);
>> m = 5:5:80;
>> err = nan(size(m));
>> for k = 1:length(m)
     qs = samplecheb(g,[a b],m(k));
     ps = antiderivcheb(0,qs,[a b]);
     p = interpcheb(ps,[a b]);
     err(k) = infnorm(@(x) f(x)-p(x),[a b],100);
   end
```
The errors are plotted in Figure 18.2.
```
>> newfig;
>> ylog; ylim([1e-18 1e2]);
>> plot(m,err,'*','displayname','abs. error');
>> xlabel('m');
```
The only singularities of the integrand are at $\pm 1i$, which lie equally far from the integration interval. The integrand is analytic in the Bernstein ellipse $E_\rho(0, 5)$ with the following value of ρ:
```
>> z = 1i;
>> t = 2/(b-a)*(z-(a+b)/2);
>> rho = max(abs(t+sqrt(t^2-1)),abs(t-sqrt(t^2-1)))
rho =
    1.918317725836914
```
The errors $\|f - p\|_\infty$ are guaranteed to converge to zero as fast as ρ_*^{-m} for any $\rho_* < \rho$. The curve 1.9^{-m} is added to the plot.

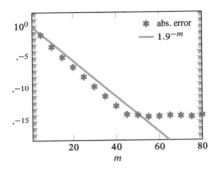

Figure 18.2. *Rate of convergence for antidifferentiation with an analytic integrand and a Chebyshev grid.*

```
>> plotfun(@(m) 1.9^(-m),[1 80],'displayname','1.9^{-m}');
```
The convergence is as fast as expected. ■

In most real problems, the integral being sought is not known in advance and therefore the error cannot be measured directly. In this case, the right-hand side of

$$\left| \int_a^b g(x)\,dx - \int_a^b q(x)\,dx \right| \le (b-a)\|g-q\|_\infty \tag{18.3}$$

(from Theorem 14.14) or the right-hand side of

$$\|f - p\|_\infty \le (b-a)\|g-q\|_\infty \tag{18.4}$$

(from Theorem 14.15) may be used as an a posteriori error estimate.

Example 18.5. Compute the incomplete elliptic integral of the second kind

$$E(\phi, 0.9) = \int_0^\phi \sqrt{1 - 0.9\sin^2\theta}\,d\theta, \quad 0 \le \phi \le \pi/2,$$

using Clenshaw–Curtis quadrature. First, predict the rate of convergence. Then, compute the integral with increasing degree m and estimate the absolute error by (18.4). Compare.

Solution. The integrand is analytic on the interval $[0, \pi/2]$, but it has singularities in the complex plane. The complex derivative of the integrand equals

$$-\frac{0.9\sin\theta\cos\theta}{\sqrt{1 - 0.9\sin^2\theta}}$$

where it is defined. The integrand is not complex differentiable at points where $1 - 0.9\sin^2\theta$ is a negative real number or zero. The closest such points to the interval of interest are a pair of numbers z with $1 - 0.9\sin^2 z = 0$, specifically

```
>> z = asin(sqrt(1/0.9))
z =
   1.570796326794897 - 0.327450150237259i
```

and its complex conjugate. The integrand is analytic in the Bernstein ellipse with the following value of ρ:

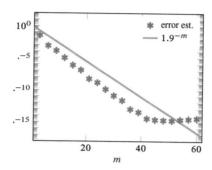

Figure 18.3. *An a posteriori error estimate for a computation of $E(\phi, 0.9)$ and a predicted rate of convergence.*

```
>> a = 0; b = pi/2;
>> t = 2/(b-a)*(z-(a+b)/2);
>> rho = max(abs(t+sqrt(t^2-1)),abs(t-sqrt(t^2-1)))
rho =
      1.946056844148527
```

We expect Clenshaw–Curtis quadrature to converge at a rate close to 1.9^{-m}, or possibly faster.

Let's test this prediction by computing the integral on finer and finer grids. With each computation, the error is estimated by (18.4).

```
>> g = @(th) sqrt(1-0.9*sin(th)^2);
>> pa = 0;
>> m = 3:3:60;
>> err = nan(size(m));
>> for k = 1:length(m)
     qs = samplecheb(g,[a b],m(k));
     ps = antiderivcheb(pa,qs,[a b]);
     q = interpcheb(qs,[a b]);
     err(k) = (b-a)*infnorm(@(x) g(x)-q(x),[a b],100);
   end
```

The error estimates and the predicted rate of convergence are graphed in Figure 18.3.

```
>> newfig;
>> ylog; ylim([1e-18 1e2]);
>> plot(m,err,'*','displayname','error est.');
>> plotfun(@(m) 1.9^(-m),[1 62],'displayname','1.9^{-m}');
>> xlabel('m');
```

The theory predicts that the absolute error should converge to zero as fast as the orange curve in the figure. The experimental estimates are consistent with this prediction. ∎

18.3 • Accuracy analysis for the indefinite integral

Finally, we consider the indefinite integral. Suppose that f is analytic in a Bernstein ellipse $E_\rho(a, b)$. Then its derivative $g = f'$ is necessarily analytic in the same ellipse (Appendix B). In addition, suppose that f and g are bounded in absolute value by some number M in the ellipse.

Theorem 18.6. *Suppose f and $g = f'$ are analytic and bounded as above. Sample $\mathbf{f} = f|_{\mathbf{x}}$ on a Chebyshev grid $\mathbf{x}$ of degree $m + 1$ and $\mathbf{g} = g|_{\mathbf{x}'}$ on a Chebyshev grid $\mathbf{x}'$*

of degree m, both over $[a, b]$. *Then*

$$\left\| \mathbf{J}_m \mathbf{f} - \tfrac{b-a}{2} \hat{\mathbf{K}}_m \mathbf{g} \right\|_\infty \le \frac{4M(2 + b - a)}{\rho - 1} \rho^{-m}, \tag{18.5}$$

in which $\mathbf{J}_m$ *and* $\hat{\mathbf{K}}_m$ *are defined as in the previous chapter.*

Proof. Let p be the degree-$(m + 1)$ Chebyshev interpolant for f, and let q be the degree-m Chebyshev interpolant for g. By Theorem 14.16,

$$\left\| \mathbf{J}_m \mathbf{f} - \tfrac{b-a}{2} \hat{\mathbf{K}}_m \mathbf{g} \right\|_\infty \le 2\| f - p \|_\infty + (b - a)\| g - q \|_\infty.$$

By Theorem 11.3,

$$\left\| \mathbf{J}_m \mathbf{f} - \tfrac{b-a}{2} \hat{\mathbf{K}}_m \mathbf{g} \right\|_\infty \le 2\frac{4M}{\rho - 1}\rho^{-m} + (b - a)\frac{4M}{\rho - 1}\rho^{-m}. \qquad \square$$

This bound is used in Chapter 28.

Notes

A bound nearly identical to (18.1) for analytic integrands was derived by Chawla in 1968 [13]. The better-than-expected accuracy of definite integration on a Chebyshev grid was investigated by Trefethen in 2008 [74]. A class of continuous but nonanalytic integrands was analyzed by Riess and Johnson in 1972 [63], and in 2012, their result was extended to a class of differentiable functions by Xiang and Bornemann [88].

    Exercise 16 is based on an example in the Trefethen paper mentioned above [74].

Exercises

Exercises 1–4: Consider computing the given integral with Clenshaw–Curtis quadrature. Using the theory of this chapter, classify the rate of convergence as algebraic, geometric, or supergeometric. If it is geometric, find an appropriate value for ρ in the bound $C\rho^{-m}$. Then test your prediction experimentally, either measuring the absolute error directly or estimating it by (18.3).

1. $\int_0^{1/\sqrt{2}} \sqrt{1 - x^2}\, dx$
2. $\int_0^{\pi/4} \log(2 \cos x)\, dx$
3. $\int_0^{e} x^2 \log x\, dx$
4. $\int_0^{2} (2\pi)^{-1/2} e^{-x^2/2}\, dx$

Exercises 5–8: Consider computing the given antiderivative on a Chebyshev grid. Using the theory of this chapter, predict the rate of convergence. Is it algebraic, geometric, or supergeometric? If it is geometric, find a numerical value for ρ in the bound $C\rho^{-m}$. Then test your prediction experimentally, either measuring the error $\| f - p \|_\infty$ directly or estimating it by (18.4).

5. $\int_0^{x} (1 + \xi^2)^{-1/2}\, d\xi,\ 0 \le x \le 10$

6. $\int_0^x ((\sin \xi)/\xi) \, d\xi$, $0 \le x \le 8\pi$. *Hint.* Be careful at $\xi = 0$.

7. $\int_0^x \cos(\pi \xi^2/2) \, d\xi$, $0 \le x \le 4$

8. $\int_{-5}^x \log(1 - e^{\xi-1}) \, d\xi$, $-5 \le x \le 0$

9. Let $g(x) = 0$ for $x < 1/\sqrt{2}$ and $g(x) = 1$ for $x \ge 1/\sqrt{2}$. Provide numerical evidence that the Clenshaw–Curtis approximation for $\int_0^1 g(x) \, dx$ converges to the correct value as $m \to \infty$, albeit slowly. Explain how $\int_0^1 g(x) \, dx - \int_0^1 q(x) \, dx$ can converge to zero even though $\|g - q\|_\infty$ does not. (Compare with Example 11.13.)

10. Consider the integral $\int_0^{\pi/4} \log x \tan x \, dx$.

    (a) What is the value of the integrand at $x = 0$? What is the limit as $x \to 0^+$?

    (b) Compute the integral using Clenshaw–Curtis quadrature. Based on experimental evidence, does the rate of convergence appear to be algebraic, geometric, or supergeometric?

    (c) Perform integration by parts on the integral with $u = \log x$ and $dv = \tan x \, dx$. Repeat parts (a) and (b) with the new expression.

11. Develop an adaptive routine for computing $\int_a^b g(x) \, dx$ using the following method. First, compute the integral on a Chebyshev grid of degree 32. Second, compute $\int_a^c g(x) \, dx + \int_c^b g(x) \, dx$, with $c = (a + b)/2$, using Chebyshev grids of degree 32 for each integral. If the two computations agree within a provided tolerance, then terminate. Otherwise, recursively compute $\int_a^c g(x) \, dx$ and $\int_c^b g(x) \, dx$ using the same adaptive procedure and return their sum. Cap the depth of the subdivision process at a provided limit to avoid unreasonably expensive computation. Demonstrate your routine on $\int_{-1}^2 |1 - x^2| \, dx$.

12. Develop an adaptive routine for computing $f_a + \int_a^x g(\xi) \, d\xi$ on $[a, b]$ using the following method. First, compute the antiderivative on $[a, b]$ using a Chebyshev grid of degree 32. Second, compute the antiderivative on $[a, (a + b)/2]$ and then on $[(a + b)/2, b]$, using a Chebyshev grid of degree 32 in each case. If the two computations agree within a provided tolerance, then terminate. Otherwise, recursively compute the antiderivative on $[a, (a + b)/2]$ and then on $[(a + b)/2, b]$, using the same adaptive procedure, and join the two results to produce a piecewise-Chebyshev interpolant. Cap the depth of the subdivision procedure at a provided limit to avoid unreasonably expensive computation. Demonstrate your routine on $5 + \int_{-1}^x |1 - \xi^2| \, d\xi$, $-1 \le x \le 2$.

13. Compute $\int_{-1}^1 \sqrt{1 - x^2} \, dx$ to an absolute error below 10^{-12} using the adaptive method of Exercise 11. How many subintervals are used?

14. Compute the incomplete gamma function $\gamma(5/3, x) = \int_0^x \xi^{2/3} e^{-\xi} \, d\xi$ for $0 \le x \le 10$ using the adaptive procedure of Exercise 12. Make sure the absolute error is everywhere below 10^{-12}. You may compare with `gamma(5/3)*gammainc(x,5/3)`. How many subintervals are used by the adaptive method?

15. Suppose g is r-times weakly differentiable on $[a, b]$ for some $r \ge 1$, with an rth derivative of bounded variation. Prove that for any $m > r$, the degree-m Clenshaw–Curtis approximation I_m for $I = \int_a^b g(x) \, dx$ satisfies $|I - I_m| \le C m^{-r}$ for a coefficient C that is constant with respect to m.

16. Compute $\int_{-1}^{1}(1/(1 + 16x^2))\, dx$ by Clenshaw–Curtis quadrature and plot the absolute error against the degree m on log-linear axes. You should see a "kink" in the error plot where the rate of convergence changes. Over what regime does the error bound (18.1) match the observed rate of convergence? Estimate the slope of the error curve above and below the kink and make a conjecture about the relationship between the two rates.

Part IV

Systems of linear equations

Chapter 19

Gaussian elimination

This chapter is concerned with the system of linear equations

$$a_{11}x_1 + a_{12}x_2 + \cdots + a_{1n}x_n = b_1,$$
$$a_{21}x_1 + a_{22}x_2 + \cdots + a_{2n}x_n = b_2,$$
$$\cdots$$
$$a_{n1}x_1 + a_{n2}x_2 + \cdots + a_{nn}x_n = b_n,$$

$$(19.1)$$

which is equivalent to the matrix-vector equation

$$\mathbf{Ax} = \mathbf{b} \tag{19.2}$$

with

$$\mathbf{A} = \begin{bmatrix} a_{11} & a_{12} & \cdots & a_{1n} \\ a_{21} & a_{22} & \cdots & a_{2n} \\ \vdots & \vdots & \ddots & \vdots \\ a_{n1} & a_{n2} & \cdots & a_{nn} \end{bmatrix}, \quad \mathbf{x} = \begin{bmatrix} x_1 \\ x_2 \\ \vdots \\ x_n \end{bmatrix}, \quad \mathbf{b} = \begin{bmatrix} b_1 \\ b_2 \\ \vdots \\ b_n \end{bmatrix}.$$

The *solution set* is the set of vectors $\mathbf{x}$ that satisfy the equation.

In the MATLAB language, the linear system can be solved with the backslash operator: x = A\b. We have seen the backslash operator before, in the implementation of intmatrix_ in Chapter 14, and we will make heavy use of it in the last several chapters of this book. In the next few chapters, we develop the underlying numerical method, called *Gaussian elimination*, and discuss its numerical stability.

19.1 ▪ Back substitution

An *upper-triangular* matrix $\mathbf{U}$ is a square matrix with zeros below its main diagonal:

$$\mathbf{U} = \begin{bmatrix} u_{11} & u_{12} & \cdots & u_{1,n-1} & u_{1n} \\ & u_{22} & \cdots & u_{2,n-1} & u_{2n} \\ & & \ddots & \vdots & \vdots \\ & & & u_{n-1,n-1} & u_{n-1,n} \\ & & & & u_{nn} \end{bmatrix}.$$

(The blank entries are understood to equal zero.) The matrix-vector equation $\mathbf{U}\mathbf{x} = \mathbf{b}$ is particularly easy to solve. It is equivalent to the system

$$
\begin{aligned}
u_{11}x_1 + u_{12}x_2 + \cdots + u_{1,n-1}x_{n-1} + u_{1n}x_n &= b_1, \\
u_{22}x_2 + \cdots + u_{2,n-1}x_{n-1} + u_{2n}x_n &= b_2, \\
&\;\;\vdots \\
u_{n-1,n-1}x_{n-1} + u_{n-1,n}x_n &= b_{n-1}, \\
u_{nn}x_n &= b_n,
\end{aligned}
$$

and the equations can be solved from the bottom up to find x_n, then x_{n-1}, then x_{n-2}, and so on, as long as no diagonal entry u_{ii} equals zero. The complete solution is

$$
x_i = \frac{1}{u_{ii}} \left(b_i - \sum_{k=i+1}^{n} u_{ik}x_k \right), \quad i = n, n-1, n-2, \ldots, 1. \tag{19.3}
$$

In vector notation, this is

$$
x_i = \frac{1}{u_{ii}} \left(b_i - \begin{bmatrix} u_{i,i+1} & \cdots & u_{in} \end{bmatrix} \begin{bmatrix} x_{i+1} \\ \vdots \\ x_n \end{bmatrix} \right), \quad i = n, n-1, n-2, \ldots, 1.
$$

Computing the solution from these equations is known as *back substitution*. The name of the method refers to the process of working backward from the bottom equation to the top, solving for one variable at a time and substituting the computed value into the remaining equations.

Back substitution is not intended for use when any u_{ii} equals zero. Such a linear system may have zero or infinitely many solutions and is outside our discussion.

Back substitution is implemented as `backsubstitute` in Listing 19.1.

Listing 19.1. Solve an upper-triangular linear system by back substitution.

```
function x = backsubstitute(U,b)

n = size(U,1);
x = nan(n,1);
for i = n:-1:1
  x(i) = (b(i)-U(i,i+1:n)*x(i+1:n))/U(i,i);
end
```

Example 19.1. Solve the linear system

$$
\begin{bmatrix} 6 & 8 & -4 \\ 0 & 3 & 1 \\ 0 & 0 & 9 \end{bmatrix} \begin{bmatrix} x_1 \\ x_2 \\ x_3 \end{bmatrix} = \begin{bmatrix} 9 \\ 0 \\ 6 \end{bmatrix}
$$

using `backsubstitute`. Check the solution.

Solution. The system is constructed.

```
>> U = [  6  8 -4 ;
          0  3  1 ;
          0  0  9 ];
```

```
>> b = [   9 ;
           0 ;
           6 ];
```

It is solved with backsubstitute.

```
>> x = backsubstitute(U,b)
x =
    2.240740740740741
   -0.222222222222222
    0.666666666666667
```

The solution is checked by substituting it into the original equation.

```
>> U*x
ans =
    9.000000000000002
                    0
    6.000000000000000
```

Because U*x equals b up to roundoff error, the solution is correct. ■

A lower-triangular system $\mathbf{Lx} = \mathbf{b}$, in which all of the nonzero entries of $\mathbf{L}$ lie on or below the main diagonal, can be solved using an analogous procedure called *forward substitution*. Whereas backward substitution starts at the bottom and works its way up, forward substitution starts at the top and works its way down (Exercise 9).

19.2 ▪ Gaussian elimination without pivoting

Gaussian elimination is a method for reducing a system of linear equations to upper-triangular form. The resulting system can be solved quickly by back substitution.

It is important that the transformation to upper-triangular form is accomplished by invertible operations. A matrix $\mathbf{B}$ is an *inverse matrix* for a matrix $\mathbf{A}$ if $\mathbf{BA} = \mathbf{AB} = \mathbf{I}$, the identity matrix. It can be proved that if $\mathbf{A}$ and $\mathbf{B}$ are square matrices, then $\mathbf{BA} = \mathbf{I}$ if and only if $\mathbf{AB} = \mathbf{I}$, so that only one of the two products must be checked. A matrix with an inverse matrix is called *invertible*. When $\mathbf{A}$ is invertible, its inverse matrix is denoted $\mathbf{A}^{-1}$.

Theorem 19.2. *If* $\mathbf{M}$ *is an invertible matrix, then the solution set for*

$$\mathbf{Ax} = \mathbf{b}$$

is identical to the solution set for

$$\mathbf{MAx} = \mathbf{Mb}.$$

Proof. If $\mathbf{Ax} = \mathbf{b}$, then also $\mathbf{MAx} = \mathbf{Mb}$. Conversely, if $\mathbf{MAx} = \mathbf{Mb}$, then also $\mathbf{M}^{-1}\mathbf{MAx} = \mathbf{M}^{-1}\mathbf{Mb}$, which is equivalent to $\mathbf{Ax} = \mathbf{b}$.   □

Gaussian elimination is illustrated in the next example. The goal is to produce an equivalent linear system that is upper triangular.

Example 19.3. Consider the matrix-vector equation

$$\begin{bmatrix} 2 & 6 & 4 \\ 1 & 4 & 6 \\ -1 & 1 & 9 \end{bmatrix} \begin{bmatrix} x_1 \\ x_2 \\ x_3 \end{bmatrix} = \begin{bmatrix} 4 \\ 7 \\ 8 \end{bmatrix}. \tag{19.4}$$

Find an equivalent matrix-vector equation in which the matrix is upper triangular.

Solution. The process of Gaussian elimination or "row reduction" may already be familiar. For the system (19.4), the elementary operation of adding a multiple of one row to another row is sufficient to accomplish the entire transformation. Forming the augmented matrix

$$\left[\begin{array}{ccc|c} 2 & 6 & 4 & 4 \\ 1 & 4 & 6 & 7 \\ -1 & 1 & 9 & 8 \end{array}\right]$$

from the left- and right-hand sides of the given equation, the reduction proceeds as follows:

$$\left[\begin{array}{ccc|c} 2 & 6 & 4 & 4 \\ 1 & 4 & 6 & 7 \\ -1 & 1 & 9 & 8 \end{array}\right]$$

$$\xrightarrow[\substack{-(1/2)R_1+R_2\to R_2 \\ (1/2)R_1+R_3\to R_3}]{} \left[\begin{array}{ccc|c} 2 & 6 & 4 & 4 \\ 0 & 1 & 4 & 5 \\ 0 & 4 & 11 & 10 \end{array}\right]$$

$$\xrightarrow[-4R_2+R_3\to R_3]{} \left[\begin{array}{ccc|c} 2 & 6 & 4 & 4 \\ 0 & 1 & 4 & 5 \\ 0 & 0 & -5 & -10 \end{array}\right].$$

The notation $cR_j + R_k \to R_k$ represents the operation of multiplying row j by the scalar c and adding the result to row k, replacing row k. The final augmented matrix represents an upper-triangular system that is equivalent to the original system:

$$\left[\begin{array}{ccc} 2 & 6 & 4 \\ 0 & 1 & 4 \\ 0 & 0 & -5 \end{array}\right] \left[\begin{array}{c} x_1 \\ x_2 \\ x_3 \end{array}\right] = \left[\begin{array}{c} 4 \\ 5 \\ -10 \end{array}\right]. \tag{19.5}$$

The equivalence of these systems can be established with matrix multiplication. The matrix

$$\mathbf{L}_1 = \left[\begin{array}{ccc} 1 & 0 & 0 \\ -1/2 & 1 & 0 \\ 1/2 & 0 & 1 \end{array}\right]$$

encodes the row operations $(-1/2)R_1 + R_2 \to R_2$ and $(1/2)R_1 + R_3 \to R_3$, and

$$\mathbf{L}_2 = \left[\begin{array}{ccc} 1 & 0 & 0 \\ 0 & 1 & 0 \\ 0 & -4 & 1 \end{array}\right]$$

encodes the row operation $-4R_2 + R_3 \to R_3$. Let $\mathbf{A}$ be the matrix in (19.4), and check that

$$\mathbf{L}_1\mathbf{A} = \left[\begin{array}{ccc} 1 & 0 & 0 \\ -1/2 & 1 & 0 \\ 1/2 & 0 & 1 \end{array}\right] \left[\begin{array}{ccc} 2 & 6 & 4 \\ 1 & 4 & 6 \\ -1 & 1 & 9 \end{array}\right] = \left[\begin{array}{ccc} 2 & 6 & 4 \\ 0 & 1 & 4 \\ 0 & 4 & 11 \end{array}\right]$$

and

$$\mathbf{L}_2\mathbf{L}_1\mathbf{A} = \left[\begin{array}{ccc} 1 & 0 & 0 \\ 0 & 1 & 0 \\ 0 & -4 & 1 \end{array}\right] \left[\begin{array}{ccc} 2 & 6 & 4 \\ 0 & 1 & 4 \\ 0 & 4 & 11 \end{array}\right] = \left[\begin{array}{ccc} 2 & 6 & 4 \\ 0 & 1 & 4 \\ 0 & 0 & -5 \end{array}\right],$$

which is the upper-triangular matrix found earlier. At the same time, with $\mathbf{b}$ denoting the vector on the right-hand side of (19.4),

$$\mathbf{L_1 b} = \begin{bmatrix} 1 & 0 & 0 \\ -1/2 & 1 & 0 \\ 1/2 & 0 & 1 \end{bmatrix} \begin{bmatrix} 4 \\ 7 \\ 8 \end{bmatrix} = \begin{bmatrix} 4 \\ 5 \\ 10 \end{bmatrix}$$

and

$$\mathbf{L_2 L_1 b} = \begin{bmatrix} 1 & 0 & 0 \\ 0 & 1 & 0 \\ 0 & -4 & 1 \end{bmatrix} \begin{bmatrix} 4 \\ 5 \\ 10 \end{bmatrix} = \begin{bmatrix} 4 \\ 5 \\ -10 \end{bmatrix},$$

which is the right-hand side for the triangular system.

Because $\mathbf{L_1}$ and $\mathbf{L_2}$ are invertible (Proposition 19.5 below), their product $\mathbf{L_2 L_1}$ is also invertible (Exercise 10), and we see that the original system

$$\mathbf{Ax = b}$$

is equivalent to

$$\mathbf{L_2 L_1 Ax = L_2 L_1 b},$$

i.e., (19.5). ∎

After a linear system is reduced to upper-triangular form, it can be solved by back substitution.

Example 19.4. Solve the matrix-vector equation (19.4) using Gaussian elimination followed by back substitution.

Solution. The system is reduced to triangular form in the previous example, producing the equivalent equation

$$\begin{bmatrix} 2 & 6 & 4 \\ 0 & 1 & 4 \\ 0 & 0 & -5 \end{bmatrix} \begin{bmatrix} x_1 \\ x_2 \\ x_3 \end{bmatrix} = \begin{bmatrix} 4 \\ 5 \\ -10 \end{bmatrix}.$$

Back substitution produces

$$x_3 = \frac{1}{-5}(-10) = 2,$$

$$x_2 = \frac{1}{1}(5 - 4x_3) = 5 - 4(2) = -3,$$

$$x_1 = \frac{1}{2}(4 - 6x_2 - 4x_3) = \frac{1}{2}(4 - 6(-3) - 4(2)) = 7.$$

The solution is

$$\begin{bmatrix} x_1 \\ x_2 \\ x_3 \end{bmatrix} = \begin{bmatrix} 7 \\ -3 \\ 2 \end{bmatrix}.$$

This should be checked by plugging the solution vector into the original equation (19.4). ∎

The reduction to triangular form in Example 19.3 illustrates *Gaussian elimination without pivoting*. (Pivoting is introduced later.) Entries below the diagonal of $\mathbf{A}$ are eliminated, i.e., transformed to zero, using square matrices of the form

$$
\mathbf{L}_j = \begin{bmatrix}
1 & & & & & & \\
 & \ddots & & & & & \\
 & & 1 & & & & \\
 & & & 1 & & & \\
 & & & -l_{j+1,j} & 1 & & \\
 & & & \vdots & & \ddots & \\
 & & & -l_{nj} & & & 1
\end{bmatrix}, \tag{19.6}
$$

which are called *elimination matrices*. After $j-1$ steps, the first $j-1$ columns are reduced:

$$
\mathbf{L}_{j-1}\cdots\mathbf{L}_1\mathbf{A} = \begin{bmatrix}
u_{11} & \cdots & u_{1,j-1} & u_{1j} & u_{1,j+1} & \cdots & u_{1n} \\
 & \ddots & \vdots & \vdots & \vdots & \ddots & \vdots \\
 & & u_{j-1,j-1} & u_{j-1,j} & u_{j-1,j+1} & \cdots & u_{j-1,n} \\
 & & & u_{jj} & u_{j,j+1} & \cdots & u_{jn} \\
 & & & \bar{a}_{j+1,j} & \bar{a}_{j+1,j+1} & \cdots & \bar{a}_{j+1,n} \\
 & & & \vdots & \vdots & \ddots & \vdots \\
 & & & \bar{a}_{nj} & \bar{a}_{n,j+1} & \cdots & \bar{a}_{nn}
\end{bmatrix}. \tag{19.7}
$$

The next elementary matrix $\mathbf{L}_j$ of the form (19.6) is constructed by setting

$$
l_{ij} = \frac{\bar{a}_{ij}}{u_{jj}}, \quad i = j+1, \ldots, n.
$$

The elimination matrix can be written more compactly as

$$
\mathbf{L}_j = \begin{bmatrix}
\mathbf{I}_{j-1} & & \\
 & 1 & \\
 & -\mathbf{v}_j & \mathbf{I}_{n-j}
\end{bmatrix},
$$

in which $\mathbf{v}_j = (l_{j+1,j}, \ldots, l_{nj})$ and $\mathbf{I}_k$ denotes a k-by-k identity matrix. Multiplying the matrix (19.7) on the left by $\mathbf{L}_j$ leaves the first j rows unchanged, zeros the entries below u_{jj}, and replaces the lower-right submatrix

$$
\mathbf{A}^{(j-1)} = \begin{bmatrix}
\bar{a}_{j+1,j+1} & \cdots & \bar{a}_{j+1,n} \\
\vdots & \ddots & \vdots \\
\bar{a}_{n,j+1} & \cdots & \bar{a}_{nn}
\end{bmatrix}
$$

by

$$
\mathbf{A}^{(j-1)} - \mathbf{v}_j \begin{bmatrix} u_{j,j+1} & \cdots & u_{jn} \end{bmatrix}.
$$

By chaining these elimination steps, the algorithm eliminates entries below the main diagonal in columns $1, 2, \ldots, n-1$, ultimately leading to the upper-triangular matrix

$$
\mathbf{L}_{n-1}\cdots\mathbf{L}_1\mathbf{A} = \mathbf{U}.
$$

The same transformations are applied to the right-hand side $\mathbf{b}$.

Note that for the above scheme to work, the elimination matrices must be invertible. Invertibility can be proved by showing that the determinant of $\mathbf{L}_j$ equals 1. Even better, though, the inverse matrix can be found explicitly.

Proposition 19.5. *An elimination matrix* $\mathbf{L}_j$ *of the form* (19.6) *is invertible, and its inverse is*

$$
\mathbf{L}_j^{-1} = \begin{bmatrix}
1 & & & & & & \\
& \ddots & & & & & \\
& & 1 & & & & \\
& & & 1 & & & \\
& & & l_{j+1,j} & 1 & & \\
& & & \vdots & & \ddots & \\
& & & l_{nj} & & & 1
\end{bmatrix}. \tag{19.8}
$$

The proof is Exercise 11.

The routine genopivot solves a linear system using Gaussian elimination without pivoting followed by back substitution. See Listing 19.2.

Listing 19.2. Solve a linear system using Gaussian elimination without pivoting.

```
function x = genopivot(A,b)

n = size(A,1);
U = A;
y = b;
% for each column,
for j = 1:n-1
  % eliminate entries below pivot
  v = U(j+1:end,j)/U(j,j);
  U(j+1:end,j) = 0;
  U(j+1:end,j+1:end) = U(j+1:end,j+1:end)-v*U(j,j+1:end);
  % apply same operation to right-hand side
  y(j+1:end) = y(j+1:end)-y(j)*v;
end
% solve upper-triangular system
x = backsubstitute(U,y);
```

Example 19.6. Solve the matrix-vector equation (19.4) with genopivot.

Solution.

```
>> A = [  2  6  4 ;
          1  4  6 ;
         -1  1  9 ];
>> b = [4; 7; 8];
>> x = genopivot(A,b)
x =
     7
    -3
     2
```

The solution agrees with the previous example. ∎

19.3 ▪ Possible failure

There is a clear possibility for failure in Gaussian elimination without pivoting. Suppose, for example, that it is time to eliminate the entries in the second column below the main diagonal:

$$
\begin{bmatrix}
\times & \times & \times & \times & \times \\
 & \times & \times & \times & \times \\
 & \times & \times & \times & \times \\
 & \times & \times & \times & \times \\
 & \times & \times & \times & \times
\end{bmatrix}.
$$

If it happens that the $2, 2$-entry equals zero,

$$
\begin{bmatrix}
\times & \times & \times & \times & \times \\
 & 0 & \times & \times & \times \\
 & \times & \times & \times & \times \\
 & \times & \times & \times & \times \\
 & \times & \times & \times & \times
\end{bmatrix},
$$

then there is no leverage. No matter what multiple of row 2 is added to the lower rows, the entries below the zero will remain unchanged.

An ill-placed zero can bring Gaussian elimination without pivoting to a standstill. But even when the algorithm runs to completion, it can produce poor results, as the next example illustrates.

Example 19.7. Let $\delta = 10^{-7}$. Solve

$$
\begin{bmatrix}
\delta & 0 & 1 \\
1 & \delta & 1 \\
0 & 1 & 1
\end{bmatrix}
\begin{bmatrix}
x_1 \\
x_2 \\
x_3
\end{bmatrix}
=
\begin{bmatrix}
1 \\
0 \\
0
\end{bmatrix}
\tag{19.9}
$$

using Gaussian elimination without pivoting. Assess the accuracy.

Solution. It is straightforward to check that the correct solution is

$$
\begin{bmatrix}
x_1 \\
x_2 \\
x_3
\end{bmatrix}
=
\frac{1}{1 - \delta + \delta^2}
\begin{bmatrix}
\delta - 1 \\
-1 \\
1
\end{bmatrix}
\approx
\begin{bmatrix}
-0.999999999999990 \\
-1.000000100000000 \\
1.000000100000000
\end{bmatrix}.
\tag{19.10}
$$

When Gaussian elimination is executed, the $2, 1$-entry of the matrix is first eliminated to produce the equivalent system

$$
\begin{bmatrix}
\delta & 0 & 1 \\
0 & \delta & 1 - 1/\delta \\
0 & 1 & 1
\end{bmatrix}
\begin{bmatrix}
x_1 \\
x_2 \\
x_3
\end{bmatrix}
=
\begin{bmatrix}
1 \\
-1/\delta \\
0
\end{bmatrix}.
$$

Then the $3, 2$-entry is eliminated to produce the following:

$$
\begin{bmatrix}
\delta & 0 & 1 \\
0 & \delta & 1 - 1/\delta \\
0 & 0 & 1 - 1/\delta + 1/\delta^2
\end{bmatrix}
\begin{bmatrix}
x_1 \\
x_2 \\
x_3
\end{bmatrix}
=
\begin{bmatrix}
1 \\
-1/\delta \\
1/\delta^2
\end{bmatrix}.
$$

The entries range in magnitude from $\delta = 10^{-7}$ to $1/\delta^2 = 10^{14}$. This is a lot to ask of floating-point arithmetic!

In fact, the computation suffers from instability:

```
>> de = 1e-7;
>> A = [ de   0   1 ;
          1  de   1 ;
          0   1   1 ];
>> b = [1; 0; 0];
>> x = genopivot(A,b)
x =
   -1.000000000583867
   -0.987201929092407
    1.000000100000000
```

Comparing with (19.10), we see that the second entry is in error by more than 0.01. ∎

During Gaussian elimination, a diagonal entry used to eliminate other entries is called a *pivot*. A zero pivot causes division by zero, and a small pivot can produce large numerical error.

19.4 ▪ Gaussian elimination with partial pivoting

A pivot is an entry of a matrix used to eliminate other entries with a row operation. The act of *pivoting* is the permutation, or rearrangement, of entries to reassign the pivot. This may be done to avoid division by zero or division by a small quantity.

The rows of a matrix may be rearranged with the help of a special kind of matrix called a *permutation matrix*. This is any square matrix that has exactly one 1 in each row and column and 0's for all other entries. A permutation matrix acts by rearranging the entries of a vector, e.g.,

$$
\begin{bmatrix} 0 & 1 & 0 & 0 \\ 0 & 0 & 1 & 0 \\ 1 & 0 & 0 & 0 \\ 0 & 0 & 0 & 1 \end{bmatrix}
\begin{bmatrix} x \\ y \\ z \\ w \end{bmatrix}
=
\begin{bmatrix} y \\ z \\ x \\ w \end{bmatrix}.
$$

A *transposition matrix* is a permutation matrix that differs from the identity matrix in exactly two columns. For example, the transposition matrix

$$
\begin{bmatrix} 0 & 0 & 1 & 0 \\ 0 & 1 & 0 & 0 \\ 1 & 0 & 0 & 0 \\ 0 & 0 & 0 & 1 \end{bmatrix}
$$

differs from the identity matrix in only the first and third columns. A transposition matrix has the effect of swapping two rows. For example, the transposition matrix above swaps the first and third rows of a matrix:

$$
\begin{bmatrix} 0 & 0 & 1 & 0 \\ 0 & 1 & 0 & 0 \\ 1 & 0 & 0 & 0 \\ 0 & 0 & 0 & 1 \end{bmatrix}
\begin{bmatrix} a_{11} & a_{12} & a_{13} & a_{14} \\ a_{21} & a_{22} & a_{23} & a_{24} \\ a_{31} & a_{32} & a_{33} & a_{34} \\ a_{41} & a_{42} & a_{43} & a_{44} \end{bmatrix}
=
\begin{bmatrix} a_{31} & a_{32} & a_{33} & a_{34} \\ a_{21} & a_{22} & a_{23} & a_{24} \\ a_{11} & a_{12} & a_{13} & a_{14} \\ a_{41} & a_{42} & a_{43} & a_{44} \end{bmatrix}.
$$

A transposition matrix is invertible and is its own inverse (Exercise 12).

A commonly practiced pivoting scheme is *partial pivoting*. Just before any entries in the jth column are eliminated, the column is inspected. If any entry below the main diagonal has a larger magnitude than the diagonal entry, then the row with the largest-magnitude entry is swapped with the jth row. In the case of a tie, we may choose the first row with a largest-magnitude entry.

Example 19.8. Demonstrate partial pivoting on (19.9).

Solution. The entry of largest magnitude in the first column of

$$\mathbf{A} = \begin{bmatrix} \delta & 0 & 1 \\ 1 & \delta & 1 \\ 0 & 1 & 1 \end{bmatrix}$$

is in the 2, 1 position. The first and second rows are swapped to make this entry the pivot using a transposition matrix $\mathbf{S}_1$:

$$\mathbf{S}_1\mathbf{A} = \begin{bmatrix} 0 & 1 & 0 \\ 1 & 0 & 0 \\ 0 & 0 & 1 \end{bmatrix} \begin{bmatrix} \delta & 0 & 1 \\ 1 & \delta & 1 \\ 0 & 1 & 1 \end{bmatrix} = \begin{bmatrix} 1 & \delta & 1 \\ \delta & 0 & 1 \\ 0 & 1 & 1 \end{bmatrix}.$$

Next, the 2, 1-entry is eliminated by adding a multiple of the first row to the second row:

$$\mathbf{L}_1\mathbf{S}_1\mathbf{A} = \begin{bmatrix} 1 & 0 & 0 \\ -\delta & 1 & 0 \\ 0 & 0 & 1 \end{bmatrix} \begin{bmatrix} 1 & \delta & 1 \\ \delta & 0 & 1 \\ 0 & 1 & 1 \end{bmatrix} = \begin{bmatrix} 1 & \delta & 1 \\ 0 & -\delta^2 & 1-\delta \\ 0 & 1 & 1 \end{bmatrix}.$$

In the second column, the potential pivots are the $-\delta^2$ on the diagonal and the 1 below the diagonal. Because 1 has the larger magnitude, it is swapped into place:

$$\mathbf{S}_2\mathbf{L}_1\mathbf{S}_1\mathbf{A} = \begin{bmatrix} 1 & 0 & 0 \\ 0 & 0 & 1 \\ 0 & 1 & 0 \end{bmatrix} \begin{bmatrix} 1 & \delta & 1 \\ 0 & -\delta^2 & 1-\delta \\ 0 & 1 & 1 \end{bmatrix} = \begin{bmatrix} 1 & \delta & 1 \\ 0 & 1 & 1 \\ 0 & -\delta^2 & 1-\delta \end{bmatrix}.$$

Eliminating the 3, 2-entry makes the matrix upper triangular:

$$\mathbf{L}_2\mathbf{S}_2\mathbf{L}_1\mathbf{S}_1\mathbf{A} = \begin{bmatrix} 1 & 0 & 0 \\ 0 & 1 & 0 \\ 0 & \delta^2 & 1 \end{bmatrix} \begin{bmatrix} 1 & \delta & 1 \\ 0 & 1 & 1 \\ 0 & -\delta^2 & 1-\delta \end{bmatrix} = \begin{bmatrix} 1 & \delta & 1 \\ 0 & 1 & 1 \\ 0 & 0 & 1-\delta+\delta^2 \end{bmatrix}.$$

Applying the same four transformations to the right-hand side

$$\mathbf{b} = \begin{bmatrix} 1 \\ 0 \\ 0 \end{bmatrix}$$

of (19.9) gives

$$\mathbf{L}_2\mathbf{S}_2\mathbf{L}_1\mathbf{S}_1\mathbf{b} = \begin{bmatrix} 0 \\ 0 \\ 1 \end{bmatrix}.$$

Thus, the triangularized system is

$$\begin{bmatrix} 1 & \delta & 1 \\ 0 & 1 & 1 \\ 0 & 0 & 1-\delta+\delta^2 \end{bmatrix} \begin{bmatrix} x_1 \\ x_2 \\ x_3 \end{bmatrix} = \begin{bmatrix} 0 \\ 0 \\ 1 \end{bmatrix}.$$

Have we avoided the instability seen in Example 19.7? Quite possibly, since we do not see entries on the order of $1/\delta$ or $1/\delta^2$. We return to this question in Example 19.10 below. ■

Theorem 19.9. *If* **A** *is invertible, then Gaussian elimination with partial pivoting runs to completion, and the diagonal entries of* **U** *are nonzero.*

Proof. In Gaussian elimination, the only potential illegal operation is division by zero. This can only occur if a pivot is zero, and the diagonal entries of **U** are the pivots. Hence, our task is to prove that every pivot is nonzero.

Suppose for the sake of contradiction that the jth pivot equals zero. Let

$$\bar{\mathbf{A}} = \mathbf{L}_{j-1}\mathbf{S}_{j-1}\cdots\mathbf{L}_1\mathbf{S}_1\mathbf{A}$$

be the matrix produced by eliminating entries in the first $j-1$ columns. Because the jth pivot is assumed to be zero, the only nonzero entries in the jth column of $\bar{\mathbf{A}}$ are in its first $j-1$ entries. Hence, the only nonzero entries in the first j columns are in the first $j-1$ rows, and this forces the first j columns to be linearly dependent. The partially reduced matrix $\bar{\mathbf{A}}$ is not invertible, and thus the original matrix $\mathbf{A} = \mathbf{S}_1\mathbf{L}_1^{-1}\cdots\mathbf{S}_{j-1}\mathbf{L}_{j-1}^{-1}\bar{\mathbf{A}}$ is not invertible either, a contradiction. □

The theorem does not apply if **A** is not invertible. In this case, $\mathbf{A}\mathbf{x} = \mathbf{b}$ may have zero or infinitely many solutions. Our implementation of Gaussian elimination is not intended for such systems.

A linear system can be solved by Gaussian elimination with partial pivoting using the routine gepp of Listing 19.3.

Example 19.10. Solve (19.9) again, this time using gepp. Assess the accuracy.

Solution. The matrix-vector equation is solved.

```
>> de = 1e-7;
>> A = [ de   0   1 ;
          1  de   1 ;
          0   1   1 ];
>> b = [1; 0; 0];
>> x = gepp(A,b)
x =
  -0.99999999999990
  -1.000000100000000
   1.000000100000000
```

The solution agrees with (19.10) in all reported digits. ■

With the addition of partial pivoting, Gaussian elimination is guaranteed to run to completion and is remarkably stable, as Example 19.10 suggests. The numerical stability is discussed further in Chapter 21.

Notes

Although Gaussian elimination with partial pivoting is remarkably stable, the algorithm can fail on certain diabolically constructed examples. See [86, p. 212] and Exercises 16–17. Fortunately, these matrices are rare. In 1997, Trefethen and Bau wrote, "In fifty years of computing,

Listing 19.3. Solve a linear system by Gaussian elimination with partial pivoting.

```
function x = gepp(A,b)

n = size(A,1);
U = A;
y = b;
% for each column,
for j = 1:n-1
  % find pivot
  [~,k] = max(abs(U(j:n,j)));
  k = j+k-1;
  if k>j
    % swap rows
    U([j k],j:n) = U([k j],j:n);
    y([j k]) = y([k j]);
  end
  % eliminate entries below pivot
  l = U(j+1:n,j)/U(j,j);
  U(j+1:n,j) = 0;
  U(j+1:n,j+1:n) = U(j+1:n,j+1:n)-l*U(j,j+1:n);
  % apply same operation to right-hand side
  y(j+1:n) = y(j+1:n)-l*y(j);
end
% solve upper-triangular system
x = backsubstitute(U,y);
```

no matrix problems that excite an explosive instability are known to have arisen under natural circumstances" [77, p. 166].

Exercise 22, on the stability of Gauss–Jordan elimination, is based on an example in a 1975 article by Peters and Wilkinson [60].

Exercises

Exercises 1–2: Solve the linear system by hand using back substitution. Report the solution to four decimal places of precision.

$$1. \quad \begin{bmatrix} 1.1 & 1.7 & -0.9 \\ 0 & 0.6 & 0.7 \\ 0 & 0 & 7.2 \end{bmatrix} \mathbf{x} = \begin{bmatrix} 5.5 \\ -2.7 \\ 6.1 \end{bmatrix}$$

$$2. \quad \begin{bmatrix} -4.8 & 2.0 & -1.2 \\ 0 & 4.1 & 1.2 \\ 0 & 0 & -3.1 \end{bmatrix} \mathbf{x} = \begin{bmatrix} 1.8 \\ -2.3 \\ -2.1 \end{bmatrix}$$

Exercises 3–4: By hand, reduce the linear system to upper-triangular form using Gaussian elimination without pivoting. Report the entries of the reduced system to four decimal places of precision.

$$3. \quad \begin{bmatrix} 1.3 & -0.1 & 4.4 \\ 1.3 & -0.7 & 4.4 \\ -3.5 & 2.5 & -3.5 \end{bmatrix} \mathbf{x} = \begin{bmatrix} 0.2 \\ -2.4 \\ -2.2 \end{bmatrix}$$

$$4. \quad \begin{bmatrix} -0.1 & 1.5 & -4.4 \\ 6.1 & -0.9 & 0.1 \\ -3.1 & 4.5 & 2.2 \end{bmatrix} \mathbf{x} = \begin{bmatrix} 2.2 \\ 3.1 \\ 0.2 \end{bmatrix}$$

Exercises 5–8: By hand, reduce the linear system to upper-triangular form using Gaussian elimination with partial pivoting. Report the entries of the reduced system to four decimal places of precision.

$$5. \quad \begin{bmatrix} 4.2 & -3.6 & 1.5 \\ 4.3 & 2.2 & 3.1 \\ 2.0 & 4.9 & 2.2 \end{bmatrix} \mathbf{x} = \begin{bmatrix} -0.9 \\ 0.9 \\ -2.4 \end{bmatrix}$$

$$6. \quad \begin{bmatrix} 2.7 & -2.4 & 1.0 \\ -3.2 & 4.3 & 4.1 \\ -3.4 & -8.8 & -2.3 \end{bmatrix} \mathbf{x} = \begin{bmatrix} -5.1 \\ -0.7 \\ -0.3 \end{bmatrix}$$

$$7. \quad \begin{bmatrix} 3.3 & -2.1 & 3.7 \\ 0.9 & -1.2 & -0.2 \\ -7.6 & 7.3 & 5.2 \end{bmatrix} \mathbf{x} = \begin{bmatrix} -4.2 \\ -1.4 \\ 5.7 \end{bmatrix}$$

$$8. \quad \begin{bmatrix} 6.4 & 1.8 & -0.5 \\ 0.6 & -1.2 & 1.8 \\ -3.6 & 0.4 & -1.1 \end{bmatrix} \mathbf{x} = \begin{bmatrix} 5.4 \\ 1.8 \\ -0.1 \end{bmatrix}$$

9. Let **L** be a square lower-triangular matrix, whose only nonzero entries lie on or below the main diagonal. Write a MATLAB routine analogous to backsubstitute that solves $\mathbf{Lx} = \mathbf{b}$ for $\mathbf{x}$. Demonstrate it.

10. Let **A** and **B** be invertible matrices with the same numbers of rows and columns. Prove that **AB** is invertible, and find a formula for the inverse matrix.

11. Prove Proposition 19.5.

12. Prove that a transposition matrix is its own inverse.

13. Develop a MATLAB function for solving the matrix-matrix equation $\mathbf{AX} = \mathbf{B}$ for **X** using Gaussian elimination with partial pivoting. (A convenient but wasteful approach is to run gepp(A,B(:,j)) for each column j. Be more efficient than this.) Demonstrate your routine.

14. Prove that backsubstitute performs $(1/2)n^2 + (1/2)n$ scalar multiplications/divisions on an n-by-n system.

15. Prove that genopivot performs $(1/3)n^3 + n^2 - (1/3)n$ scalar multiplications/divisions on an n-by-n system, including the call to backsubstitute. (See the previous exercise.)

16. Consider running Gaussian elimination with exact arithmetic on the n-by-n matrix

$$\mathbf{A}_n = \begin{bmatrix} 1 & 0 & 0 & \cdots & 0 & 1 \\ -1 & 1 & 0 & \cdots & 0 & 1 \\ -1 & -1 & 1 & \cdots & 0 & 1 \\ \vdots & \vdots & \vdots & \ddots & \vdots & \vdots \\ -1 & -1 & -1 & \cdots & 1 & 1 \\ -1 & -1 & -1 & \cdots & -1 & 1 \end{bmatrix}. \tag{19.11}$$

Find the resulting upper-triangular matrix. Give an exact answer.

17. Let $\mathbf{A}_n$ be the n-by-n matrix (19.11), and consider the matrix-vector equation

$$
\mathbf{A}_n \mathbf{x} =
\begin{bmatrix}
1 \\
1 \\
\vdots \\
1 \\
2 \\
0
\end{bmatrix}.
\tag{19.12}
$$

Find the exact solution for any $n \geq 2$. Then solve using gepp for $n = 10, 20, 30,$ $\ldots, 100$. Something goes wrong. What? How? *Comment.* Fortunately, systems that induce this behavior are vanishingly rare in real-world practice.

18. A more cautious method than partial pivoting is *complete pivoting*. Before eliminating entries in the jth column, the entry of largest magnitude in $\mathbf{A}(j : n, j : n)$, the lower-right submatrix extending from the j, j-entry to the n, n-entry, is found. Rows and columns are swapped to bring this entry to the pivot position j, j. Implement Gaussian elimination with complete pivoting and demonstrate your code on an example. *Hint.* Each column of $\mathbf{A}$ is associated with a variable x_j. By swapping two columns, the order of the variables is also changed.

19. Solve the linear system (19.12) with $n = 100$ using your implementation of complete pivoting from Exercise 18. Is the computed solution accurate?

20. Introductory linear algebra courses often emphasize *Gauss–Jordan elimination* for solving $\mathbf{Ax} = \mathbf{b}$. This method uses the same elimination operations as Gaussian elimination, but it reduces the matrix to diagonal form rather than triangular form. With partial pivoting, the method proceeds as follows:

    - for $j = 1, \ldots, n$,
        - if necessary, swap row j with some row $i > j$ so that $|a_{jj}|$ is the largest among $|a_{ij}|, i = j, \ldots, n$;
        - eliminate all entries above and below the j, j-entry using elementary elimination operations, and apply the same operations to $\mathbf{b}$;
    - set $x_j = b_j / a_{jj}, j = 1, \ldots, n$.

    Implement Gauss–Jordan elimination as a MATLAB function and demonstrate it on a 3-by-3 example.

21. Show that Gauss–Jordan elimination, as specified in Exercise 20, performs $(1/2)n^3 + n^2 - (1/2)n$ scalar multiplications/additions.

22. Consider the following linear system:

$$
\begin{bmatrix}
0.815 & 0.632 & 0.958 & 0.957 \\
0 & 10^{-8} & 0.965 & 0.485 \\
0 & 0 & 0.158 & 0.800 \\
0 & 0 & 0 & 0.142
\end{bmatrix}
\begin{bmatrix}
x_1 \\
x_2 \\
x_3 \\
x_4
\end{bmatrix}
=
\begin{bmatrix}
0.422 \\
-24.886 \\
0.792 \\
0.959
\end{bmatrix}.
$$

Solve the system in two ways: (1) by back substitution and (2) by Gauss–Jordan elimination. (See Exercise 20.) Does either method achieve a smaller residual? If so, which one?

Chapter 20

LU factorization

This chapter introduces the *LU factorization*, which expresses a matrix $\mathbf{A}$ as a product of a lower-triangular matrix $\mathbf{L}$ and an upper-triangular matrix $\mathbf{U}$:

$$\mathbf{A} = \mathbf{LU}.$$

When this factorization exists, it can be computed using Gaussian elimination. The upper-triangular factor is the result of Gaussian elimination, and the lower-triangular factor records the sequence of elimination operations used in the reduction.

The LU factorization is useful for proving theorems because it expresses the relationship between $\mathbf{A}$ and $\mathbf{U}$ in one succinct equation, rather than through a long sequence of row operations.

The LU factorization is also useful for solving the matrix-vector equation

$$\mathbf{Ax} = \mathbf{b}$$

when the matrix $\mathbf{A}$ is known in advance but the right-hand side $\mathbf{b}$ is measured later. The matrix may be replaced by its LU factorization immediately:

$$\mathbf{LUx} = \mathbf{b}.$$

Then, when $\mathbf{b}$ is revealed, the system can be solved quickly in two steps: (1) Solve $\mathbf{Lw} = \mathbf{b}$ for $\mathbf{w}$ and (2) solve $\mathbf{Ux} = \mathbf{w}$ for $\mathbf{x}$. (The original system is solved because $\mathbf{Ax} = \mathbf{LUx} = \mathbf{Lw} = \mathbf{b}$.) The final steps can be completed very quickly by forward substitution and back substitution.

20.1 ▪ LU factorization without pivoting

A *unit lower-triangular* matrix $\mathbf{L}$ is a square matrix in which every entry above the main diagonal equals zero and every entry on the main diagonal equals 1:

$$\mathbf{L} = \begin{bmatrix} 1 & & & & & & \\ l_{21} & 1 & & & & & \\ l_{31} & l_{32} & 1 & & & & \\ \vdots & \vdots & \ddots & \ddots & & & \\ l_{n-1,1} & l_{n-1,2} & \cdots & l_{n-1,n-2} & 1 & \\ l_{n1} & l_{n2} & \cdots & l_{n,n-2} & l_{n,n-1} & 1 \end{bmatrix}. \tag{20.1}$$

An *LU factorization* expresses a matrix $\mathbf{A}$ as a product of a unit lower-triangular matrix $\mathbf{L}$ and an upper-triangular matrix $\mathbf{U}$:

$$\mathbf{A} = \mathbf{LU}. \tag{20.2}$$

When Gaussian elimination runs to completion without pivoting, the initial matrix $\mathbf{A}$ is related to the final upper-triangular matrix $\mathbf{U}$ by

$$\mathbf{L}_{n-1} \cdots \mathbf{L}_2 \mathbf{L}_1 \mathbf{A} = \mathbf{U}, \tag{20.3}$$

in which $\mathbf{L}_1, \ldots, \mathbf{L}_{n-1}$ are elimination matrices. The lower-triangular factor $\mathbf{L}$ in the LU factorization is constructed from $\mathbf{L}_1, \ldots, \mathbf{L}_{n-1}$ in a very simple way, as the following theorem shows.

Theorem 20.1. *Suppose Gaussian elimination runs to completion without pivoting, computing the upper-triangular factor $\mathbf{U}$ by (20.3) using elimination matrices*

$$\mathbf{L}_j = \begin{bmatrix} 1 & & & & & & \\ & \ddots & & & & & \\ & & 1 & & & & \\ & & & 1 & & & \\ & & & -l_{j+1,j} & 1 & & \\ & & & \vdots & & \ddots & \\ & & & -l_{nj} & & & 1 \end{bmatrix}, \quad j = 1, \ldots, n-1.$$

Let

$$\mathbf{L} = \begin{bmatrix} 1 & & & & & \\ l_{21} & 1 & & & & \\ l_{31} & l_{32} & 1 & & & \\ \vdots & \vdots & \ddots & \ddots & & \\ l_{n-1,1} & l_{n-1,2} & \cdots & l_{n-1,n-2} & 1 & \\ l_{n1} & l_{n,2} & \cdots & l_{n,n-2} & l_{n,n-1} & 1 \end{bmatrix}. \tag{20.4}$$

Then

$$\mathbf{A} = \mathbf{LU}.$$

Proof. The triangular reduction (20.3) is equivalent to

$$\mathbf{A} = \mathbf{L}_1^{-1} \mathbf{L}_2^{-1} \cdots \mathbf{L}_{n-1}^{-1} \mathbf{U}.$$

Thus, the major task is to prove

$$\mathbf{L}_1^{-1} \mathbf{L}_2^{-1} \cdots \mathbf{L}_{n-1}^{-1} = \mathbf{L}.$$

Let $\mathbf{M}_j$ be the submatrix sitting in the lower-right corner of $\mathbf{L}$ starting at row j and column j:

$$\mathbf{M}_j = \begin{bmatrix} 1 & & & \\ l_{j+1,j} & 1 & & \\ \vdots & & \ddots & \\ l_{n,j} & \cdots & l_{n,n-1} & 1 \end{bmatrix}.$$

We prove by induction that

$$\mathbf{L}_j^{-1}\cdots\mathbf{L}_{n-1}^{-1} = \begin{bmatrix} \mathbf{I}_{j-1} & \mathbf{0} \\ \mathbf{0} & \mathbf{M}_j \end{bmatrix} \tag{20.5}$$

for $j = n - 1, n - 2, \ldots, 1$, in which $\mathbf{I}_k$ denotes the k-by-k identity matrix. The base case is $j = n - 1$. We find

$$\mathbf{M}_{n-1} = \begin{bmatrix} 1 & 0 \\ l_{n,n-1} & 1 \end{bmatrix}.$$

By Proposition 19.5,

$$\mathbf{L}_{n-1}^{-1} = \begin{bmatrix} \mathbf{I}_{n-2} & \mathbf{0} & \mathbf{0} \\ \mathbf{0} & 1 & 0 \\ \mathbf{0} & l_{n,n-1} & 1 \end{bmatrix} = \begin{bmatrix} \mathbf{I}_{n-2} & \mathbf{0} \\ \mathbf{0} & \mathbf{M}_{n-1} \end{bmatrix}.$$

The base case is complete.

For the induction step, assume that (20.5) has been proved for some value of $j > 1$. Then

$$\mathbf{L}_j^{-1}\cdots\mathbf{L}_{n-1}^{-1} = \begin{bmatrix} \mathbf{I}_{j-1} & \mathbf{0} \\ \mathbf{0} & \mathbf{M}_j \end{bmatrix} = \begin{bmatrix} \mathbf{I}_{j-2} & \mathbf{0} & \mathbf{0} \\ \mathbf{0} & 1 & 0 \\ \mathbf{0} & \mathbf{0} & \mathbf{M}_j \end{bmatrix}.$$

Let

$$\mathbf{v} = \begin{bmatrix} l_{j,j-1} \\ \vdots \\ l_{n,j-1} \end{bmatrix}.$$

Then

$$\mathbf{L}_{j-1}^{-1}\mathbf{L}_j^{-1}\cdots\mathbf{L}_{n-1}^{-1} = \begin{bmatrix} \mathbf{I}_{j-2} & \mathbf{0} & \mathbf{0} \\ \mathbf{0} & 1 & 0 \\ \mathbf{0} & \mathbf{v} & \mathbf{I} \end{bmatrix}\begin{bmatrix} \mathbf{I}_{j-2} & \mathbf{0} & \mathbf{0} \\ \mathbf{0} & 1 & 0 \\ \mathbf{0} & \mathbf{0} & \mathbf{M}_j \end{bmatrix}$$

$$= \begin{bmatrix} \mathbf{I}_{j-2} & \mathbf{0} & \mathbf{0} \\ \mathbf{0} & 1 & 0 \\ \mathbf{0} & \mathbf{v} & \mathbf{M}_j \end{bmatrix} = \begin{bmatrix} \mathbf{I}_{j-2} & \mathbf{0} \\ \mathbf{0} & \mathbf{M}_{j-1} \end{bmatrix}.$$

The induction step is complete.

By induction, (20.5) is true for $j = 1, \ldots, n - 1$. In the case $j = 1$, the statement thus proved is $\mathbf{L}_1^{-1}\cdots\mathbf{L}_{n-1}^{-1} = \mathbf{M}_1 = \mathbf{L}$. $\square$

Example 20.2. Find an LU factorization for the matrix

$$\mathbf{A} = \begin{bmatrix} 2 & 6 & 4 \\ 1 & 4 & 6 \\ -1 & 1 & 9 \end{bmatrix},$$

seen earlier in Example 19.3.

Solution. The upper-triangular factor is the result of Gaussian elimination, which was found earlier:

$$\mathbf{U} = \begin{bmatrix} 2 & 6 & 4 \\ 0 & 1 & 4 \\ 0 & 0 & -5 \end{bmatrix}.$$

For the lower-triangular factor, Theorem 20.1 instructs us to negate and combine the lower-triangular entries of $\mathbf{L}_1$ and $\mathbf{L}_2$ from the earlier example:

$$\mathbf{L} = \begin{bmatrix} 1 & 0 & 0 \\ 1/2 & 1 & 0 \\ -1/2 & 4 & 1 \end{bmatrix}.$$

Let's check that $\mathbf{A}$ equals $\mathbf{LU}$.

```
>> L = [   1   0   0   ;
          1/2   1   0   ;
         -1/2   4   1   ];
>> U = [   2   6   4   ;
           0   1   4   ;
           0   0  -5   ];
>> L*U
ans =
      2    6    4
      1    4    6
     -1    1    9
```

Yes, $\mathbf{A} = \mathbf{LU}$. ∎

The computer code for Gaussian elimination without pivoting can be modified to produce an LU factorization (when the routine runs to completion without dividing by zero). The resulting routine is called lunopivot, and it is shown in Listing 20.1.

Listing 20.1. Compute an LU factorization without pivoting.

```
function [L,U] = lunopivot(A)

n = size(A,1);
L = eye(n);
U = A;
% for each column,
for j = 1:n-1
  v = U(j+1:n,j)/U(j,j);
  % store elimination operation
  L(j+1:n,j) = v;
  % apply elimination operation
  U(j+1:n,j) = 0;
  U(j+1:n,j+1:n) = U(j+1:n,j+1:n)-v*U(j,j+1:n);
end
```

Example 20.3. Verify that lunopivot computes the LU factorization of Example 20.2.

Solution.

```
>> A = [   2  6  4  ;
           1  4  6  ;
          -1  1  9  ];
>> [L,U] = lunopivot(A)
L =
    1.000000000000000                   0                   0
    0.500000000000000   1.000000000000000                   0
   -0.500000000000000   4.000000000000000   1.000000000000000
U =
```

$$\begin{array}{rrr} 2 & 6 & 4 \\ 0 & 1 & 4 \\ 0 & 0 & -5 \end{array}$$

Yes, this agrees with the previous example. ∎

20.2 ▪ LU factorization with pivoting

Not every matrix has an LU factorization.

Example 20.4. Prove that

$$\mathbf{A} = \begin{bmatrix} 0 & 0 \\ 1 & 0 \end{bmatrix}$$

does not have an LU factorization.

Solution. Suppose there were an LU factorization

$$\mathbf{A} = \begin{bmatrix} 1 & 0 \\ l_{21} & 1 \end{bmatrix} \begin{bmatrix} u_{11} & u_{12} \\ 0 & u_{22} \end{bmatrix}.$$

Expanding the matrix product produces the equivalent equation

$$\begin{bmatrix} 0 & 0 \\ 1 & 0 \end{bmatrix} = \begin{bmatrix} u_{11} & u_{12} \\ l_{21}u_{11} & l_{21}u_{12} + u_{22} \end{bmatrix}.$$

From the top-left entry we see $u_{11} = 0$, but from the bottom-left entry we see $l_{21}u_{11} = 1$. This is a contradiction. ∎

Because an LU factorization may not exist, the routine lunopivot may produce matrices containing Inf, -Inf, or NaN.

This failure is not the end of the story, however. Every matrix admits an LU factorization after its rows are permuted appropriately, just as Gaussian elimination always runs to completion when rows are pivoted carefully. Recall from the previous chapter that a permutation matrix $\mathbf{P}$ is a square matrix with a single 1 in each row and column and 0's everywhere else. We need two important facts about permutation matrices. First, the product of two permutation matrices is a permutation matrix (Exercise 10). Second, every permutation matrix is invertible, and its inverse is its transpose (Exercise 12):

$$\mathbf{P}^T \mathbf{P} = \mathbf{I}.$$

Lemma 20.5. *Let $\mathbf{P}$ be a permutation matrix of the form*

$$\mathbf{P} = \left[\begin{array}{c|c|c} \mathbf{I}_{k-1} & & \\ \hline & 1 & \\ \hline & & \mathbf{Q} \end{array} \right],$$

in which $\mathbf{I}_{k-1}$ is the $(k-1)$-by-$(k-1)$ identity matrix and $\mathbf{Q}$ is an $(n-k)$-by-$(n-k)$ permutation matrix. Let $\mathbf{L}_k$ be an elimination matrix of the form

$$\mathbf{L}_k = \left[\begin{array}{c|c|c} \mathbf{I}_{k-1} & & \\ \hline & 1 & \\ \hline & -\mathbf{v} & \mathbf{I}_{n-k} \end{array} \right].$$

Then

$$\mathbf{PL}_k = \tilde{\mathbf{L}}_k \mathbf{P},$$ (20.6)

in which $\tilde{\mathbf{L}}_k$ *is the elimination matrix*

$$\tilde{\mathbf{L}}_k = \left[\begin{array}{c|c|c} \mathbf{I}_{k-1} & & \\ \hline & 1 & \\ \hline & -\mathbf{Qv} & \mathbf{I}_{n-k} \end{array}\right].$$

Proof. Both sides of (20.6) equal

$$\left[\begin{array}{c|c|c} \mathbf{I}_{k-1} & & \\ \hline & 1 & \\ \hline & -\mathbf{Qv} & \mathbf{Q} \end{array}\right].$$ □

Theorem 20.6. *Let* $\mathbf{A}$ *be a square matrix. There exists a permutation matrix* $\mathbf{P}$ *such that* $\mathbf{PA}$ *has an LU factorization*

$$\mathbf{PA} = \mathbf{LU},$$ (20.7)

and $\mathbf{P}$ *can be chosen so that no entry of* $\mathbf{L}$ *has absolute value greater than 1.*

Proof. Recall Gaussian elimination with partial pivoting: (1) pivot, (2) eliminate entries below the pivot, and (3) repeat on the next column. In a single equation, this is

$$\mathbf{L}_{n-1}\mathbf{S}_{n-1}\cdots\mathbf{L}_1\mathbf{S}_1\mathbf{A} = \mathbf{U},$$ (20.8)

in which each $\mathbf{S}_j$ is a transposition matrix or the identity matrix and each $\mathbf{L}_j$ is an elimination matrix. (Compare with Example 19.8.) Essentially, we want to move the elimination matrices in (20.8) to the right-hand side, leaving the permutation matrices in place.

We prove by induction that

$$(\mathbf{L}_{n-1}\mathbf{S}_{n-1})(\mathbf{L}_{n-2}\mathbf{S}_{n-2})\cdots(\mathbf{L}_j\mathbf{S}_j) = (\mathbf{L}_{n-1}\tilde{\mathbf{L}}_{n-2}\cdots\tilde{\mathbf{L}}_j)\mathbf{P}_j,$$

in which $\tilde{\mathbf{L}}_j$ is related to $\mathbf{L}_j$ as in Lemma 20.5 and $\mathbf{P}_j$ is a permutation matrix whose top-left $(j-1)$-by-$(j-1)$ submatrix equals an identity matrix. The base case $j = n-1$ is trivial: the statement reduces to $\mathbf{L}_{n-1}\mathbf{S}_{n-1} = \mathbf{L}_{n-1}\mathbf{P}_{n-1}$, so just take $\mathbf{P}_{n-1} = \mathbf{S}_{n-1}$. Now suppose that the statement has been proved for some value of $j > 1$. The goal for the induction step is to prove that it must also be true for $j-1$. The induction hypothesis implies

$$(\mathbf{L}_{n-1}\mathbf{S}_{n-1})(\mathbf{L}_{n-2}\mathbf{S}_{n-2})\cdots(\mathbf{L}_j\mathbf{S}_j)(\mathbf{L}_{j-1}\mathbf{S}_{j-1}) = (\mathbf{L}_{n-1}\tilde{\mathbf{L}}_{n-2}\cdots\tilde{\mathbf{L}}_j)\mathbf{P}_j\mathbf{L}_{j-1}\mathbf{S}_{j-1}.$$

By the lemma, there exists an elimination matrix $\tilde{\mathbf{L}}_{j-1}$ that facilitates moving $\mathbf{P}_j$ to the right:

$$(\mathbf{L}_{n-1}\mathbf{S}_{n-1})(\mathbf{L}_{n-2}\mathbf{S}_{n-2})\cdots(\mathbf{L}_j\mathbf{S}_j)(\mathbf{L}_{j-1}\mathbf{S}_{j-1}) = (\mathbf{L}_{n-1}\tilde{\mathbf{L}}_{n-2}\cdots\tilde{\mathbf{L}}_j)\tilde{\mathbf{L}}_{j-1}\mathbf{P}_j\mathbf{S}_{j-1}.$$

Set $\mathbf{P}_{j-1} = \mathbf{P}_j\mathbf{S}_{j-1}$ and note that the $(j-2)$-by-$(j-2)$ top-left submatrix of $\mathbf{P}_{j-1}$ equals the identity since $\mathbf{S}_{j-1}$ does not modify the first $j-2$ columns of $\mathbf{P}_j$. The induction proof is complete.

In particular, when $j = 1$, we have

$$\mathbf{L}_{n-1}\mathbf{S}_{n-1}\cdots\mathbf{L}_1\mathbf{S}_1 = \mathbf{L}_{n-1}\tilde{\mathbf{L}}_{n-2}\cdots\tilde{\mathbf{L}}_1\mathbf{P}_1.$$

Combining with (20.8) and writing $\mathbf{P} = \mathbf{P}_1$ gives

$$\mathbf{L}_{n-1}\tilde{\mathbf{L}}_{n-2}\cdots\tilde{\mathbf{L}}_1\mathbf{PA} = \mathbf{U}.$$

From this, Theorem 20.1 produces the LU factorization

$$\mathbf{PA} = \mathbf{LU}.$$

Note that no entry in any elimination matrix is greater than 1 in magnitude by the partial pivoting strategy, and thus the entries of $\mathbf{L}$ obey the same bound. □

Example 20.7. By hand, find a $\mathbf{PA} = \mathbf{LU}$ factorization for

$$\mathbf{A} = \begin{bmatrix} 0 & 3 & -2 \\ 1 & -2 & 1 \\ -2 & 2 & 0 \end{bmatrix}.$$

Solution. Gaussian elimination with partial pivoting proceeds as follows. Rows 1 and 3 are swapped:

$$\mathbf{S}_1\mathbf{A} = \begin{bmatrix} 0 & 0 & 1 \\ 0 & 1 & 0 \\ 1 & 0 & 0 \end{bmatrix}\begin{bmatrix} 0 & 3 & -2 \\ 1 & -2 & 1 \\ -2 & 2 & 0 \end{bmatrix} = \begin{bmatrix} -2 & 2 & 0 \\ 1 & -2 & 1 \\ 0 & 3 & -2 \end{bmatrix}.$$

The $2, 1$-entry is eliminated:

$$\mathbf{L}_1\mathbf{S}_1\mathbf{A} = \begin{bmatrix} 1 & 0 & 0 \\ 1/2 & 1 & 0 \\ 0 & 0 & 1 \end{bmatrix}\begin{bmatrix} -2 & 2 & 0 \\ 1 & -2 & 1 \\ 0 & 3 & -2 \end{bmatrix} = \begin{bmatrix} -2 & 2 & 0 \\ 0 & -1 & 1 \\ 0 & 3 & -2 \end{bmatrix}.$$

Rows 2 and 3 are swapped:

$$\mathbf{S}_2\mathbf{L}_1\mathbf{S}_1\mathbf{A} = \begin{bmatrix} 1 & 0 & 0 \\ 0 & 0 & 1 \\ 0 & 1 & 0 \end{bmatrix}\begin{bmatrix} -2 & 2 & 0 \\ 0 & -1 & 1 \\ 0 & 3 & -2 \end{bmatrix} = \begin{bmatrix} -2 & 2 & 0 \\ 0 & 3 & -2 \\ 0 & -1 & 1 \end{bmatrix}.$$

The $3, 2$-entry is eliminated:

$$\mathbf{L}_2\mathbf{S}_2\mathbf{L}_1\mathbf{S}_1\mathbf{A} = \begin{bmatrix} 1 & 0 & 0 \\ 0 & 1 & 0 \\ 0 & 1/3 & 1 \end{bmatrix}\begin{bmatrix} -2 & 2 & 0 \\ 0 & 3 & -2 \\ 0 & -1 & 1 \end{bmatrix} = \begin{bmatrix} -2 & 2 & 0 \\ 0 & 3 & -2 \\ 0 & 0 & 1/3 \end{bmatrix}.$$

The final matrix is the upper-triangular factor $\mathbf{U}$ in the desired factorization.

Following the proof of Theorem 20.6, we want to push the transposition matrix $\mathbf{S}_2$ to the right of $\mathbf{L}_1$ in order to find the $\mathbf{L}$ and $\mathbf{P}$ factors. We find

$$\mathbf{S}_2\mathbf{L}_1 = \begin{bmatrix} 1 & 0 & 0 \\ 0 & 0 & 1 \\ 0 & 1 & 0 \end{bmatrix}\begin{bmatrix} 1 & 0 & 0 \\ 1/2 & 1 & 0 \\ 0 & 0 & 1 \end{bmatrix} = \begin{bmatrix} 1 & 0 & 0 \\ 0 & 0 & 1 \\ 1/2 & 1 & 0 \end{bmatrix}$$

$$= \begin{bmatrix} 1 & 0 & 0 \\ 0 & 1 & 0 \\ 1/2 & 0 & 1 \end{bmatrix}\begin{bmatrix} 1 & 0 & 0 \\ 0 & 0 & 1 \\ 0 & 1 & 0 \end{bmatrix} = \tilde{\mathbf{L}}_1\mathbf{S}_2.$$

Thus,

$$\mathbf{L}_2\tilde{\mathbf{L}}_1\mathbf{S}_2\mathbf{S}_1\mathbf{A} = \mathbf{U}.$$

Finally, let

$$\mathbf{P} = \mathbf{S}_2\mathbf{S}_1 = \begin{bmatrix} 0 & 0 & 1 \\ 1 & 0 & 0 \\ 0 & 1 & 0 \end{bmatrix}$$

and

$$\mathbf{L} = \tilde{\mathbf{L}}_1^{-1}\mathbf{L}_2^{-1} = \begin{bmatrix} 1 & 0 & 0 \\ 0 & 1 & 0 \\ -1/2 & -1/3 & 1 \end{bmatrix}.$$

The factorization $\mathbf{PA} = \mathbf{LU}$ is

$$\begin{bmatrix} 0 & 0 & 1 \\ 1 & 0 & 0 \\ 0 & 1 & 0 \end{bmatrix}\begin{bmatrix} 0 & 3 & -2 \\ 1 & -2 & 1 \\ -2 & 2 & 0 \end{bmatrix} = \begin{bmatrix} 1 & 0 & 0 \\ 0 & 1 & 0 \\ -1/2 & -1/3 & 1 \end{bmatrix}\begin{bmatrix} -2 & 2 & 0 \\ 0 & 3 & -2 \\ 0 & 0 & 1/3 \end{bmatrix}. \blacksquare$$

Routine lupp in Listing 20.2 computes a factorization $\mathbf{PA} = \mathbf{LU}$.

Listing 20.2. Compute an LU factorization with partial pivoting.

```
function [L,U,P] = lupp(A)

n = size(A,1);
sigma = 1:n;
L = eye(n);
U = A;
% for each column,
for j = 1:n-1
  % find pivot
  [~,k] = max(abs(U(j:n,j)));
  k = j+k-1;
  if k>j
    % swap rows
    sigma([j k]) = sigma([k j]);
    L([j k],1:j-1) = L([k j],1:j-1);
    U([j k],j:n) = U([k j],j:n);
  end
  l = U(j+1:n,j)/U(j,j);
  % store and apply elimination operation
  L(j+1:n,j) = l;
  U(j+1:n,j) = 0;
  U(j+1:n,j+1:n) = U(j+1:n,j+1:n)-l*U(j,j+1:n);
end
% construct permutation matrix
P = eye(n);
P = P(sigma,:);
```

Example 20.8. Compute the $\mathbf{PA} = \mathbf{LU}$ factorization of the previous example using lupp.

Solution.

```
>> A = [   0   3  -2 ;
           1  -2   1 ;
          -2   2   0 ];
>> [L,U,P] = lupp(A)
L =
    1.000000000000000                    0                    0
                    0    1.000000000000000                    0
   -0.500000000000000   -0.333333333333333    1.000000000000000
U =
   -2.000000000000000    2.000000000000000                    0
                    0    3.000000000000000   -2.000000000000000
                    0                    0    0.333333333333333
P =
        0        0        1
        1        0        0
        0        1        0
```

This agrees with the previous example. ■

Exercises

Exercises 1–2: Compute an LU factorization by hand without pivoting. Report the entries to four decimal places of precision.

1. $\begin{bmatrix} -6.0 & -4.2 \\ -3.0 & 9.4 \end{bmatrix}$

2. $\begin{bmatrix} -2.5 & 3.6 & -5.7 \\ 3.0 & -3.1 & 2.0 \\ -0.8 & -5.6 & -0.7 \end{bmatrix}$

Exercises 3–6: For the given matrix $\mathbf{A}$, compute by hand a $\mathbf{PA} = \mathbf{LU}$ factorization having the properties guaranteed by Theorem 20.6. Report the entries of the matrices to four decimal places of precision.

3. $\begin{bmatrix} -0.8 & 1.2 \\ 5.7 & 0.8 \end{bmatrix}$

4. $\begin{bmatrix} 6.4 & 3.3 & -4.7 \\ 2.8 & 0.9 & 0.4 \\ -3.2 & -1.0 & -4.6 \end{bmatrix}$

5. $\begin{bmatrix} 1.8 & -1.8 & -5.1 \\ 6.6 & 0.9 & 2.5 \\ -8.1 & -3.3 & 2.5 \end{bmatrix}$

6. $\begin{bmatrix} -2.0 & 1.3 & 2.6 \\ -2.8 & -1.9 & 5.4 \\ -0.4 & 9.5 & -4.1 \end{bmatrix}$

7. Prove that the product of lower-triangular matrices is lower triangular. Prove that the product of unit lower-triangular matrices is unit lower triangular.

8. Let $\mathbf{L}$ be an invertible lower-triangular matrix. Prove that $\mathbf{L}^{-1}$ is lower triangular. Also prove that if $\mathbf{L}$ is unit lower triangular, then $\mathbf{L}^{-1}$ is also unit lower triangular.

9. Prove that a matrix is a permutation matrix if and only if (1) it is square, (2) every entry equals 0 or 1, (3) every column sum equals 1, and (4) every row sum equals 1. (A column sum is the sum of all entries in a given column, and a row sum is the sum of all entries in a given row.)

10. Prove that the product of two permutation matrices is a permutation matrix.

11. Prove that any permutation matrix can be represented as a product of transposition matrices. (We can say that the identity matrix equals the empty product.)

12. Let $\mathbf{P}$ be a permutation matrix. Prove that $\mathbf{P}$ is invertible and that $\mathbf{P}^{-1} = \mathbf{P}^T$.

13. Solve

$$\begin{bmatrix} 1.4 & -1.2 & 0.5 \\ 1.4 & 0.7 & 1.0 \\ 0.7 & 1.6 & 0.7 \end{bmatrix} \begin{bmatrix} x_1 \\ x_2 \\ x_3 \end{bmatrix} = \begin{bmatrix} -0.3 \\ 0.3 \\ -0.8 \end{bmatrix}$$

as follows: (1) compute a $\mathbf{PA} = \mathbf{LU}$ factorization for the matrix; (2) solve $\mathbf{Lw} = \mathbf{Pb}$ using forward substitution (see Exercise 9 in Chapter 19); and (3) solve $\mathbf{Ux} = \mathbf{w}$ using back substitution.

14. Consider the matrix-matrix equation $\mathbf{AX} = \mathbf{B}$. Let $\mathbf{b}_j$ denote the jth column of $\mathbf{B}$ and $\mathbf{x}_j$ the jth column of $\mathbf{X}$. Write a MATLAB function that solves the matrix-matrix equation as follows: (1) compute $\mathbf{PA} = \mathbf{LU}$; (2) solve $\mathbf{Lw}_j = \mathbf{Pb}_j$ by forward substitution (see Exercise 9 in Chapter 19); and (3) solve $\mathbf{Ux}_j = \mathbf{w}_j$ by back substitution. Demonstrate your implementation on an example.

15. Suppose that an LU factorization $\mathbf{PA} = \mathbf{LU}$ is computed using partial pivoting and that s row swaps are performed during the reduction. Prove that $\det \mathbf{A} = (-1)^s \prod_{i=1}^{n} u_{ii}$, in which u_{ii} is the ith diagonal entry of $\mathbf{U}$.

16. Complete pivoting is introduced in Exercise 18 of Chapter 19. Let $\mathbf{A}$ be a square matrix and let $\mathbf{U}$ be the upper-triangular matrix that results from Gaussian elimination with complete pivoting. By collecting the row and column swaps and the elimination steps, it is possible to construct permutation matrices $\mathbf{P}$ and $\mathbf{Q}$ and a unit lower-triangular matrix $\mathbf{L}$ such that

$$\mathbf{PAQ} = \mathbf{LU}.$$

Develop a MATLAB routine to compute this factorization.

Chapter 21

Conditioning of linear systems

Many numerical methods suffer from two types of error: truncation error and numerical roundoff error. For example, when a definite integral $\int_a^b f(x)\,dx$ is approximated by a Riemann sum $\sum_i f(x_i)\Delta x$, the replacement of the continuous integral by the finite sum incurs truncation error, and the subsequent multiplications and additions incur roundoff error when performed on a finite-precision computer. Many of the earlier error bounds in this book concentrate on truncation error as opposed to roundoff error.

The present context is different. The matrix-vector equation $\mathbf{Ax} = \mathbf{b}$ is already finite, so there is no truncation error associated with Gaussian elimination. However, for some linear systems known as *ill-conditioned* systems, the tiny roundoff errors committed by computer hardware can result in much larger errors in the final computed solution. This chapter explores this phenomenon.

We consider two measures of accuracy. The first is the residual

$$\mathbf{b} - \mathbf{A}\hat{\mathbf{x}}, \tag{21.1}$$

which measures the difference between the left- and right-hand sides of the given equation $\mathbf{Ax} = \mathbf{b}$ when the numerically computed solution $\hat{\mathbf{x}}$ is substituted for $\mathbf{x}$. (Ideally, this difference would be zero.) The second is the error

$$\mathbf{x} - \hat{\mathbf{x}}, \tag{21.2}$$

which measures the difference between the exact solution and the numerically computed solution. These are different measures! For an ill-conditioned system, a small residual does not necessarily imply a small error.

21.1 ▪ Vector and matrix infinity norms

A norm is a one-number summary of magnitude. The vector norm $\|\mathbf{x}\|_\infty$ and the matrix norm $\|\mathbf{A}\|_\infty$ are defined below. These norms are different from (but related to) the infinity norm $\|f\|_\infty$ of a function f, which plays a prominent role in interpolation and integration.

The *vector infinity norm* of a column vector $\mathbf{x} = (x_1, \ldots, x_n)$ is the maximum absolute value of any entry:

$$\|\mathbf{x}\|_\infty = \max_{i=1,\ldots,n} |x_i|. \tag{21.3}$$

The *matrix infinity norm* of an *m*-by-*n* matrix $\mathbf{A} = [a_{ij}]$ is

$$\|\mathbf{A}\|_\infty = \max_{i=1,\ldots,m} \sum_{j=1}^n |a_{ij}|. \tag{21.4}$$

The matrix infinity norm is often referred to as the "maximum row sum," but this phrase elides the absolute value.

The MATLAB routine `norm` computes the infinity norm when its second argument is `Inf`, as demonstrated in the next example.

Example 21.1. Compute $\|\mathbf{x}\|_\infty$ and $\|\mathbf{A}\|_\infty$ for

$$\mathbf{x} = \begin{bmatrix} 3 \\ -5 \end{bmatrix}, \quad \mathbf{A} = \begin{bmatrix} 1 & 2 \\ -3 & 4 \end{bmatrix}.$$

Solution. The infinity norm of $\mathbf{x}$ is just the largest magnitude of any entry, in this case 5. For the matrix, the infinity norm is the maximum sum of the absolute values of any row: $\|\mathbf{A}\|_\infty = |-3| + |4| = 7$. The `norm` routine concurs:

```
>> x = [3; -5];
>> norm(x,Inf)
ans =
     5
>> A = [ 1 2; -3 4 ];
>> norm(A,Inf)
ans =
     7
```

∎

The matrix infinity norm is important in error analyses because of the following theorem.

Theorem 21.2. *Let* $\mathbf{A} = [a_{ij}]$ *be an m-by-n matrix.*

1. *If* $\mathbf{x}$ *is any n-by-1 vector, then*

$$\|\mathbf{A}\mathbf{x}\|_\infty \leq \|\mathbf{A}\|_\infty \|\mathbf{x}\|_\infty. \tag{21.5}$$

2. *There exists a nonzero vector* $\mathbf{x}$ *for which*

$$\|\mathbf{A}\mathbf{x}\|_\infty = \|\mathbf{A}\|_\infty \|\mathbf{x}\|_\infty. \tag{21.6}$$

Proof. First, we prove (21.5) for any vector $\mathbf{x} = (x_1, \ldots, x_n)$. The ith entry of $\mathbf{A}\mathbf{x}$ has absolute value

$$\left| \sum_{j=1}^n a_{ij} x_j \right| \leq \sum_{j=1}^n |a_{ij} x_j| = \sum_{j=1}^n |a_{ij}| |x_j|$$

$$\leq \sum_{j=1}^n |a_{ij}| \|\mathbf{x}\|_\infty = \left(\sum_{j=1}^n |a_{ij}| \right) \|\mathbf{x}\|_\infty \leq \|\mathbf{A}\|_\infty \|\mathbf{x}\|_\infty.$$

Because every entry of $\mathbf{Ax}$ is bounded by $\|\mathbf{A}\|_\infty\|\mathbf{x}\|_\infty$ in absolute value, the same bound applies to $\|\mathbf{Ax}\|_\infty$.

Now, we prove that the inequality is "tight," i.e., that there exists a nonzero vector $\mathbf{x}$ for which equality is attained: $\|\mathbf{Ax}\|_\infty = \|\mathbf{A}\|_\infty\|\mathbf{x}\|_\infty$. Suppose row k of $\mathbf{A}$ has the largest sum, i.e., $\sum_{j=1}^n |a_{kj}| = \|\mathbf{A}\|_\infty$. Let $\mathbf{x}$ be the n-by-1 vector whose jth entry equals 1 if $a_{kj} \geq 0$ or -1 if $a_{kj} < 0$. Then $\|\mathbf{x}\|_\infty = 1$ and the kth entry of $\mathbf{Ax}$ equals

$$\sum_{j=1}^n a_{kj} x_j = \sum_{j=1}^n |a_{kj}| = \|\mathbf{A}\|_\infty = \|\mathbf{A}\|_\infty\|\mathbf{x}\|_\infty.$$

Therefore, $\|\mathbf{Ax}\|_\infty \geq \|\mathbf{A}\|_\infty\|\mathbf{x}\|_\infty$. Considering (21.5), the quantities must in fact be equal. □

21.2 ▪ Residual from Gaussian elimination

When $\hat{\mathbf{x}}$ is an approximate solution to $\mathbf{Ax} = \mathbf{b}$, the residual is $\mathbf{b} - \mathbf{A}\hat{\mathbf{x}}$. If the entries of the residual are small, then $\hat{\mathbf{x}}$ nearly solves the matrix-vector equation.

Gaussian elimination with partial pivoting computes solutions with small residuals. A good rule of thumb is

$$\|\mathbf{b} - \mathbf{A}\hat{\mathbf{x}}\|_\infty \approx 10^{-16}\, \|\mathbf{A}\|_\infty\|\hat{\mathbf{x}}\|_\infty. \tag{21.7}$$

(See, however, the notes at the end of this chapter.) The factor of 10^{-16} comes from the precision of the IEEE 754 double-precision format. Justifying this rule of thumb requires a close examination of roundoff error, which we avoid in keeping with the approach of this book.

21.3 ▪ Condition number

Although the residual $\mathbf{b} - \mathbf{A}\hat{\mathbf{x}}$ provides a way of measuring the accuracy of a computed solution $\hat{\mathbf{x}}$, the term *error* is reserved for the difference $\mathbf{x} - \hat{\mathbf{x}}$, in which $\mathbf{x}$ is the exact solution to $\mathbf{Ax} = \mathbf{b}$. What does a small residual imply about error? The relationship is delicate, and ignoring this subtlety is dangerous.

The norm of the error

$$\|\mathbf{x} - \hat{\mathbf{x}}\|_\infty$$

can be bounded in terms of the norm of the residual

$$\|\mathbf{b} - \mathbf{A}\hat{\mathbf{x}}\|_\infty$$

using the *absolute condition number*

$$\kappa_{\mathrm{abs}} = \|\mathbf{A}^{-1}\|_\infty \tag{21.8}$$

or the *relative condition number*

$$\kappa_{\mathrm{rel}} = \|\mathbf{A}\|_\infty\|\mathbf{A}^{-1}\|_\infty. \tag{21.9}$$

If $\mathbf{A}$ is not invertible, then both condition numbers are defined to be ∞. In the following theorem, think of $\hat{\mathbf{x}}$ as an approximation to $\mathbf{x}$ as usual.

Theorem 21.3. *Suppose* $\mathbf{Ax} = \mathbf{b}$. *Then*

$$\|\mathbf{x} - \hat{\mathbf{x}}\|_\infty \le \kappa_{\mathrm{abs}} \|\mathbf{b} - \mathbf{A}\hat{\mathbf{x}}\|_\infty \tag{21.10}$$

and

$$\frac{\|\mathbf{x} - \hat{\mathbf{x}}\|_\infty}{\|\mathbf{x}\|_\infty} \le \kappa_{\mathrm{rel}} \frac{\|\mathbf{b} - \mathbf{A}\hat{\mathbf{x}}\|_\infty}{\|\mathbf{b}\|_\infty}. \tag{21.11}$$

Proof. If $\mathbf{A}$ is not invertible, then the right-hand sides of the bounds are infinite, so assume for the rest of the proof that $\mathbf{A}$ is invertible.

Observe that

$$\mathbf{x} - \hat{\mathbf{x}} = \mathbf{A}^{-1}(\mathbf{b} - \mathbf{A}\hat{\mathbf{x}}).$$

By (21.5),

$$\|\mathbf{x} - \hat{\mathbf{x}}\|_\infty = \|\mathbf{A}^{-1}(\mathbf{b} - \mathbf{A}\hat{\mathbf{x}})\|_\infty \le \|\mathbf{A}^{-1}\|_\infty \|\mathbf{b} - \mathbf{A}\hat{\mathbf{x}}\|_\infty$$

This proves (21.10). Furthermore, since $\|\mathbf{b}\|_\infty = \|\mathbf{Ax}\|_\infty \le \|\mathbf{A}\|_\infty \|\mathbf{x}\|_\infty$,

$$\frac{\|\mathbf{x} - \hat{\mathbf{x}}\|_\infty}{\|\mathbf{x}\|_\infty} \le \frac{\|\mathbf{A}^{-1}\|_\infty \|\mathbf{b} - \mathbf{A}\hat{\mathbf{x}}\|_\infty}{\|\mathbf{b}\|_\infty / \|\mathbf{A}\|_\infty} = \|\mathbf{A}\|_\infty \|\mathbf{A}^{-1}\|_\infty \frac{\|\mathbf{b} - \mathbf{A}\hat{\mathbf{x}}\|_\infty}{\|\mathbf{b}\|_\infty}.$$

This proves (21.11).   □

When $\mathbf{A}$ has a large condition number, $\hat{\mathbf{x}}$ may be far from $\mathbf{x}$ even when the residual is small. Such a system is called *ill conditioned*. Systems with smaller condition numbers are better behaved and called *well conditioned*.

The next two examples consider well-conditioned and ill-conditioned linear systems.

Example 21.4. The solution to the linear system

$$\begin{bmatrix} 1 & 1 & 2 \\ -2 & 1 & -1 \\ 1 & 1 & 1 \end{bmatrix} \begin{bmatrix} x_1 \\ x_2 \\ x_3 \end{bmatrix} = \begin{bmatrix} 0 \\ 2 \\ 0 \end{bmatrix}$$

is approximated by

$$\hat{\mathbf{x}} = \begin{bmatrix} -0.66667 \\ 0.66667 \\ 0.00000 \end{bmatrix}.$$

What does (21.11) predict about the error? What actually happens?

Solution. The provided approximation produces a small residual, considering the precision of the figures.

```
>> A = [  1   1   2 ;
         -2   1  -1 ;
          1   1   1 ];
>> b = [0; 2; 0];
>> xapprox = [-0.66667; 0.66667; 0.00000];
>> format short;
>> res = norm(b-A*xapprox,Inf)
res =
   1.0000e-05
```

The norm of the residual is about 10^{-5}. To bound the error, it is necessary to know the condition number of the matrix. The MATLAB library provides a function condest for estimating the relative condition number.

```
>> kappa = condest(A')
kappa =
     8
```

(See Exercise 21 to understand why it's condest(A') instead of just condest(A).) This is a small relative condition number, and it leads to the bound

$$\frac{\|\mathbf{x} - \hat{\mathbf{x}}\|_\infty}{\|\mathbf{x}\|_\infty} \leq \kappa_{\text{rel}} \frac{\|\mathbf{b} - \mathbf{A}\hat{\mathbf{x}}\|_\infty}{\|\mathbf{b}\|_\infty} \approx 8 \frac{10^{-5}}{2} = 4 \times 10^{-5}.$$

What actually happens? It's easy to check that the exact solution is

$$\mathbf{x} = \begin{bmatrix} -2/3 \\ 2/3 \\ 0 \end{bmatrix}.$$

The actual relative error is

```
>> x = [-2/3; 2/3; 0];
>> norm(x-xapprox,Inf)/norm(x,Inf)
ans =
    5.0000e-06
>> format long;
```

The actual relative error of $\approx 5 \times 10^{-6}$ is smaller than the bound of $\approx 4 \times 10^{-5}$ as expected. ∎

Example 21.5. The solution to the linear system

$$\begin{bmatrix} 1 & 1 & 2 \\ -2 & 1 & -1 \\ 1.00001 & 1 & 2 \end{bmatrix} \begin{bmatrix} x_1 \\ x_2 \\ x_3 \end{bmatrix} = \begin{bmatrix} 0 \\ 1 \\ 0 \end{bmatrix}$$

is approximated by

$$\hat{\mathbf{x}} = \begin{bmatrix} -0.33333 \\ 0.33333 \\ 0.00000 \end{bmatrix}.$$

What does (21.11) predict about the error? What actually happens?

Solution. The norm of the residual and the relative condition number are measured.

```
>> A = [  1         1  2 ;
         -2         1 -1 ;
          1.00001   1  2 ];
>> b = [0; 1; 0];
>> xapprox = [-0.33333; 0.33333; 0.00000];
>> format short;
>> res = norm(b-A*xapprox,Inf)
res =
    1.0000e-05
>> kappa = condest(A')
kappa =
    8.0000e+05
```

The residual is small, but the relative condition number is over 10^5. Inequality (21.11) gives

$$\frac{\|\mathbf{x} - \hat{\mathbf{x}}\|_\infty}{\|\mathbf{x}\|_\infty} \leq \kappa_{\text{rel}} \frac{\|\mathbf{b} - A\hat{\mathbf{x}}\|_\infty}{\|\mathbf{b}\|_\infty} \approx (8 \times 10^5) \frac{10^{-5}}{1} = 8.$$

It is possible for the error to be much larger than the residual.

In fact, the exact solution is

$$\mathbf{x} = \begin{bmatrix} 0 \\ 2/3 \\ -1/3 \end{bmatrix}.$$

There is not even a superficial resemblance between $\hat{\mathbf{x}}$ and $\mathbf{x}$, even though $\hat{\mathbf{x}}$ produces a small residual. Quantitatively, the difference is as follows.

```
>> x = [0; 2/3; -1/3];
>> norm(x-xapprox,Inf)/norm(x,Inf)
ans =
    0.5000
>> format long;
```

The relative error is indeed quite large. ∎

In Example 21.4, a small residual and a small condition number together guarantee a small error. In Example 21.5, the larger condition number allows $\hat{\mathbf{x}}$ to wander farther from $\mathbf{x}$, even though the residual is small.

What condition numbers qualify as "large?" Inequality (21.11) is equivalent to

$$\log_{10} \frac{\|\mathbf{x} - \hat{\mathbf{x}}\|_\infty}{\|\mathbf{x}\|_\infty} \leq \log_{10} \kappa_{\text{rel}} + \log_{10} \frac{\|\mathbf{b} - A\mathbf{x}\|_\infty}{\|\mathbf{b}\|_\infty}. \tag{21.12}$$

The left-hand side measures the number of correct significant digits in the entries of $\hat{\mathbf{x}}$. The right-hand side shows that $\log_{10} \kappa_{\text{rel}}$ significant digits can be lost to ill conditioning. For example, a relative condition number of $\kappa_{\text{rel}} = 10^8$ can be responsible for a loss of eight significant digits of accuracy in $\hat{\mathbf{x}}$—approximately half of the digits available in the common IEEE 754 standard. Hence, a relative condition number of 10^8 may be large enough to cause concern. By contrast, a relative condition number of 10^3 and the potential loss of three significant digits of accuracy may not be so bad.

21.4 ▪ Distance from noninvertibility

A matrix has a large condition number if and only if it is nearly noninvertible. Rather than state this fact as a formal theorem, we illustrate with an example.

Example 21.6. From Example 21.5, we know that the matrix

$$\begin{bmatrix} 1 & 1 & 2 \\ -2 & 1 & -1 \\ 1.00001 & 1 & 2 \end{bmatrix} \tag{21.13}$$

has a condition number of 8×10^5. Find a nearby matrix that is not invertible.

Solution. The matrix

$$\begin{bmatrix} 1 & 1 & 2 \\ -2 & 1 & -1 \\ 1 & 1 & 2 \end{bmatrix} \tag{21.14}$$

is not invertible because its third column equals the sum of its first two columns—the columns are linearly dependent. This matrix differs from the original matrix by just 0.00001 in its lower-left entry. ∎

Matrix (21.14) has an infinite condition number and is not invertible. The nearby matrix (21.13), while invertible, has a fairly large condition number that poses difficulties for numerical computation.

21.5 ▪ Scaling and units of measurement

Conditioning is affected by the units of measurement chosen for a problem.

Example 21.7. An athlete designs a meal consisting of chicken breast, low-fat milk, and almonds to have 15 g carbohydrates, 60 g protein, and 1.7 mg iron. She consults the following nutrition table:

	Chicken breast (1 g)	Low-fat milk (1 cL)	Almonds (1 nut)
Carbohydrates (g)	0.00	0.49	0.27
Protein (g)	0.31	0.34	0.27
Iron (mg)	0.0094	0.0021	0.011

An overzealous science-major teammate converts all units to SI units. Find the linear systems for both choices of units and compare their condition numbers.

Solution. The matrix-vector equation encoding the athlete's nutritional requirements in the original units is

$$\begin{bmatrix} 0.00 & 0.49 & 0.27 \\ 0.31 & 0.34 & 0.27 \\ 0.0094 & 0.0021 & 0.011 \end{bmatrix} \begin{bmatrix} x_1 \\ x_2 \\ x_3 \end{bmatrix} = \begin{bmatrix} 15 \\ 60 \\ 1.7 \end{bmatrix}.$$

The SI units for mass and volume are kg and m^3, respectively. The mass of an almond is considered to be 1.1 g. In SI units, the table is as follows:

	Chicken breast (1 kg)	Low-fat milk (1 m³)	Almonds (1 kg)
Carbohydrates (kg)	0.00	49	0.25
Protein (kg)	0.31	34	0.25
Iron (kg)	0.0094	0.21	0.010

The dietary requirements are 0.015 kg carbohydrates, 0.060 kg protein, and 1.7×10^{-6} kg iron. The linear system in SI units is

$$\begin{bmatrix} 0.00 & 49 & 0.25 \\ 0.31 & 34 & 0.25 \\ 0.0094 & 0.21 & 0.010 \end{bmatrix} \begin{bmatrix} x_1 \\ x_2 \\ x_3 \end{bmatrix} = \begin{bmatrix} 0.015 \\ 0.060 \\ 0.0000017 \end{bmatrix}.$$

The relative condition numbers of the matrices are computed.

```
>> A = [ 0.00 0.49 0.27; 0.31 0.34 0.27; 0.0094 0.0021 0.011 ];
>> kappaA = condest(A')
kappaA =
     1.313020288398554e+02
```

```
>> B = [ 0.00 49 0.25; 0.31 34 0.25; 0.0094 0.21 0.010 ];
>> kappaB = condest(B')
kappaB =
       7.803996762141967e+03
```

The second system has a larger condition number. ∎

The difference in condition numbers in the example is not large, but the point is that the condition numbers are different, even though the underlying scientific problem is unchanged. A more malign choice of units could cause more severe conditioning problems.

A linear system $\mathbf{Ax} = \mathbf{b}$ may be *rescaled* by applying the following substitutions, in which $\mathbf{D}$ and $\mathbf{E}$ are invertible matrices:

$$\mathbf{x} \mapsto \mathbf{E}^{-1}\mathbf{x},$$

$$\mathbf{b} \mapsto \mathbf{D}^{-1}\mathbf{b},$$

$$\mathbf{A} \mapsto \mathbf{D}^{-1}\mathbf{AE}.$$

This approach reproduces the change of units in the previous example when $\mathbf{D}$ is the diagonal matrix with 1000 g/kg, 1000 g/kg, and 10^6 mg/kg along its diagonal and $\mathbf{E}$ is the diagonal matrix with 1000 g/kg, 10^5 cL/m^3, and 910 nuts/kg along its diagonal. Note that the substitutions do not change the solution in theory since $\mathbf{Ax} = \mathbf{b}$ is equivalent to

$$(\mathbf{D}^{-1}\mathbf{AE})(\mathbf{E}^{-1}\mathbf{x}) = \mathbf{D}^{-1}\mathbf{b}.$$

However, a careful rescaling may improve the condition number. That is, a carefully rescaled matrix $\mathbf{D}^{-1}\mathbf{AE}$ may have a smaller condition number than the original matrix $\mathbf{A}$.

In the remainder of the book, we usually do not mention scaling in order to simplify the discussion.

21.6 ▪ Conclusion

For an ill-conditioned system, a small residual does not guarantee a small error. A tiny error in the problem statement or an intermediate computation can potentially be amplified into a huge error in the solution. Once an ill-conditioned matrix is delivered, there is unfortunately little that can be done in many circumstances because ill conditioning is a property of the problem rather than any particular solution method. However, if the matrix comes from a physical model, then a modification of the model may produce a better-conditioned mathematical problem.

With this danger acknowledged, we should emphasize that ill conditioning is not an ever-present plague. Plenty of interesting and natural systems are perfectly well conditioned. Also, we are helped by the fact that standard computer hardware stores far more digits than are needed in many applications; often a loss of several digits is tolerable.

Notes

Rule of thumb (21.7) appears in the text of Golub and Van Loan [35, p. 138]. It relies on a bound by Wilkinson [87, p. 108]. More details and historical notes can be found in Higham's text [43, pp. 163–166, 183–187].

Although the rule of thumb (21.7) is reliable in practice, it fails to hold on some pathological examples. See the notes at the end of Chapter 19.

The condition number $\|A^{-1}\|_\infty$ is just one example of an absolute condition number. The general definition is stated in terms of computing a function value $y = f(x)$ and requires norms $\|\cdot\|_X$ and $\|\cdot\|_Y$ for the domain and codomain, respectively. The absolute condition number for the computation of $y = f(x)$ is defined to be

$$\lim_{\delta \to 0} \sup_{\|\Delta x\|_X \leq \delta} \frac{\|f(x + \Delta x) - f(x)\|_Y}{\|\Delta x\|_X}.$$

Notice that this is a kind of derivative; the purpose is to measure how a small change in the input can perturb the output. The specific problem of solving $Ax = b$, given input b, can be expressed as computing $f(b) = A^{-1}b$. Using the infinity norm for the input and output spaces, the absolute condition number with respect to perturbations in b is

$$\lim_{\delta \to 0} \sup_{\|\Delta b\|_\infty \leq \delta} \frac{\|A^{-1}(b + \Delta b) - A^{-1}b\|_\infty}{\|\Delta b\|_\infty} = \lim_{\delta \to 0} \sup_{\|\Delta b\|_\infty \leq \delta} \frac{\|A^{-1}\Delta b\|_\infty}{\|\Delta b\|_\infty}$$

$$= \lim_{\delta \to 0} \|A^{-1}\|_\infty = \|A^{-1}\|_\infty,$$

in which we used both conclusions of Theorem 21.2. Another example of a condition number is the Lebesgue constant introduced in Chapter 6. It quantifies the possible error in a polynomial interpolant caused by perturbations of the interpolation points (Theorem 6.11). In the language of conditioning, we can say that interpolation on a uniform grid of high degree is ill conditioned (Theorem 9.4), while interpolation on Chebyshev grids is well conditioned (Theorem 11.14).

Exercises

1. Let c be a scalar and x a column vector. Prove that $\|cx\|_\infty = |c|\,\|x\|_\infty$.

2. Let x and y be column vectors with the same number of entries. Prove that $\|x + y\|_\infty \leq \|x\|_\infty + \|y\|_\infty$.

3. Let
$$A = \begin{bmatrix} 8 & 8 & -3 \\ -9 & -6 & 5 \\ 1 & 7 & 10 \end{bmatrix}.$$

   Find $\|A\|_\infty$ and a nonzero vector x that realizes $\|Ax\|_\infty = \|A\|_\infty\|x\|_\infty$.

4. Let A and B be m-by-n matrices. Prove that $\|A + B\|_\infty \leq \|A\|_\infty + \|B\|_\infty$.

5. Another common vector norm is the Euclidean norm, also known as the 2-norm. Let $x = (x_1, \ldots, x_n)$ be an n-by-1 vector. Its Euclidean norm is

$$\|x\|_2 = \sqrt{x_1^2 + \cdots + x_n^2}.$$

   Prove that $\|x\|_\infty \leq \|x\|_2$ and that $\|x\|_2 \leq \sqrt{n}\|x\|_\infty$.

6. Another common vector norm is the 1-norm. Let $x = (x_1, \ldots, x_n)$ be an n-by-1 vector. Its 1-norm is $\|x\|_1 = |x_1| + \cdots + |x_n|$. Prove that $\|x\|_\infty \leq \|x\|_1$ and that $\|x\|_1 \leq n\|x\|_\infty$.

7. Is the rule of thumb (21.7) always valid? Test on the linear system defined by (19.11)–(19.12) for $n = 10, 20, 30, \ldots, 100$.

8. Construct a matrix whose relative condition number is exactly 10^{10}.

Exercises 9–12: The solution to the linear system $\mathbf{A}\mathbf{x} = \mathbf{b}$ is approximated by $\hat{\mathbf{x}}$. Measure the residual $\|\mathbf{b} - \mathbf{A}\hat{\mathbf{x}}\|_\infty$ and bound the relative error using (21.11).

9. $\mathbf{A} = \begin{bmatrix} 8 & 5 & 7 \\ 6 & 5 & 9 \\ 6 & 1 & 4 \end{bmatrix}, \mathbf{b} = \begin{bmatrix} 9 \\ 3 \\ 9 \end{bmatrix}; \hat{\mathbf{x}} = \begin{bmatrix} 2.1429 \\ -0.4286 \\ -0.8571 \end{bmatrix}$

10. $\mathbf{A} = \begin{bmatrix} 8 & 2 & -5 \\ -9 & 2 & 5 \\ -2 & -8 & -1 \end{bmatrix}, \mathbf{b} = \begin{bmatrix} 36 \\ -33 \\ -20 \end{bmatrix}; \hat{\mathbf{x}} = \begin{bmatrix} 3.6667 \\ 1.6667 \\ -0.6667 \end{bmatrix}$

11. $\mathbf{A} = \begin{bmatrix} 6.250 & 3.750 & -1.768 \\ 3.750 & 6.250 & 1.768 \\ -1.768 & 1.768 & 2.500 \end{bmatrix}, \mathbf{b} = \begin{bmatrix} 2.500 \\ -2.500 \\ -3.536 \end{bmatrix}; \hat{\mathbf{x}} = \begin{bmatrix} 0.5000 \\ -0.5000 \\ -0.7071 \end{bmatrix}$

12. $\mathbf{A} = \begin{bmatrix} -6.663 & 6.364 & -1.670 \\ -8.981 & -4.191 & -7.776 \\ 1.900 & -5.758 & -1.230 \end{bmatrix}, \mathbf{b} = \begin{bmatrix} -6.197 \\ -11.954 \\ 0.655 \end{bmatrix}; \hat{\mathbf{x}} = \begin{bmatrix} 0.825 \\ 0.038 \\ 0.564 \end{bmatrix}$

13. The matrix

$$\begin{bmatrix} 15 & 5 & -7.0711 \\ 5 & 15 & 7.0711 \\ -7.0711 & 7.0711 & 10 \end{bmatrix}$$

has a relative condition number of approximately 7×10^5. Find a nearby matrix that is noninvertible by changing no entry by more than 10^{-4}. Prove that your answer is not invertible.

14. Let

$$\mathbf{A} = \begin{bmatrix} 500000 & 9000 & -400 & 2000 \\ 9000 & 3000 & 100 & 200 \\ -400 & 100 & -2 & -6 \\ 2000 & 200 & -6 & 6 \end{bmatrix}.$$

Measure the relative condition number. Then find a diagonal matrix $\mathbf{D}$ for which the relative condition number of $\mathbf{D}\mathbf{A}\mathbf{D}$ is below 100.

15. For a matrix $\mathbf{C} = [c_{ij}]$, define $\max |\mathbf{C}| = \max_{i,j} |c_{ij}|$. (Note that this is different from the matrix infinity norm.) Let $\mathbf{A}$ be an n-by-n matrix with absolute condition number $\kappa_{\text{abs}} < \infty$, and let $\mathbf{B}$ be an n-by-p matrix. Suppose that $\mathbf{X} = [x_{ij}]$ is the solution to $\mathbf{A}\mathbf{X} = \mathbf{B}$ and that $\hat{\mathbf{X}} = [\hat{x}_{ij}]$ is an approximation to $\mathbf{X}$. Prove

$$\max |\mathbf{X} - \hat{\mathbf{X}}| \le \kappa_{\text{abs}} \max |\mathbf{B} - \mathbf{A}\hat{\mathbf{X}}|.$$

16. As discussed in Chapter 14, an integration matrix $\mathbf{K}_{x,x'}$ can be computed by solving the equation

$$\begin{bmatrix} \mathbf{D}_{x',x} \\ \mathbf{E}_{a,x} \end{bmatrix} \mathbf{K}_{x,x'} = \begin{bmatrix} \mathbf{I} \\ \mathbf{0} \end{bmatrix}. \tag{21.15}$$

This could produce an inaccurate result if

$$\begin{bmatrix} \mathbf{D}_{x',x} \\ \mathbf{E}_{a,x} \end{bmatrix} \tag{21.16}$$

had a large condition number. Let $\mathbf{x}$ and $\mathbf{x}'$ be uniform grids of degrees $m + 1$ and m, respectively, on $[0, 1]$, and let $a = 0$ be the lower limit of integration. Measure the absolute condition number for (21.16) with $m = 0, 1, 2, 3$. *Hint.* Because $\|\mathbf{A}^{-1}\|_\infty = (\|\mathbf{A}\|_\infty \|\mathbf{A}^{-1}\|_\infty)/\|\mathbf{A}\|_\infty$, the absolute condition number of a matrix $\mathbf{A}$ can be estimated with condest(A')/norm(A,Inf).

17. Repeat the previous exercise but with Chebyshev grids on $[-1, 1]$ and a lower limit of integration at $a = -1$. Measure the absolute condition number for $m = 10, 20, 30, \ldots, 100$.

18. Define $\mathbf{A}_n$ as in (19.11). Measure the relative condition number for $n = 10, 20, 30, \ldots, 100$. Is the failure observed in Exercise 17 of Chapter 19 the result of ill conditioning? Explain.

19. The n-by-n *Hilbert matrix* $\mathbf{H} = [h_{ij}]$ has entries $h_{ij} = 1/(i + j - 1)$, $i, j = 1, \ldots, n$. It can be constructed in code by evaluating hilb(n). Let $\mathbf{b} = [b_i]$ be the n-by-1 column vector with $b_i = (1/i) + (1/(i + 1))$. Solve $\mathbf{Hx} = \mathbf{b}$ by hand and then numerically for $n = 3, 6, 9, 12, 15$. What phenomenon is observed?

20. The 1-norm of a vector $\mathbf{x}$, denoted $\|\mathbf{x}\|_1$, is the sum of the absolute values of its entries. (See Exercise 6.) The 1-norm of a matrix $\mathbf{A}$, denoted $\|\mathbf{A}\|_1$, is the largest 1-norm of any of its columns. Prove that $\|\mathbf{Ax}\|_1 \leq \|\mathbf{A}\|_1 \|\mathbf{x}\|_1$ for any $\mathbf{x}$ and that there exists a nonzero $\mathbf{x}$ for which $\|\mathbf{Ax}\|_1 = \|\mathbf{A}\|_1 \|\mathbf{x}\|_1$.

21. The 1-norm of a matrix is defined in Exercise 20. Prove that $\|\mathbf{A}\|_\infty \|\mathbf{A}^{-1}\|_\infty = \|\mathbf{A}^T\|_1 \|(\mathbf{A}^T)^{-1}\|_1$. (This equation helps to explain the use of condest(A') instead of condest(A) in, e.g., Example 21.4. The condest function is designed to estimate the condition number in the 1-norm but can be manipulated to compute the condition number in the infinity norm by applying a transpose.)

Part V

Linear differential equations

Chapter 22

Introduction to linear differential equations

In many mathematical models, a system evolves over time according to a *differential equation* that specifies the rate and direction of change.

In this chapter, we introduce differential equations and discuss what it means to solve them. We focus on a particular type called a linear differential equation, which is important in theory and in numerical practice. The more general class of nonlinear differential equations is the subject of Chapters 32–34.

22.1 ▪ First-order linear differential equations

A *first-order linear differential equation* is an equation that can be written in the form

$$\alpha(t)u(t) + \beta(t)u'(t) = g(t). \tag{22.1}$$

Typically, functions α, β, and g are known, and u is an unknown function to be found. The equation is called a differential equation because it involves a derivative; it is first order because the highest derivative is a first derivative; and it is linear because the left-hand side is a linear combination of $u(t)$ and $u'(t)$.

Example 22.1. Carbon-14 is a radioactive isotope. If a mass of it is left alone, its atoms spontaneously decay, emitting particles and transforming into nitrogen. Over time, the amount of carbon-14 decreases.

On average, the number of radioactive emissions in a given period of time is proportional to the total number of atoms present. Therefore, the rate of change of the mass u is modeled by

$$u'(t) = -\lambda u(t) \tag{22.2}$$

for some constant of proportionality λ. For carbon-14, the value of λ is about 0.00012/ year, i.e., 0.012% per year. (This determines the isotope's half-life, which is often mentioned in the context of radiometric dating.)

Verify that the differential equation above is a first-order linear differential equation and then solve it by guessing and checking.

Solution. The equation is a first-order linear differential equation because it can be written in the form

$$\lambda u(t) + u'(t) = 0, \tag{22.3}$$

i.e., as (22.1) with $\alpha(t) = \lambda$, $\beta(t) = 1$, and $g(t) = 0$.

To solve the equation, we want to find a function whose derivative is nearly equal to itself but differs by a factor of $-\lambda$. The exponential function e^t, whose derivative is $\frac{d}{dt}[e^t] = e^t$, comes to mind. Through trial and error, we find that $e^{-\lambda t}$ works just right. For this function, the left-hand side of (22.2) is $\frac{d}{dt}[e^{-\lambda t}] = e^{-\lambda t}\frac{d}{dt}[-\lambda t] = e^{-\lambda t}(-\lambda)$, and the right-hand side is $-\lambda e^{-\lambda t}$. Because these are equal, the function is a solution to the differential equation.

In fact, the *general solution* to the differential equation is the infinite family $u(t) = Ce^{-\lambda t}$, in which C can be any constant coefficient. Again, plug the function into both sides of (22.2) and check that they are equal to verify.

Finally, note that the solution $u(t) = Ce^{-\lambda t}$ seems physically reasonable as long as $C \geq 0$ because it approaches zero monotonically from above in the $t \to +\infty$ limit. In other words, it predicts that the mass of carbon-14 always decreases and would disappear given unlimited time. ■

It is common for a differential equation to have infinitely many solutions. A single solution can be pinned down by adding an *initial condition* that specifies the value of the solution at a specific moment in time. The pair

$$\alpha(t)u(t) + \beta(t)u'(t) = g(t), \quad u(a) = u_a \tag{22.4}$$

is an *initial-value problem* (IVP). The initial time is $t = a$, and the initial value at that time is u_a.

Example 22.2. Continuing the previous example, suppose that a sample of carbon-14 has mass 0.1 mg at time $t = 0$. Then the mass of carbon-14 obeys the IVP

$$\lambda u(t) + u'(t) = 0, \quad u(0) = 0.1. \tag{22.5}$$

Solve the IVP.

Solution. Among the infinite family of solutions $Ce^{-\lambda t}$ from the previous example, only one has the correct initial value:

$$u(0) = 0.1 \iff Ce^{-\lambda \times 0} = 0.1 \iff C = 0.1.$$

Hence, the IVP has the unique solution

$$u(t) = 0.1e^{-\lambda t}. \qquad ■$$

IVPs are common in physical situations. The initial condition specifies the original state of a system, and the differential equation specifies how it evolves over time.

An additional example is useful for testing upcoming numerical methods.

Example 22.3. Find a solution to the IVP

$$tu(t) + u'(t) = 0, \quad u(0) = 1.$$

Solution. At $t = 0$, the solution is positive. Let's proceed for now under the assumption that the solution stays positive. Then the differential equation is equivalent to

$$\frac{u'(t)}{u(t)} = -t.$$

The left-hand side equals $\frac{d}{dt}[\log u(t)]$, and the right-hand side equals $\frac{d}{dt}[-t^2/2]$. Thus,

$$\log u(t) = -\frac{t^2}{2} + C,$$

so

$$u(t) = e^{-t^2/2+C} = ke^{-t^2/2},$$

in which $k = e^C$. The initial condition forces $k = 1$:

$$u(t) = e^{-t^2/2}.$$

To have complete confidence, we check the solution $u(t) = e^{-t^2/2}$ from scratch. Note that this function satisfies the differential equation because $te^{-t^2/2} + \frac{d}{dt}[e^{-t^2/2}] = te^{-t^2/2} - te^{-t^2/2} = 0$, and it satisfies the initial condition because $e^{-0^2/2} = 1$. Hence, $u(t) = e^{-t^2/2}$ is a solution to the IVP. ∎

22.2 ▪ Second-order linear differential equations

A *second-order linear differential equation* is an equation that can be written in the form

$$\alpha(t)u(t) + \beta(t)u'(t) + \gamma(t)u''(t) = g(t). \tag{22.6}$$

Of course, there are third- and fourth- and higher-order equations as well, but we choose to stop at second order.

We can thank Isaac Newton for the ubiquity of second-order differential equations in classical mechanics. His second law for rectilinear motion—force equals mass times acceleration—is the second-order differential equation $F = Mu''(t)$. For example, when applied to an idealized spring obeying Hooke's law $F(u) = -Ku$, the differential equation is

$$Ku(t) + Mu''(t) = 0,$$

which is second order and linear. (Lowercase letters k and m are more commonly used, but since this book reserves m for the degree of a polynomial, we use capital letters instead.)

The behavior of a second-order system is often fully determined once the initial position *and* initial velocity are given. The resulting IVP has the form

$$\alpha(t)u(t) + \beta(t)u'(t) + \gamma(t)u''(t) = g(t), \quad u(a) = u_a, \quad u'(a) = v_a. \tag{22.7}$$

In the next few examples, second-order linear differential equations are solved by hand. The solutions are useful for testing numerical methods later.

Example 22.4. A spring obeys Hooke's law, exerting a force $F(u) = -Ku$ on an attached mass, in which u is displacement from equilibrium and K is the stiffness coefficient. In addition, the position of the mass obeys Newton's second law of motion. If the mass starts at position u_0 with zero velocity, then its position satisfies the IVP

$$Ku(t) + Mu''(t) = 0, \quad u(0) = u_0, \quad u'(0) = 0.$$

Show that this IVP has solution $u(t) = u_0 \cos(\lambda t)$, in which $\lambda = \sqrt{K/M}$.

Solution. Because the function $u_0 \cos(\lambda t)$ has not yet been verified to be a solution, we call it a candidate solution and write

$$u_c(t) = u_0 \cos(\lambda t).$$

Its first two derivatives are the following:

$$u_c'(t) = -u_0 \lambda \sin(\lambda t),$$
$$u_c''(t) = -u_0 \lambda^2 \cos(\lambda t).$$

The differential equation is satisfied because

$$Ku_c(t) + Mu_c''(t) = Ku_0 \cos(\lambda t) - Mu_0 \lambda^2 \cos(\lambda t) = 0.$$

Also, the initial conditions are satisfied because $u_c(0) = u_0 \cos(0) = u_0$ and $u_c'(0) = -u_0 \lambda \sin(0) = 0$. The IVP is solved. ∎

Example 22.5. The damped harmonic oscillator is a model for a mass-spring system that loses energy over time. The position u, relative to the equilibrium position, of a mass started from rest is determined by the IVP

$$Ku(t) + Bu'(t) + Mu''(t) = 0, \quad u(0) = u_0, \quad u'(0) = 0.$$

As in the previous example, K is the stiffness coefficient and M is the mass. The new parameter B, called the damping coefficient, controls the loss of energy due to effects such as drag. Show that the IVP has solution

$$u(t) = u_0 e^{-(B/(2M))t} \left(\cos(\lambda t) + \frac{B}{2\lambda M} \sin(\lambda t) \right),$$

in which $\lambda = \sqrt{4KM - B^2}/(2M)$.

Solution. The candidate solution satisfies

$$u_c(t) = u_0 e^{-(B/(2M))t} \left(\cos(\lambda t) + \frac{B}{2\lambda M} \sin(\lambda t) \right),$$
$$u_c'(t) = -\frac{u_0 K}{\lambda M} e^{-(B/(2M))t} \sin(\lambda t),$$
$$u_c''(t) = \frac{u_0 K}{2\lambda M^2} e^{-(B/(2M))t} (-2M\lambda \cos(\lambda t) + B \sin(\lambda t)).$$

The differential equation is satisfied because

$$Ku_c(t) + Bu_c'(t) + Mu_c''(t)$$
$$= u_0 e^{-(B/(2M))t} \left(\left(K - M\tfrac{K}{M} \right) \cos(\lambda t) + \left(K\tfrac{B}{2\lambda M} - B\tfrac{K}{\lambda M} + M\tfrac{BK}{2\lambda M^2} \right) \sin(\lambda t) \right)$$
$$= u_0 e^{-(B/(2M))t} \times 0 = 0.$$

Also, the initial conditions are satisfied because

$$u_c(0) = u_0 e^0 (\cos(0) + (B/(2\lambda M)) \sin(0)) = u_0$$

and $u_c'(0) = -\frac{u_0 K}{\lambda M} e^0 \sin(0) = 0$. ∎

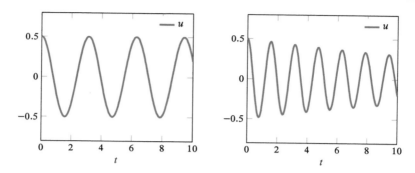

Figure 22.1. *The position of the mass in an undamped (left) or damped (right) mass-spring system.*

Example 22.6. A 1-kg mass is attached to a spring with stiffness coefficient 4 N/m. It is released from rest at a distance of 0.5 m from the equilibrium position in the positive-u direction. Using the solutions to the previous two examples, plot trajectories with no damping and with damping of 0.1 N·s/m.

Solution. First, no damping is applied.

```
>> M = 1;
>> K = 4;
>> ua = 0.5;
>> u1 = @(t) ua*cos(sqrt(K/M)*t);
>> ax = newfig(1,2); plotfun(u1,[0 10]);
```

As seen in the left plot in Figure 22.1, the mass oscillates back and forth about the equilibrium position with a constant amplitude.

Next, the damping coefficient is set to 0.1 N·s/m.

```
>> B = 0.1;
>> u2 = @(t) ua*exp(-(B/(2*M))*t)*(cos(K*t)+B/(2*K*M)*sin(K*t));
>> subplot(ax(2)); plotfun(u2,[0 10]);
```

See the plot on the right, and notice how energy is lost because of damping. ∎

Example 22.7. The Bessel function J_0, plotted in Figure 22.2, is useful for describing the shape of a vibrating drum head. It can be defined by

$$J_0(x) = \frac{1}{\pi} \int_0^\pi \cos(x \sin \theta) \, d\theta.$$

Verify that J_0 satisfies the IVP

$$x u(x) + u'(x) + x u''(x) = 0, \quad u(0) = 1, \quad u''(0) = -1/2. \tag{22.8}$$

(Note that the second initial condition constrains u'', not u'. This is somewhat unusual but valid.)

Solution. The Bessel function and its first two derivatives, which can be found by

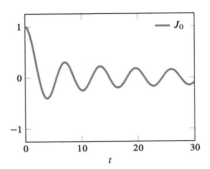

Figure 22.2. *The Bessel function J_0.*

differentiating under the integral, are as follows:

$$J_0(x) = \frac{1}{\pi} \int_0^\pi \cos(x \sin \theta) \, d\theta,$$

$$J_0'(x) = \frac{1}{\pi} \int_0^\pi \sin(x \sin \theta)(-\sin \theta) \, d\theta,$$

$$J_0''(x) = \frac{1}{\pi} \int_0^\pi \cos(x \sin \theta)(-\sin^2 \theta) \, d\theta.$$

At $x = 0$, we have

$$J_0(0) = \frac{1}{\pi} \int_0^\pi (\cos 0) \, d\theta = \frac{1}{\pi} \int_0^\pi 1 \, d\theta = 1,$$

$$J_0''(0) = \frac{1}{\pi} \int_0^\pi (-\sin^2 \theta) \, d\theta = \frac{1}{\pi} \int_0^\pi \frac{\cos(2\theta) - 1}{2} \, d\theta = -\frac{1}{2}.$$

The initial conditions are satisfied.

Apply integration by parts to $J_0'(x)$:

$$J_0'(x) = \frac{1}{\pi} [\sin(x \sin \theta) \cos(\theta)]_{\theta=0}^\pi - \frac{1}{\pi} \int_0^\pi \cos(\theta) \cos(x \sin \theta)(x \cos \theta) \, d\theta$$

$$= \frac{1}{\pi} \int_0^\pi \cos(x \sin \theta)(-x \cos^2 \theta) \, d\theta.$$

Then substitute into the left-hand side of the differential equation:

$$x J_0(x) + J_0'(x) + x J_0''(x) = \frac{1}{\pi} \int_0^\pi x(1 - \cos^2 \theta - \sin^2 \theta) \cos(x \sin \theta) \, d\theta.$$

Because $1 - \cos^2 \theta - \sin^2 \theta = 0$, the integrand equals zero, and therefore $x J_0(x) + J_0'(x) + x J_0''(x) = 0$. The differential equation is verified. ∎

Some readers may know that initial condition $u''(0) = -1/2$ in (22.8) is redundant. This is true (Exercise 13), but the computer code we develop in the next few chapters is not smart enough to deduce the value of $u''(0)$ automatically.

22.3 ▪ Existence and uniqueness

A linear IVP may have zero, one, or infinitely many solutions.

The following theorem establishes the existence of a unique solution for a common problem, a particular kind of IVP on a closed interval $[a, b]$. At the endpoints of the interval, the derivative should be interpreted as a one-sided derivative: $u'(a) = \lim_{t \to a+} (u(t) - u(a))/(t - a)$ and $u'(b) = \lim_{t \to b-} (u(t) - u(b))/(t - b)$.

Theorem 22.8. *If $\alpha(t)$, $\beta(t)$, and $g(t)$ are continuous on $[a, b]$ and $\beta(t)$ equals zero nowhere in the interval, then the IVP*

$$\alpha(t)u(t) + \beta(t)u'(t) = g(t), \quad u(a) = u_a$$

has a unique solution on the interval.

The proof is omitted.

Although the theorem applies to a great many problems, there are also interesting problems that do not satisfy the hypotheses of the theorem.

Example 22.9. Count the number of solutions to each of the following IVPs:

$$u(t) - tu'(t) = 0, \quad u(0) = 1;$$
$$u(t) - tu'(t) = 0, \quad u(0) = 0;$$
$$u(t) - tu'(t) = 0, \quad u'(0) = 1.$$

Solution. The differential equation $u(t) - tu'(t) = 0$ has general solution

$$u(t) = kt,$$

in which k is an arbitrary coefficient. (Derive this by separating variables or using an integrating factor. Check it by plugging it into the differential equation.) The first IVP has no solution since $u(0) = k \times 0 = 0$ regardless of the value of k. The second IVP, on the other hand, has infinitely many solutions because every function $u(t) = kt$ satisfies $u(0) = 0$. The third IVP has a unique solution: only $u(t) = t$ satisfies $u'(0) = 1$. ∎

The differential equation $u(t) - tu'(t) = 0$ in the example has a "regular singular point" at $t = 0$, where the coefficient function $\beta(t) = t$ vanishes and thus the hypotheses of Theorem 22.8 are not satisfied. We see that in a neighborhood of a singular point, the number of solutions can depend delicately on the initial condition.

The theory for second-order linear differential equations is similar.

Theorem 22.10. *If $\alpha(t)$, $\beta(t)$, $\gamma(t)$, and $g(t)$ are continuous on $[a, b]$ and $\gamma(t)$ equals zero nowhere in the interval, then the IVP*

$$\alpha(t)u(t) + \beta(t)u'(t) + \gamma(t)u''(t) = g(t), \quad u(a) = u_a, \quad u'(a) = v_a$$

has a unique solution on the interval.

If $\gamma(t) = 0$ somewhere, then the differential equation may have a singular point, and the IVP may have zero, one, or infinitely many solutions depending on the initial conditions. The Bessel differential equation of Example 22.7 has a coefficient function $\gamma(x) = x$ that vanishes at $x = 0$. It is no coincidence that an unusual initial condition, specifying an initial value for the *second* derivative, accompanies the differential equation in that example.

Notes

Differentiating under the integral sign as in Example 22.7 is valid under certain conditions. See, for example, the text by Whittaker and Watson [85, p. 67].

Theorems 22.8 and 22.10 are proved, for example, in the differential equations text by Coddington [17, p. 260].

Exercises

Exercises 1–6: Verify that the IVP has the given solution.

1. $u + u' = t$, $u(0) = 7$; $u(t) = -1 + t + 8e^{-t}$
2. $2u + 5u' = t^2$, $u(0) = 25/4$; $u(t) = \frac{1}{4}(2t^2 - 10t + 25)$
3. $5u + u' = \sin t$, $u(0) = -1/26$; $u(t) = \frac{1}{26}(-\cos t + 5 \sin t)$
4. $(\cos t)u + u' = \cos t$, $u(0) = 4$; $u(t) = 3e^{-\sin t} + 1$
5. $u + 2u' + u'' = 0$, $u(0) = 6$, $u'(0) = -2$; $u(t) = 6e^{-t} + 4te^{-t}$
6. $2u + u' + u'' = 0$, $u(0) = 0$, $u'(0) = \sqrt{7}$; $u(t) = 2e^{-t/2} \sin((\sqrt{7}/2)t)$

Exercises 7–10: Solve the IVP by hand.

7. $-2u + u' = 0$, $u(0) = 10$
8. $tu + u' = 0$, $u(0) = -1$
9. $u + tu' = 0$, $u(1) = 5$
10. $u + u'' = 0$, $u(0) = 2$, $u'(0) = -1$

11. After one half-life, the mass of a radioactive isotope is half of its original amount. Find the half-life of carbon-14 using the decay coefficient $\lambda = 0.000121/\text{year}$. Justify.

12. An aquarium contains 150 liters of too-salty water, with 6 kg of dissolved salt. To adjust the salinity, a drain is opened at the bottom of the tank, allowing salt water to flow out at 0.5 liters per minute, and pure water is added to the tank at the same rate to keep it full. The salt water is kept well mixed throughout. Write an IVP for the amount of salt in the tank. You do not need to solve it.

13. The Bessel function $J_0(x)$ satisfies three properties: (1) it obeys the differential equation $xJ_0(x) + J_0'(x) + xJ_0''(x) = 0$; (2) it can be expanded in a power series $J_0(x) = \sum_{i=0}^{\infty} c_i x^i$ that converges for all x; and (3) it satisfies the initial condition $J_0(0) = 1$.

    (a) Prove that the power series begins $J_0(x) = 1 - (1/4)x^2 + (1/64)x^4 + \cdots$.
    (b) Deduce that $J''(0) = -1/2$, which agrees with the second initial condition in (22.8).

14. Suppose $\alpha(t)$ and $\beta(t)$ are continuous on a closed interval and that $\beta(t)$ equals zero nowhere in the interval. Let $u_h(t)$ be one solution to $\alpha(t)u + \beta(t)u' = 0$ that is not identically zero. Prove that every scalar multiple $cu_h(t)$ is also a solution to the differential equation and that there are no other solutions.

15. Suppose $\alpha(t)$, $\beta(t)$, and $g(t)$ are continuous on a closed interval and that $\beta(t)$ equals zero nowhere in the interval. Also, suppose $u_h(t)$ is a solution to $\alpha(t)u +$

$\beta(t)u' = 0$ that is not identically zero and $u_p(t)$ is a solution to $\alpha(t)u + \beta(t)u' = g(t)$. Show that the solutions to $\alpha(t)u + \beta(t)u' = g(t)$ are precisely the functions of the form $u_p(t) + cu_h(t)$, with c a scalar.

16. Consider the vector-valued IVP

$$\begin{bmatrix} 1/2 & -1 \\ 1 & 1/2 \end{bmatrix} \begin{bmatrix} u_1 \\ u_2 \end{bmatrix} + \begin{bmatrix} 1 & 0 \\ 0 & 1 \end{bmatrix} \begin{bmatrix} u_1' \\ u_2' \end{bmatrix} = \begin{bmatrix} 0 \\ 0 \end{bmatrix}, \quad \begin{bmatrix} u_1(0) \\ u_2(0) \end{bmatrix} = \begin{bmatrix} 1 \\ 0 \end{bmatrix}.$$

Show that $u_1(t) = e^{-t/2} \cos t$, $u_2(t) = -e^{-t/2} \sin t$ is a solution.

Exercises 17–18: A *boundary-value problem* (BVP) of the form

$$\alpha(x)u(x) + \beta(x)u'(x) + \gamma(x)u''(x) = g(x), \quad u(a) = u_a, \quad u(b) = u_b$$

is given. The *boundary conditions* $u(a) = u_a, u(b) = u_b$ constrain the solution at the endpoints of an interval $[a, b]$. Verify that the BVP has the given solution.

17. $u + u'' = 0$, $u(0) = 1$, $u(\pi/2) = 2$; $u(t) = \cos t + 2 \sin t$
18. $u - u'' = 0$, $u(-1) = 0$, $u(1) = 1$; $u(t) = \frac{1}{e^4-1}(e^{3+t} - e^{1-t})$

Chapter 23

Collocation

The next several chapters are devoted to the numerical solution of IVPs involving the first-order linear differential equation

$$\alpha(t)u(t) + \beta(t)u'(t) = g(t)$$

and the second-order linear differential equation

$$\alpha(t)u(t) + \beta(t)u'(t) + \gamma(t)u''(t) = g(t).$$

We cannot hope to solve an arbitrary IVP exactly and everywhere with limited resources. Below, we compute solutions using an approach known as *collocation*. A collocation method enforces an equation at a finite set of points rather than an entire interval, leading to a finite linear system that can be solved mechanically.

23.1 ▪ First-order problems

We start with the first-order IVP

$$\alpha(t)u(t) + \beta(t)u'(t) = g(t), \quad c_0 u(a) + c_1 u'(a) = d. \tag{23.1}$$

The most common form of initial condition, $u(a) = u_a$, is recovered by setting $c_0 = 1$, $c_1 = 0$, and $d = u_a$.

Definition 23.1. *Given distinct points $s_0, \ldots, s_m$ on the real line, a* polynomial *collocation solution for*

$$\alpha(t)u(t) + \beta(t)u'(t) = g(t)$$

is a polynomial $p(t)$ of degree at most $m + 1$ that satisfies

$$\alpha(s_i)p(s_i) + \beta(s_i)p'(s_i) = g(s_i), \quad i = 0, \ldots, m. \tag{23.2}$$

The points $s_0, \ldots, s_m$ are called collocation nodes.

A collocation solution satisfies a differential equation at a finite set of collocation nodes rather than at every point in a continuous interval. Equivalently, the *residual function*

$$R(t) = g(t) - \alpha(t)p(t) - \beta(t)p'(t) \tag{23.3}$$

equals zero at the collocation nodes:

$$R(s_i) = 0, \quad i = 0, \dots, m.$$

When the nodes are packed tightly together, the residual function may stay small between the nodes as well, and thus the collocation solution may nearly satisfy the differential equation.

If a collocation solution $p(t)$ satisfying (23.2) also satisfies the initial condition

$$c_0 p(a) + c_1 p'(a) = d, \tag{23.4}$$

then it is called a collocation solution for the IVP (23.1).

A collocation solution can be found in a finite number of computational steps. Suppose p is a collocation solution, and let $q = p'$. Sample $\mathbf{p} = p|_{\mathbf{t}}$ and $\mathbf{q} = q|_{\mathbf{t}'}$ on grids $\mathbf{t}$ and $\mathbf{t}'$ of degrees $m + 1$ and m, respectively. The collocation solution is characterized by three matrix-vector equations involving $\mathbf{p}$ and $\mathbf{q}$, which are developed below.

First, the condition that $q = p'$ is equivalent to

$$\mathbf{J}_{\mathbf{t}}\mathbf{p} = \hat{\mathbf{K}}_{\mathbf{t},\mathbf{t}'}\mathbf{q} \tag{23.5}$$

by Theorem 14.8. (Alternatively, we could write $\mathbf{q} = \mathbf{D}_{\mathbf{t}',\mathbf{t}}\mathbf{p}$, using the differentiation matrix of (13.1)–(13.2), but we prefer the integration matrices because of the associated error bound in Theorem 14.16.)

Second, the collocation equations (23.2) can be expressed in terms of $\mathbf{p}$ and $\mathbf{q}$ by resampling. Note that

$$\begin{bmatrix} p(s_0) \\ \vdots \\ p(s_m) \end{bmatrix} = \mathbf{E}_{\mathbf{s},\mathbf{t}}\mathbf{p} \quad \text{and} \quad \begin{bmatrix} q(s_0) \\ \vdots \\ q(s_m) \end{bmatrix} = \mathbf{E}_{\mathbf{s},\mathbf{t}'}\mathbf{q},$$

in which $\mathbf{s} = (s_0, \dots, s_m)$ is the grid of collocation nodes. (See Theorem 12.1.) Let

$$\mathbf{A} = \begin{bmatrix} \alpha(s_0) & & \\ & \ddots & \\ & & \alpha(s_m) \end{bmatrix} \mathbf{E}_{\mathbf{s},\mathbf{t}}$$

and

$$\mathbf{B} = \begin{bmatrix} \beta(s_0) & & \\ & \ddots & \\ & & \beta(s_m) \end{bmatrix} \mathbf{E}_{\mathbf{s},\mathbf{t}'}.$$

Then (23.2) is equivalent to

$$\mathbf{A}\mathbf{p} + \mathbf{B}\mathbf{q} = \mathbf{g}, \tag{23.6}$$

in which $\mathbf{g} = g|_{\mathbf{s}}$ is a sample of g on the collocation nodes.

Third, the initial condition can also be expressed with the help of evaluation matrices. Note that

$$p(a) = \mathbf{E}_{a,\mathbf{t}}\mathbf{p}, \quad p'(a) = \mathbf{E}_{a,\mathbf{t}'}\mathbf{q}.$$

Thus, the initial condition (23.4) is equivalent to

$$c_0 \mathbf{E}_{a,\mathbf{t}}\mathbf{p} + c_1 \mathbf{E}_{a,\mathbf{t}'}\mathbf{q} = d. \tag{23.7}$$

Theorem 23.2. *Let p and q be polynomials of degrees $m + 1$ and m, respectively, and let $\mathbf{p} = p|_{\mathbf{t}}$ and $\mathbf{q} = q|_{\mathbf{t}'}$. Then the following are equivalent:*

1. p is a collocation solution to (23.1) with collocation nodes $s_0, \ldots, s_m$, and $q = p'$.

2. The matrix-vector equation

$$
\begin{bmatrix}
\mathbf{J}_t & -\hat{\mathbf{K}}_{t,t'} \\
\mathbf{A} & \mathbf{B} \\
c_0 \mathbf{E}_{a,t} & c_1 \mathbf{E}_{a,t'}
\end{bmatrix}
\begin{bmatrix}
\mathbf{p} \\
\mathbf{q}
\end{bmatrix}
=
\begin{bmatrix}
\mathbf{0} \\
\mathbf{g} \\
d
\end{bmatrix}
\tag{23.8}
$$

holds.

Proof. By Theorem 14.8, $q = p'$ if and only if (23.5) holds. By Theorem 12.1, the collocation equations (23.2) are equivalent to (23.6), and the initial condition (23.4) is equivalent to (23.7). The system of equations consisting of (23.5), (23.6), and (23.7) is equivalent to the block-matrix equation (23.8). □

Example 23.3. Let $\mathbf{s} = (0, 0.5)$, $\mathbf{t} = (0, 0.25, 0.5)$, and $\mathbf{t}' = (0, 0.5)$. Find the collocation solution for

$$(3 + t)u(t) + u'(t) = 4 - 2t, \quad u(0) = 2.$$

Solution. The interpolation grids $\mathbf{t}$, $\mathbf{t}'$ are the same as the grids $\mathbf{x}$, $\mathbf{x}'$ of Example 14.9, so the integration matrix pair is

$$
\mathbf{J}_t = \begin{bmatrix} -1 & 1 & 0 \\ -1 & 0 & 1 \end{bmatrix}, \quad
\hat{\mathbf{K}}_{t,t'} = \begin{bmatrix} 3/16 & 1/16 \\ 1/4 & 1/4 \end{bmatrix}.
$$

Because $t_0 = s_0$ and $t_2 = s_1$ and because $t_0' = s_0$ and $t_1' = s_1$, the resampling matrices are

$$
\mathbf{E}_{s,t} = \begin{bmatrix} 1 & 0 & 0 \\ 0 & 0 & 1 \end{bmatrix}, \quad
\mathbf{E}_{s,t'} = \begin{bmatrix} 1 & 0 \\ 0 & 1 \end{bmatrix}.
$$

The coefficient functions are $\alpha(t) = 3 + t$ and $\beta(t) = 1$. We have $\alpha(s_0) = 3$, $\alpha(s_1) = 3.5$ and $\beta(s_0) = 1$, $\beta(s_1) = 1$, so

$$
\mathbf{A} = \begin{bmatrix} 3 & 0 \\ 0 & 3.5 \end{bmatrix} \mathbf{E}_{s,t} = \begin{bmatrix} 3 & 0 & 0 \\ 0 & 0 & 3.5 \end{bmatrix}, \quad
\mathbf{B} = \begin{bmatrix} 1 & 0 \\ 0 & 1 \end{bmatrix} \mathbf{E}_{s,t'} = \begin{bmatrix} 1 & 0 \\ 0 & 1 \end{bmatrix}.
$$

Because the initial condition and the first node in each interpolation grid are placed at the same location $t = 0$, the matrices involved in the initial condition are simply

$$
\mathbf{E}_{a,t} = \begin{bmatrix} 1 & 0 & 0 \end{bmatrix}, \quad \mathbf{E}_{a,t'} = \begin{bmatrix} 1 & 0 \end{bmatrix}.
$$

The right-hand side of the differential equation is sampled at the collocation nodes to find $g(s_0) = 4$ and $g(s_1) = 3$. The initial condition is of the form $c_0 u(a) + c_1 u'(a) = d$, with $c_0 = 1$, $c_1 = 0$, and $d = 2$.

The entire linear system is

$$
\left[\begin{array}{ccc|cc}
-1 & 1 & 0 & -3/16 & -1/16 \\
-1 & 0 & 1 & -1/4 & -1/4 \\
\hline
3 & 0 & 0 & 1 & 0 \\
0 & 0 & 3.5 & 0 & 1 \\
\hline
1 & 0 & 0 & 0 & 0
\end{array}\right]
\begin{bmatrix}
p(t_0) \\
p(t_1) \\
p(t_2) \\
\hline
q(t_0') \\
q(t_1')
\end{bmatrix}
=
\begin{bmatrix}
0 \\
0 \\
4 \\
3 \\
2
\end{bmatrix},
$$

and its solution, found by Gaussian elimination, is

$$
\begin{bmatrix}
p(t_0) \\
p(t_1) \\
p(t_2) \\
\hline
q(t_0') \\
q(t_1')
\end{bmatrix}
=
\begin{bmatrix}
2 \\
1.55 \\
1.2 \\
\hline
-2 \\
-1.2
\end{bmatrix}.
$$

The collocation solution is the quadratic interpolating polynomial

$$
\begin{aligned}
p(t) &= \frac{2(t-0.25)(t-0.5)}{(0-0.25)(0-0.5)} + \frac{1.55(t-0)(t-0.5)}{(0.25-0)(0.25-0.5)} + \frac{1.2(t-0)(t-0.25)}{(0.5-0)(0.5-0.25)} \\
&= 0.8t^2 - 2t + 2.
\end{aligned}
$$

Note that p is a polynomial of degree 2; it satisfies the collocation equations $(3 + s_i)p(s_i) + p'(s_i) = 4 - 2s_i$ for $i = 0, 1$; and it satisfies the initial condition $p(0) = 2$. ∎

Collocation for the first-order linear IVP is implemented by ivpl1gen and ivpl1_ in Listing 23.1. The name ivpl1gen is short for "IVP (linear, first-order) on a general grid." In the code, the matrix in (23.8) is constructed as L, and the matrix-vector equation is solved by Gaussian elimination using the MATLAB backslash operator in the line sol = L\rhs.

We adopt a convention of two-letter abbreviations for Greek letters in code: al for α, be for β, ga for γ, la for λ, and so on.

Listing 23.1. Solve a first-order linear IVP by collocation.

```
function [ps,qs] = ivpl1gen(al,be,g,a,c,d,ss,ts,ts_)

% sample coefficient functions and right-hand side of DE
als = arrayfun(al,ss);
bes = arrayfun(be,ss);
gs = arrayfun(g,ss);
% construct integration, resampling, and evaluation matrices
[J,K] = indefinitegen(ts,ts_);
E1 = evalmatrixgen(ss,ts);
E2 = evalmatrixgen(ss,ts_);
Ea1 = evalmatrixgen(a,ts);
Ea2 = evalmatrixgen(a,ts_);
% solve collocation system
[ps,qs] = ivpl1_(als,bes,gs,c,d,J,K,E1,E2,Ea1,Ea2);
```

```
function [ps,qs] = ivpl1_(als,bes,gs,c,d,J,K,E1,E2,Ea1,Ea2)

m = length(als)-1;
A = diag(als)*E1; B = diag(bes)*E2;
c0 = c(1); c1 = c(2);
L = [ J        -K      ;
      A         B      ;
      c0*Ea1 c1*Ea2 ];
rhs = [ zeros(m+1,1); gs; d ];
sol = L\rhs;
ps = sol(1:m+2); qs = sol(m+3:end);
```

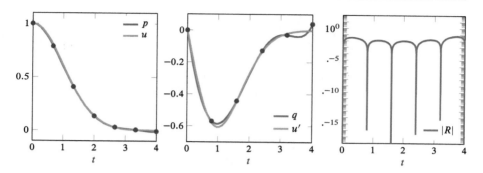

Figure 23.1. *A collocation solution (left) to a first-order IVP, its derivative (middle), and its absolute residual (right).*

The following example illustrates the usage of ivpl1gen. Beginning with this example, we omit most plot formatting options for the sake of brevity.

Example 23.4. Compute a collocation solution to

$$tu(t) + u'(t) = 0, \quad u(0) = 1$$

for $0 \le t \le 4$. Use a uniform grid of degree 5 for the collocation grid and uniform grids of appropriate degrees for **t** and **t′**. Compare the collocation solution with the exact solution. Then plot the absolute residual function and check that it equals zero at the collocation nodes.

Solution. In the usual notation, $\alpha(t) = t$, $\beta(t) = 1$, and $g(t) = 0$, and the domain is $[a, b] = [0, 4]$.

```
>> al = @(t) t; be = @(t) 1; g = @(t) 0;
>> a = 0; b = 4;
```

The initial condition can be expressed as $1 \times u(a) + 0 \times u'(a) = 1$. The coefficients in the left-hand side are recorded in a row vector, and the right-hand side is a scalar.

```
>> c = [1 0]; ua = 1;
```

The grids are placed. In the computer code, grid **t** is called ts, and grid **t′** is called ts_.

```
>> m = 5;
>> ss = griduni([a b],m);
>> ts = griduni([a b],m+1);
>> ts_ = griduni([a b],m);
```

Now, the collocation solution can be computed.

```
>> [ps,qs] = ivpl1gen(al,be,g,a,c,ua,ss,ts,ts_);
>> p = interpgen(ts,ps);
>> q = interpgen(ts_,qs);
```

The collocation solution functions p and q are graphed in the left and middle plots of Figure 23.1.

```
>> ax = newfig(1,3); plotsample(ts,ps); plotfun(p,[a b]);
>> subplot(ax(2)); plotsample(ts_,qs); plotfun(q,[a b]);
```

The exact solution is $u(t) = e^{-t^2/2}$ (see Example 22.3). This function and its derivative are added to the plots.

```
>> subplot(ax(1)); plotfun(@(t) exp(-t^2/2),[a b]);
>> subplot(ax(2)); plotfun(@(t) -t*exp(-t^2/2),[a b]);
```

Note that p and q do not agree perfectly with u and u' at the interpolation nodes. Instead, the residual $R(t) = g(t) - \alpha(t)p(t) - \beta(t)q(t)$ equals zero at the collocation nodes; see the plot on the right.

```
>> R = @(t) g(t)-al(t)*p(t)-be(t)*q(t);
>> subplot(ax(3)); ylog; ylim([1e-18 1e2]);
>> plotfun(@(t) abs(R(t)),[a b]);
```                                                                        ∎

23.2 ▪ Second-order problems

Collocation for the second-order IVP

$$\alpha(t)u(t) + \beta(t)u'(t) + \gamma(t)u''(t) = g(t), \quad \begin{aligned} c_{10}u(a) + c_{11}u'(a) + c_{12}u''(a) &= d_1, \\ c_{20}u(a) + c_{21}u'(a) + c_{22}u''(a) &= d_2 \end{aligned}$$
$$(23.9)$$

is similar.

Definition 23.5. *Given a sequence of distinct collocation nodes* $s_0, \ldots, s_m$ *on the real line, a* polynomial collocation solution *for the second-order linear differential equation*

$$\alpha(t)u(t) + \beta(t)u'(t) + \gamma(t)u''(t) = g(t)$$

is a polynomial $p(t)$ *of degree at most* $m + 2$ *for which*

$$\alpha(s_i)p(s_i) + \beta(s_i)p'(s_i) + \gamma(s_i)p''(s_i) = g(s_i), \quad i = 0, \ldots, m. \qquad (23.10)$$

For a collocation solution, the residual function

$$R(t) = g(t) - \alpha(t)p(t) - \beta(t)p'(t) - \gamma(t)p''(t) \qquad (23.11)$$

equals zero at each collocation node.

If a polynomial collocation solution p also satisfies the initial conditions

$$c_{10}p(a) + c_{11}p'(a) + c_{12}p''(a) = d_1,$$
$$c_{20}p(a) + c_{21}p'(a) + c_{22}p''(a) = d_2,$$

then it is called a collocation solution for the IVP (23.9).

The computation of a second-order collocation solution proceeds much as before. Let p be a polynomial of degree at most $m + 2$; let $q = p'$ and $r = p''$; and sample the three polynomials: $\mathbf{p} = p|_t$ on a degree-$(m + 2)$ grid $\mathbf{t}$, $\mathbf{q} = q|_{t'}$ on a degree-$(m + 1)$ grid $\mathbf{t}'$, and $\mathbf{r} = r|_{t''}$ on a degree-m grid $\mathbf{t}''$. The relations $q = p'$ and $r = q'$ are equivalent to

$$\mathbf{J}_t\mathbf{p} = \hat{\mathbf{K}}_{t,t'}\mathbf{q}$$

and

$$\mathbf{J}_{t'}\mathbf{q} = \hat{\mathbf{K}}_{t',t''}\mathbf{r}.$$

Let $\text{diag}(x_1, \ldots, x_n)$ denote the diagonal matrix with entries $x_1, \ldots, x_n$ from the top-left to the bottom-right along its main diagonal, and set

$$\mathbf{A} = \text{diag}(\alpha(s_0), \ldots, \alpha(s_m))\mathbf{E}_{s,t}, \qquad (23.12)$$
$$\mathbf{B} = \text{diag}(\beta(s_0), \ldots, \beta(s_m))\mathbf{E}_{s,t'}, \qquad (23.13)$$
$$\mathbf{C} = \text{diag}(\gamma(s_0), \ldots, \gamma(s_m))\mathbf{E}_{s,t''}, \qquad (23.14)$$

in which $\mathbf{s} = (s_0, \ldots, s_m)$. Then the collocation equations are equivalent to

$$\mathbf{Ap} + \mathbf{Bq} + \mathbf{Cr} = \mathbf{g},$$

with $\mathbf{g} = g|_\mathbf{s}$. Finally, the initial conditions are equivalent to

$$c_{10}\mathbf{E}_{a,\mathbf{t}}\mathbf{p} + c_{11}\mathbf{E}_{a,\mathbf{t}'}\mathbf{q} + c_{12}\mathbf{E}_{a,\mathbf{t}''}\mathbf{r} = d_1,$$
$$c_{20}\mathbf{E}_{a,\mathbf{t}}\mathbf{p} + c_{21}\mathbf{E}_{a,\mathbf{t}'}\mathbf{q} + c_{22}\mathbf{E}_{a,\mathbf{t}''}\mathbf{r} = d_2.$$

The proof of the following theorem is very similar to the one for Theorem 23.2.

Theorem 23.6. *Let p, q, and r be polynomials of degrees $m + 2$, $m + 1$, and m, respectively, and let $\mathbf{p} = p|_\mathbf{t}$, $\mathbf{q} = q|_{\mathbf{t}'}$, and $\mathbf{r} = r|_{\mathbf{t}''}$. Then the following are equivalent:*

1. *p is a collocation solution to the IVP (23.9) on collocation nodes $s_0, \ldots, s_m$, and $q = p'$ and $r = p''$.*

2. *The matrix-vector equation*

$$\begin{bmatrix} \mathbf{J}_\mathbf{t} & -\hat{\mathbf{K}}_{\mathbf{t},\mathbf{t}'} & \mathbf{0} \\ \mathbf{0} & \mathbf{J}_{\mathbf{t}'} & -\hat{\mathbf{K}}_{\mathbf{t}',\mathbf{t}''} \\ \mathbf{A} & \mathbf{B} & \mathbf{C} \\ c_{10}\mathbf{E}_{a,\mathbf{t}} & c_{11}\mathbf{E}_{a,\mathbf{t}'} & c_{12}\mathbf{E}_{a,\mathbf{t}''} \\ c_{20}\mathbf{E}_{a,\mathbf{t}} & c_{21}\mathbf{E}_{a,\mathbf{t}'} & c_{22}\mathbf{E}_{a,\mathbf{t}''} \end{bmatrix} \begin{bmatrix} \mathbf{p} \\ \mathbf{q} \\ \mathbf{r} \end{bmatrix} = \begin{bmatrix} \mathbf{0} \\ \mathbf{0} \\ \mathbf{g} \\ d_1 \\ d_2 \end{bmatrix} \qquad (23.15)$$

holds.

Routine `ivpl2gen` computes a collocation solution to a second-order problem. The implementation is omitted here, but the usage is illustrated by the following example.

Example 23.7. Solve the harmonic oscillator IVP

$$\pi^2 u(t) + u''(t) = 0, \quad u(0) = 1, \quad u'(0) = 0$$

on $0 \le t \le 2$ by collocation. Use a uniform grid of degree 4 for the collocation nodes and uniform grids of appropriate degrees for $\mathbf{t}$, $\mathbf{t}'$, and $\mathbf{t}''$. Plot the solution and the absolute residual. Check that the residual equals zero at the collocation nodes.

Solution. The problem is stated.

```
>> al = @(t) pi^2; be = @(t) 0; ga = @(t) 1; g = @(t) 0;
>> a = 0; b = 2;
>> ua = 1; va = 0;
```

The grids are set.

```
>> m = 4;
>> ss = griduni([a b],m);
>> ts = griduni([a b],m+2);
>> ts_ = griduni([a b],m+1);
>> ts__ = griduni([a b],m);
```

The problem is solved numerically.

```
>> [ps,qs,rs] = ivpl2gen(al,be,ga,g,a,[ 1 0 0; 0 1 0 ],[ua; va] ...
                ,ss,ts,ts_,ts__);
>> p = interpgen(ts,ps);
>> q = interpgen(ts_,qs);
>> r = interpgen(ts__,rs);
```

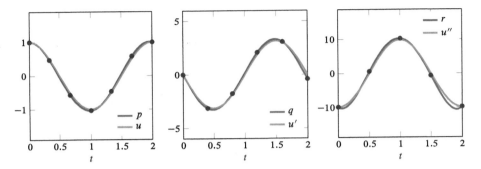

Figure 23.2. *A collocation solution (left) to a second-order IVP, its first derivative (middle), and its second derivative (right).*

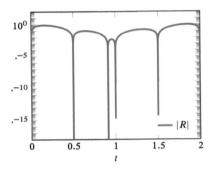

Figure 23.3. *The absolute residual for the collocation solution of Figure* 23.2.

The numerical solution is plotted and compared to the exact solution $u(t) = \cos(\pi t)$ in Figure 23.2.

```
>> ax = newfig(1,3); plotsample(ts,ps); plotfun(p,[a b]);
>> plotfun(@(t) cos(pi*t),[a b]);
>> subplot(ax(2)); plotsample(ts_,qs); plotfun(q,[a b]);
>> plotfun(@(t) -pi*sin(pi*t),[a b]);
>> subplot(ax(3)); plotsample(ts__,rs); plotfun(r,[a b]);
>> plotfun(@(t) -pi^2*cos(pi*t),[a b]);
```

The absolute residual is graphed in Figure 23.3. Notice that it falls to zero at the collocation nodes 0, 0.5, 1, 1.5, and 2 as expected. There is an additional zero near $t = 0.9$ by happenstance.

```
>> R = @(t) g(t)-al(t)*p(t)-be(t)*q(t)-ga(t)*r(t);
>> newfig; ylog; ylim([1e-18 1e2]);
>> plotfun(@(t) abs(R(t)),[a b]);
```

∎

Notes

Many numerical methods for differential equations construct a sequence of points that fall near the graph of the solution. Collocation methods by contrast produce continuous function approximations.

Collocation methods were introduced by Kantorovich in 1934 for BVPs [47, p. 427]. Atkinson's book situates collocation in the broader setting of projection methods [2], and Brunner's book provides additional history and references on collocation [10]. One reason for focusing on collocation in this book is to set up a smooth transition from IVPs to BVPs (see Exercise 14).

We should note, however, that collocation methods are not always the fastest methods for IVPs, particularly for "nonstiff" problems.

The discrete systems presented here use integration matrices instead of differentiation matrices. Greengard advocated the use of integral equations in place of differential equations in 1991 [37], emphasizing the *ill conditioning* of many differentiation matrices. On the other hand, replacing derivatives by integrals is not the only effective approach. The ultraspherical method of Olver and Townsend [59], for example, uses well-conditioned differentiation.

In the collocation method presented here, the solution p is sampled on a grid of degree $m + 1$, but the collocation nodes form a grid of degree m. This gap is bridged in part through resampling matrices, as in the method of Driscoll and Hale [25].

Collocation methods have gained additional attention with the rise of Chebyshev technology in practice. Chebyshev collocation is covered in Chapter 27. Additional references are listed at the end of that chapter.

Exercises

Exercises 1–4: A first-order linear IVP is given. A collocation solution to the IVP is determined by a linear system of the form (23.8). Construct the linear system by hand, using the specified collocation and interpolation grids.

1. $t^2 u + u' = t + 1, u(0) = 2; \mathbf{s} = \mathbf{t}' = (1/2), \mathbf{t} = (0, 1)$
2. $u + t u' = 1/t, u(1) = 5; \mathbf{s} = \mathbf{t}' = (1.25), \mathbf{t} = (1, 1.5)$
3. $u + (1 + t^2) u' = t^2, u(0) = -1; \mathbf{s} = \mathbf{t}' = (0, 1),$ and $\mathbf{t} = (0, 1/2, 1)$
4. $2u + (1 + 1/t) u' = -1, u'(1) = 1/3; \mathbf{s} = \mathbf{t}' = (1, 3),$ and $\mathbf{t} = (1, 2, 3)$

Exercises 5–8: Solve the IVP on the given domain using `ivpl1gen` or `ivpl2gen`. Use a uniform collocation grid of degree m and uniform interpolation grids of appropriate degrees. Plot the approximate solution and the absolute residual, and confirm that the residual equals zero at the collocation nodes.

5. $e^{-t} u + u' = 0, u(0) = 1; 0 \le t \le 2; m = 6$
6. $u + t u' = -1, u(1) = 1; 1 \le t \le 5; m = 8$
7. $4u + u' + 2u'' = 0, u(0) = -4, u'(0) = 0; 0 \le t \le 5; m = 8$
8. $u + u'' = \sin(2t), u(0) = 0, u'(0) = 0; 0 \le t \le 4; m = 10$

9. The IVP
$$\alpha(t) u + \beta(t) u' = g(t), \quad u(a) = u_a$$
   is solved by collocation using a single collocation node $s_0 = a$. The collocation solution $p(t)$ is sampled at $t_0 = a$ and $t_1 = b$, and its derivative is sampled at $t_0' = (a + b)/2$. Find an explicit formula for $p(b)$.

10. The IVP
$$\alpha(t) u + \beta(t) u' = g(t), \quad u(a) = u_a$$
    is solved by collocation using a single collocation node $s_0 = b$. The collocation solution $p(t)$ is sampled at $t_0 = a$ and $t_1 = b$, and its derivative is sampled at $t_0' = (a + b)/2$. Find an explicit formula for $p(b)$.

11. Find an IVP of the form
$$\alpha(t) u + u' = g(t), \quad u(a) = u_a$$

that has a unique solution on an interval $[a, b]$ but for which no collocation solution exists when the nodes are $s_0 = (a + b)/2$, $t_0 = a$, $t_1 = b$, and $t_0' = (a + b)/2$.

12. Find an IVP of the form

$$\alpha(t)u + u' = g(t), \quad u(a) = u_a$$

that has a unique solution on an interval $[a, b]$ but for which there are infinitely many collocation solutions when the nodes are $s_0 = (a + b)/2$, $t_0 = a$, $t_1 = b$, and $t_0' = (a + b)/2$.

13. The collocation system (23.8) enforces the equation $\mathbf{J}_t\mathbf{p} = \hat{\mathbf{K}}_{t,t'}\mathbf{q}$. Duplicate and modify ivpl1_ and ivpl1gen to instead enforce the equation $\mathbf{D}_{t',t}\mathbf{p} = \mathbf{q}$. Check the new implementation on the problem of Example 23.4.

14. Consider the second-order linear BVP

$$\alpha(x)u + \beta(x)u' + \gamma(x)u'' = g(x), \quad c_{10}u(a) + c_{11}u'(a) + c_{12}u''(a) = d_1,$$
$$c_{20}u(b) + c_{21}u'(b) + c_{22}u''(b) = d_2. \tag{23.16}$$

Note that there is one boundary condition at each endpoint of an interval $[a, b]$, rather than a pair of initial conditions at $x = a$. Duplicate and modify ivpl2_ and ivpl2gen to solve this BVP. Demonstrate your implementation on the hanging-cable problem $u - u'' = 0$, $u(-1) = 1$, $u(1) = 1$.

15. Develop routines ivpl3_ and ivpl3gen for solving the third-order IVP

$$\alpha(t)u + \beta(t)u' + \gamma(t)u'' + \delta(t)u''' = g(t),$$
$$c_{10}u(a) + c_{11}u'(a) + c_{12}u''(a) + c_{13}u'''(a) = d_1,$$
$$c_{20}u(a) + c_{21}u'(a) + c_{22}u''(a) + c_{23}u'''(a) = d_2,$$
$$c_{30}u(a) + c_{31}u'(a) + c_{32}u''(a) + c_{33}u'''(a) = d_3,$$

by collocation. Test your pair of routines on the problem

$$u + 2u' + u'' + 2u''' = 0, \quad u(0) = 5, \quad u'(0) = 0, \quad u''(0) = 0$$

for $0 \le t \le 4\pi$, comparing with the exact solution $u(t) = 4e^{-t/2} + \cos t + 2\sin t$.

16. Let $A(t)$ and $B(t)$ be r-by-r matrix-valued functions of t, and let $g(t)$ be an r-by-1 vector-valued function of t. Consider the vector-valued IVP

$$A(t)u(t) + B(t)u'(t) = g(t), \quad c_0u(a) + c_1u'(a) = d. \tag{23.17}$$

Let $p(t) = (p_1(t), \ldots, p_r(t))$ be a vector-valued function in which each p_k is a polynomial of degree at most $m + 1$, and let $q(t) = (q_1(t), \ldots, q_r(t))$ be a vector-valued function in which each q_k is a polynomial of degree at most m. Let $\mathbf{t} = (t_0, \ldots, t_{m+1})$ and $\mathbf{t}' = (t_0', \ldots, t_m')$ be interpolation grids and sample

$$\mathbf{p} = \begin{bmatrix} p_1(t_0) & \cdots & p_1(t_{m+1}) \\ \vdots & \ddots & \vdots \\ p_r(t_0) & \cdots & p_r(t_{m+1}) \end{bmatrix}, \quad \mathbf{q} = \begin{bmatrix} q_1(t_0') & \cdots & q_1(t_m') \\ \vdots & \ddots & \vdots \\ q_r(t_0') & \cdots & q_r(t_m') \end{bmatrix}. \tag{23.18}$$

Also, let $s_0, \ldots, s_m$ be a sequence of collocation nodes and sample

$$
\mathbf{g} = \begin{bmatrix} g_1(s_0) & \cdots & g_1(s_m) \\ \vdots & \ddots & \vdots \\ g_r(s_0) & \cdots & g_r(s_m) \end{bmatrix}.
$$

Design a matrix $\mathbf{L}$ such that

$$
\mathbf{L} \begin{bmatrix} \mathrm{vec}(\mathbf{p}) \\ \mathrm{vec}(\mathbf{q}) \end{bmatrix} = \begin{bmatrix} \mathbf{0} \\ \mathrm{vec}(\mathbf{g}) \\ u_a \end{bmatrix} \tag{23.19}
$$

if and only if (1) p satisfies the collocation equations

$$
A(s_i)p(s_i) + B(s_i)p'(s_i) = g(s_i), \quad i = 0, \ldots, m;
$$

(2) the initial condition $c_0 p(a) + c_1 p'(a) = d$ is satisfied; and (3) $p'_k = q_k$ for $k = 1, \ldots, r$. Justify. *Hint.* See Exercises 17–19 in Chapter 12 and Exercise 17 in Chapter 14.

17. Develop a MATLAB routine to solve (23.19) and thus compute a collocation solution to (23.17). The routine should return $\mathbf{p}$ and $\mathbf{q}$ of the form (23.18). Demonstrate your implementation by solving

$$
\begin{bmatrix} 1/2 & -1 \\ 1 & 1/2 \end{bmatrix} \begin{bmatrix} u_1 \\ u_2 \end{bmatrix} + \begin{bmatrix} 1 & 0 \\ 0 & 1 \end{bmatrix} \begin{bmatrix} u'_1 \\ u'_2 \end{bmatrix} = \begin{bmatrix} 0 \\ 0 \end{bmatrix}, \quad \begin{bmatrix} u_1(0) \\ u_2(0) \end{bmatrix} = \begin{bmatrix} 1 \\ 0 \end{bmatrix}
$$

on $0 \le t \le 10$. Compare with the exact solution $u_1(t) = e^{-t/2} \cos t$, $u_2(t) = -e^{-t/2} \sin t$.

Chapter 24

Collocation accuracy

When the IVP

$$\alpha(t)u(t) + \beta(t)u'(t) = g(t), \quad c_0 u(a) + c_1 u'(a) = d \qquad (24.1)$$

is solved approximately by a function $p(t)$, we ask two questions concerning accuracy:

1. Is the numerical solution $p(t)$ close to the exact solution $u(t)$ in the sense of producing a small error $u(t) - p(t)$?

2. Does the numerical solution $p(t)$ nearly satisfy the differential equation in the sense of producing a small residual $R(t) = g(t) - \alpha(t)p(t) - \beta(t)p'(t)$?

In this chapter, we develop bounds on both measures.

24.1 ▪ Accuracy analysis for first-order problems

The notation of the previous chapter for first-order collocation is used below. In particular, suppose that

$$\alpha(t)u(t) + \beta(t)u'(t) = g(t), \quad c_0 u(a) + c_1 u'(a) = d \qquad (24.2)$$

has a unique solution $u(t)$ and that $p(t)$ is a polynomial collocation solution satisfying

$$\alpha(s_i)p(s_i) + \beta(s_i)p'(s_i) = g(s_i), \quad i = 0, \ldots, m,$$

and

$$c_0 p(a) + c_1 p'(a) = d.$$

The quality of the approximation will be measured on an interval $[a, b]$ that contains all of the grid nodes.

Let

$$\mathbf{u} = u|_{\mathbf{t}}, \quad \mathbf{v} = u'|_{\mathbf{t}'}, \quad \mathbf{p} = p|_{\mathbf{t}}, \quad \mathbf{q} = p'|_{\mathbf{t}'}.$$

Stack $\mathbf{p}$ and $\mathbf{q}$ into a single column vector $\mathbf{x}$, and stack $\mathbf{u}$ and $\mathbf{v}$ into a single column vector $\mathbf{y}$:

$$\mathbf{x} = \begin{bmatrix} \mathbf{p} \\ \mathbf{q} \end{bmatrix}, \quad \mathbf{y} = \begin{bmatrix} \mathbf{u} \\ \mathbf{v} \end{bmatrix}.$$

Also, let

$$\mathbf{L} = \begin{bmatrix} \mathbf{J_t} & -\hat{\mathbf{K}}_{t,t'} \\ \mathbf{A} & \mathbf{B} \\ c_0\mathbf{E}_{a,t} & c_1\mathbf{E}_{a,t'} \end{bmatrix}, \quad \mathbf{b} = \begin{bmatrix} \mathbf{0} \\ \mathbf{g} \\ d \end{bmatrix}. \tag{24.3}$$

The collocation solution satisfies

$$\mathbf{Lx} = \mathbf{b} \tag{24.4}$$

(Theorem 23.2). Our plan, informally stated, is to show that

$$\mathbf{Ly} \approx \mathbf{b}, \tag{24.5}$$

i.e., that the exact solution to the IVP nearly solves the discretized linear system. Then we argue that, for sufficiently well-behaved problems, (24.4)–(24.5) imply $\mathbf{x} \approx \mathbf{y}$ and therefore $\mathbf{p} \approx \mathbf{u}$ and $\mathbf{q} \approx \mathbf{v}$. With this approach in mind, define

$$\boldsymbol{\varepsilon} = \mathbf{b} - \mathbf{Ly}.$$

We plan to show that $\boldsymbol{\varepsilon}$ has small entries.

The bound on $\|\boldsymbol{\varepsilon}\|_\infty$ in the following lemma involves two interpolating polynomials: the degree-$(m + 1)$ polynomial $\tilde{p}$ satisfying

$$\tilde{p}(t_i) = u(t_i), \quad i = 0, \ldots, m + 1,$$

and the degree-m polynomial $\tilde{q}$ satisfying

$$\tilde{q}(t_i') = u'(t_i'), \quad i = 0, \ldots, m.$$

The lemma relates the residual norm $\|\boldsymbol{\varepsilon}\|_\infty$ to the interpolation errors $\|u - \tilde{p}\|_\infty$ and $\|u' - \tilde{q}\|_\infty$. It shows that if the grids $\mathbf{t}$ and $\mathbf{t}'$ are sufficiently fine to enable low interpolation error, then they are also fine enough to produce a small $\|\boldsymbol{\varepsilon}\|_\infty$.

Lemma 24.1. *We have*

$$\|\boldsymbol{\varepsilon}\|_\infty \le C_1\|u - \tilde{p}\|_\infty + C_2\|u' - \tilde{q}\|_\infty, \tag{24.6}$$

in which $C_1 = 2 \vee \|\alpha\|_\infty \vee |c_0|$ *and* $C_2 = (b - a) \vee \|\beta\|_\infty \vee |c_1|$.

Proof. We have

$$\boldsymbol{\varepsilon} = \begin{bmatrix} -\mathbf{J_t}\mathbf{u} + \hat{\mathbf{K}}_{t,t'}\mathbf{v} \\ \mathbf{g} - \mathbf{Au} - \mathbf{Bv} \\ d - c_0\mathbf{E}_{a,t}\mathbf{u} - c_1\mathbf{E}_{a,t'}\mathbf{v} \end{bmatrix}.$$

By Theorem 14.16, the entries of $-\mathbf{J_t}\mathbf{u} + \hat{\mathbf{K}}_{t,t'}\mathbf{v}$ are bounded in absolute value by $2\|u - \tilde{p}\|_\infty + (b - a)\|u' - \tilde{q}\|_\infty$.

The entries of $\mathbf{g} - \mathbf{Au} - \mathbf{Bv}$ are $g(s_i) - \alpha(s_i)\tilde{p}(s_i) - \beta(s_i)\tilde{q}(s_i)$, $i = 0, \ldots, m$. Because u satisfies the differential equation at all t,

$$g(s_i) - \alpha(s_i)u(s_i) - \beta(s_i)u'(s_i) = 0.$$

Adding $\alpha(s_i)(u(s_i) - \tilde{p}(s_i)) + \beta(s_i)(u'(s_i) - \tilde{q}(s_i))$ to both sides gives

$$g(s_i) - \alpha(s_i)\tilde{p}(s_i) - \beta(s_i)\tilde{q}(s_i) = \alpha(s_i)(u(s_i) - \tilde{p}(s_i)) + \beta(s_i)(u'(s_i) - \tilde{q}(s_i)).$$

The left-hand side is the ith entry of $\mathbf{g} - \mathbf{A}u - \mathbf{B}v$, and the right-hand side is bounded in magnitude by $\|\alpha\|_\infty\|u - \tilde{p}\|_\infty + \|\beta\|_\infty\|u' - \tilde{q}\|_\infty$.

Because u satisfies the initial condition,

$$d - c_0 u(a) - c_1 u'(a) = 0.$$

Adding $c_0(u(a) - \tilde{p}(a)) + c_1(u'(a) - \tilde{q}(a))$ to both sides gives

$$d - c_0 \tilde{p}(a) - c_1 \tilde{q}(a) = c_0(u(a) - \tilde{p}(a)) + c_1(u'(a) - \tilde{q}(a)).$$

The left-hand side is the last entry of $\boldsymbol{\varepsilon}$, and the right-hand side is bounded in absolute value by $|c_0|\,\|u - \tilde{p}\|_\infty + |c_1|\,\|u' - \tilde{q}\|_\infty$. □

The following theorem proceeds to the errors $\mathbf{u} - \mathbf{p}$ and $\mathbf{v} - \mathbf{q}$. These are guaranteed to be small if both $\|\boldsymbol{\varepsilon}\|_\infty$ is small and the system of collocation equations is well conditioned. (See Chapter 21 for an introduction to condition numbers.)

Theorem 24.2. *We have*

$$\|\mathbf{u} - \mathbf{p}\|_\infty \le \kappa_{\text{abs}}\|\boldsymbol{\varepsilon}\|_\infty, \tag{24.7}$$

$$\|\mathbf{v} - \mathbf{q}\|_\infty \le \kappa_{\text{abs}}\|\boldsymbol{\varepsilon}\|_\infty, \tag{24.8}$$

in which $\kappa_{\text{abs}} = \|\mathbf{L}^{-1}\|_\infty$ *is the absolute condition number for* $\mathbf{L}$.

Proof. If $\mathbf{L}$ is not invertible, then $\kappa_{\text{abs}} = \infty$. For the rest of the proof, assume that $\mathbf{L}$ is invertible.

We find

$$\mathbf{L}(\mathbf{y} - \mathbf{x}) = \mathbf{L}\mathbf{y} - \mathbf{L}\mathbf{x} = (\mathbf{b} - \boldsymbol{\varepsilon}) - \mathbf{b} = -\boldsymbol{\varepsilon}.$$

Multiplying on the left by $\mathbf{L}^{-1}$ gives $\mathbf{y} - \mathbf{x} = -\mathbf{L}^{-1}\boldsymbol{\varepsilon}$, which implies

$$\|\mathbf{y} - \mathbf{x}\|_\infty \le \|\mathbf{L}^{-1}\|_\infty\|\boldsymbol{\varepsilon}\|_\infty = \kappa_{\text{abs}}\|\boldsymbol{\varepsilon}\|_\infty. \tag{24.9}$$

Because the entries of $\mathbf{u} - \mathbf{p}$ and $\mathbf{v} - \mathbf{q}$ are also entries of $\mathbf{y} - \mathbf{x}$, the infinity norms $\|\mathbf{u} - \mathbf{p}\|_\infty$ and $\|\mathbf{v} - \mathbf{q}\|_\infty$ are bounded by $\kappa_{\text{abs}}\|\boldsymbol{\varepsilon}\|_\infty$ as well. □

The previous theorem provides bounds on the discrete errors $\mathbf{u} - \mathbf{p}$ and $\mathbf{v} - \mathbf{q}$. Just as important are $u - p$ and $u' - p'$, which measure the collocation error over the continuous interval $[a, b]$, rather than at a finite set of nodes.

Theorem 24.3. *Let* $\Lambda_{\mathbf{t}}$ *and* $\Lambda_{\mathbf{t'}}$ *be the Lebesgue constants for* $\mathbf{t}$ *and* $\mathbf{t'}$, *respectively. Define* $\tilde{p}$ *and* $\tilde{q}$ *as above. We have*

$$\|u - p\|_\infty \le \|u - \tilde{p}\|_\infty + \Lambda_{\mathbf{t}}\|\mathbf{u} - \mathbf{p}\|_\infty \tag{24.10}$$

and

$$\|u' - p'\|_\infty \le \|u' - \tilde{q}\|_\infty + \Lambda_{\mathbf{t'}}\|\mathbf{v} - \mathbf{q}\|_\infty. \tag{24.11}$$

Proof. By the triangle inequality,

$$\|u - p\|_\infty \le \|u - \tilde{p}\|_\infty + \|\tilde{p} - p\|_\infty.$$

Apply Theorem 6.11 to $\|p - \tilde{p}\|_\infty$, taking $\delta = \|\mathbf{u} - \mathbf{p}\|_\infty$.

The proof of the second inequality is similar.   □

The continuous residual $R(t) = g(t) - \alpha(t)p(t) - \beta(t)p'(t)$ can be bounded similarly.

Theorem 24.4. *Let $u(t)$ be any solution to the differential equation*

$$\alpha(t)u(t) + \beta(t)u'(t) = g(t).$$

Then the collocation residual $R(t) = g(t) - \alpha(t)p(t) - \beta(t)p'(t)$ satisfies

$$\|R\|_\infty \leq \|\alpha\|_\infty \|u - p\|_\infty + \|\beta\|_\infty \|u' - p'\|_\infty. \qquad (24.12)$$

Proof. Because $u(t)$ solves the differential equation $\alpha(t)u(t) + \beta(t)u'(t) = g(t)$, we have

$$g(t) - \alpha(t)u(t) - \beta(t)u'(t) = 0$$

for all t. Adding $\alpha(t)(u(t) - p(t))$ and $\beta(t)(u'(t) - p'(t))$ to both sides gives

$$g(t) - \alpha(t)p(t) - \beta(t)p'(t) = \alpha(t)(u(t) - p(t)) + \beta(t)(u'(t) - p'(t)).$$

The left-hand side is the residual. Thus,

$$|R(t)| \leq |\alpha(t)||u(t) - p(t)| + |\beta(t)||u'(t) - p'(t)|. \qquad □$$

Let's summarize this section. Given a well-conditioned problem, by choosing sufficiently fine and well-behaved grids,

1. $\|\varepsilon\|_\infty$ must be small by Lemma 24.1;

2. and thus $\|\mathbf{u} - \mathbf{p}\|_\infty$ and $\|\mathbf{v} - \mathbf{q}\|_\infty$ must be small by Theorem 24.2;

3. and thus $\|u - p\|_\infty$ and $\|u' - p'\|_\infty$ must be small by Theorem 24.3;

4. and thus $\|R\|_\infty$ must be small by Theorem 24.4.

This chapter began with two questions: (1) Is the collocation solution close to the correct solution? and (2) Does the collocation solution produce a small residual? For first-order problems, we can now answer both questions with a qualified yes. When the IVP is well conditioned and the grids are sufficiently fine and well behaved, the residuals and errors are guaranteed to be small.

24.2 ▪ Accuracy analysis for second-order problems

For the second-order problem

$$\alpha(t)u(t) + \beta(t)u'(t) + \gamma(t)u''(t) = g(t), \quad c_{10}u(a) + c_{11}u'(a) + c_{12}u''(a) = d_1,$$
$$c_{20}u(a) + c_{21}u'(a) + c_{22}u''(a) = d_2,$$
$$(24.13)$$

let $\mathbf{u} = u|_{\mathbf{t}}$, $\mathbf{v} = u'|_{\mathbf{t'}}$, $\mathbf{w} = u''|_{\mathbf{t''}}$. The collocation solution p is determined by the samples $\mathbf{p} = p|_{\mathbf{t}}$, $\mathbf{q} = p'|_{\mathbf{t'}}$, $\mathbf{r} = p''|_{\mathbf{t''}}$, which satisfy the equation $\mathbf{Lx} = \mathbf{b}$ with

$$
\mathbf{L} = \begin{bmatrix} \mathbf{J}_{\mathbf{t}} & -\hat{\mathbf{K}}_{\mathbf{t},\mathbf{t'}} & \mathbf{0} \\ \mathbf{0} & \mathbf{J}_{\mathbf{t'}} & -\hat{\mathbf{K}}_{\mathbf{t'},\mathbf{t''}} \\ \mathbf{A} & \mathbf{B} & \mathbf{C} \\ c_{10}\mathbf{E}_{a,\mathbf{t}} & c_{11}\mathbf{E}_{a,\mathbf{t'}} & c_{12}\mathbf{E}_{a,\mathbf{t''}} \\ c_{20}\mathbf{E}_{a,\mathbf{t}} & c_{21}\mathbf{E}_{a,\mathbf{t'}} & c_{22}\mathbf{E}_{a,\mathbf{t''}} \end{bmatrix}, \quad \mathbf{x} = \begin{bmatrix} \mathbf{p} \\ \mathbf{q} \\ \mathbf{r} \end{bmatrix}, \quad \mathbf{b} = \begin{bmatrix} \mathbf{0} \\ \mathbf{0} \\ \mathbf{g} \\ d_1 \\ d_2 \end{bmatrix}.
$$

Let

$$
\mathbf{y} = \begin{bmatrix} \mathbf{u} \\ \mathbf{v} \\ \mathbf{w} \end{bmatrix}
$$

and

$$
\boldsymbol{\varepsilon} = \mathbf{b} - \mathbf{Ly}.
$$

The following results, analogous to earlier ones for the first-order problem, are stated without proof.

Lemma 24.5. *Suppose* $\|\alpha\|_\infty$, $\|\beta\|_\infty$, *and* $\|\gamma\|_\infty$ *are finite. Let* $\tilde{p}$ *be the degree-$(m+2)$ polynomial that interpolates u at the nodes in* $\mathbf{t}$, *let* $\tilde{q}$ *be the degree-$(m+1)$ polynomial that interpolates u' at the nodes in* $\mathbf{t'}$, *and let* $\tilde{r}$ *be the degree-m polynomial that interpolates u'' at the nodes in* $\mathbf{t''}$. *Then*

$$
\|\boldsymbol{\varepsilon}\|_\infty \le C_1 \|u - \tilde{p}\|_\infty + C_2 \|u' - \tilde{q}\|_\infty + C_3 \|u'' - \tilde{r}\|_\infty
$$

for coefficients C_1, C_2, *and* C_3 *that are independent of the choice of grids.*

Theorem 24.6. *We have*

$$
\|\mathbf{u} - \mathbf{p}\|_\infty \le \kappa_{\text{abs}} \|\boldsymbol{\varepsilon}\|_\infty, \tag{24.14}
$$

$$
\|\mathbf{v} - \mathbf{q}\|_\infty \le \kappa_{\text{abs}} \|\boldsymbol{\varepsilon}\|_\infty, \tag{24.15}
$$

$$
\|\mathbf{w} - \mathbf{r}\|_\infty \le \kappa_{\text{abs}} \|\boldsymbol{\varepsilon}\|_\infty, \tag{24.16}
$$

in which κ_{abs} *is the absolute condition number for* $\mathbf{L}$.

Theorem 24.7. *Let* $\Lambda_{\mathbf{t}}$, $\Lambda_{\mathbf{t'}}$, *and* $\Lambda_{\mathbf{t''}}$ *be the Lebesgue constants for* $\mathbf{t}$, $\mathbf{t'}$, *and* $\mathbf{t''}$, *respectively. Define* $\tilde{p}$, $\tilde{q}$, *and* $\tilde{r}$ *as in the previous theorem. We have*

$$
\|u - p\|_\infty \le \|u - \tilde{p}\|_\infty + \Lambda_{\mathbf{t}} \|\mathbf{u} - \mathbf{p}\|_\infty, \tag{24.17}
$$

$$
\|u' - p'\|_\infty \le \|u' - \tilde{q}\|_\infty + \Lambda_{\mathbf{t'}} \|\mathbf{v} - \mathbf{q}\|_\infty, \tag{24.18}
$$

$$
\|u'' - p''\|_\infty \le \|u'' - \tilde{r}\|_\infty + \Lambda_{\mathbf{t''}} \|\mathbf{w} - \mathbf{r}\|_\infty. \tag{24.19}
$$

Theorem 24.8. *The residual function* $R(t) = g(t) - \alpha(t)p(t) - \beta(t)p'(t) - \gamma(t)p''(t)$ *satisfies*

$$
\|R\|_\infty \le \|\alpha\|_\infty \|u - p\|_\infty + \|\beta\|_\infty \|u' - p'\|_\infty + \|\gamma\|_\infty \|u'' - p''\|_\infty \tag{24.20}
$$

for any solution u to the differential equation $\alpha(t)u(t) + \beta(t)u'(t) + \gamma(t)u''(t) = g(t)$.

Again, for a well-conditioned problem, the discrete error, continuous error, and residual can all be made small by using sufficiently fine and well-behaved grids.

Notes

Theorems 24.2 and 24.6 provide bounds on absolute collocation error using the absolute condition number. It is also possible to prove bounds on relative error, such as the following analogue to (24.9):

$$\frac{\|\mathbf{y} - \mathbf{x}\|_\infty}{\|\mathbf{x}\|_\infty} \leq \kappa_{\text{rel}} \frac{\|\boldsymbol{\varepsilon}\|}{\|\mathbf{b}\|_\infty}.$$

This in turn implies

$$\frac{\|\mathbf{u} - \mathbf{p}\|_\infty}{\|\mathbf{p}\|_\infty \vee \|\mathbf{q}\|_\infty} \leq \kappa_{\text{rel}} \frac{\|\boldsymbol{\varepsilon}\|_\infty}{\|\mathbf{b}\|_\infty}, \qquad \frac{\|\mathbf{v} - \mathbf{q}\|_\infty}{\|\mathbf{p}\|_\infty \vee \|\mathbf{q}\|_\infty} \leq \kappa_{\text{rel}} \frac{\|\boldsymbol{\varepsilon}\|_\infty}{\|\mathbf{b}\|_\infty}.$$

In many situations, a bound on relative error is preferred to a bound on absolute error because relative error is unitless and therefore often easier to interpret. In this situation, however, there is an additional concern about balancing. We would like to measure $\|\mathbf{u} - \mathbf{p}\|_\infty$ relative to $\|\mathbf{u}\|_\infty \approx \|\mathbf{p}\|_\infty$ and $\|\mathbf{v} - \mathbf{q}\|_\infty$ relative to $\|\mathbf{v}\|_\infty \approx \|\mathbf{q}\|_\infty$. This happens in the bounds above only if $\|\mathbf{p}\|_\infty$ and $\|\mathbf{q}\|_\infty$ are of similar magnitude. In practice, it may be necessary to change units of measurement or to rescale equations to ensure that quantities are well balanced, and the relative condition number addresses only some of these concerns. Ultimately, we choose to use the absolute condition number for the sake of simplicity. Some exercises on scaling appear in Chapter 29.

Exercises

1. True or false, and explain: Collocation is effective because the collocation solution is guaranteed to equal the exact solution at a prescribed set of nodes.

2. Modify `ivpl1_` and `ivpl1gen` to return L and rhs as additional outputs. Then repeat Example 23.4 and measure $\|\boldsymbol{\varepsilon}\|_\infty$, κ_{abs}, $\|\mathbf{u}-\mathbf{p}\|_\infty$, $\|u-p\|_\infty$, and $\|R\|_\infty$. (Estimate $\kappa_{\text{abs}} = \kappa_{\text{rel}}/\|\mathbf{L}\|_\infty$ with `condest(L')/norm(L,Inf)`.)

3. Modify `ivpl2_` and `ivpl2gen` to return L and rhs as additional outputs. Then repeat Example 23.7 and measure $\|\boldsymbol{\varepsilon}\|_\infty$, κ_{abs}, $\|\mathbf{u}-\mathbf{p}\|_\infty$, $\|u-p\|_\infty$, and $\|R\|_\infty$. (Estimate $\kappa_{\text{abs}} = \kappa_{\text{rel}}/\|\mathbf{L}\|_\infty$ with `condest(L')/norm(L,Inf)`.)

4. Explain why $u - tu' = 1, u(0) = 1$ is a poorly posed problem from the perspective of differential equations. Try solving the IVP on $0 \leq t \leq 1$ by collocation using a uniform grid of degree 1 for the collocation nodes and uniform grids of appropriate degrees for sampling p and p'. Identify where the numerical computation goes awry.

5. Explain why $u - tu' = 1, u(0) = 2$ is a poorly posed problem from the perspective of differential equations. Try solving the IVP on $0 \leq t \leq 1$ by collocation using a uniform grid of degree 1 for the collocation nodes and uniform grids of appropriate degrees for sampling p and p'. Identify where the numerical computation goes awry.

6. Suppose $u(t)$ is a solution to $e^t u + tu' = t + 1$ that can be expanded in a power series $u(t) = \sum_{i=0}^\infty c_i t^i$ with a positive radius of convergence.

   (a) Prove that $u(0) = 1$ and $u'(0) = 0$.

   (b) Solve $e^t u + tu' = t + 1, u(0) = 1$ on $0 \leq t \leq 5$ using a uniform grid of degree 4 for the collocation nodes and uniform grids of appropriate degrees for the interpolation nodes.

(c) Solve again using the same differential equation and the same grids, but this time using the initial condition $u'(0) = 0$.

(d) Which computation produces a more faithful answer? Why does the other give a poor answer?

7. In a collocation solution of the first-order IVP (24.2), it is often the case that $t_0 = t_0' = a$. That is, the first node in each interpolation grid coincides with the location of the initial condition. In this case, the error bound in Lemma 24.1 can be simplified (and possibly tightened depending on the other parameters). State the simpler error bound and justify.

8. Prove Lemma 24.5. Include expressions for C_1, C_2, and C_3.

9. Prove Theorem 24.6.

10. Prove Theorem 24.7.

11. Prove Theorem 24.8.

12. Let $p(t)$ be a collocation solution for the first-order IVP (24.1). Let $\hat{q}(t) = (g(t) - \alpha(t)p(t))/\beta(t)$ be the *iterated collocation* approximation for $u'(t)$. Prove

$$\|u' - \hat{q}\|_\infty \leq \|\alpha/\beta\|_\infty \|u - p\|_\infty.$$

13. Let $p(t)$ be a collocation solution for the second-order IVP (24.13). Let $\hat{r}(t) = (g(t) - \alpha(t)p(t) - \beta(t)p'(t))/\gamma(t)$ be the *iterated collocation* approximation for $u''(t)$. Prove

$$\|u'' - \hat{r}\|_\infty \leq \|\alpha/\gamma\|_\infty \|u - p\|_\infty + \|\beta/\gamma\|_\infty \|u' - p'\|_\infty.$$

14. Exercise 14 in Chapter 23 introduced collocation for BVPs. How and to what extent do Lemma 24.5 and Theorems 24.6–24.8 change when the pair of initial conditions in (24.13) is replaced by the boundary conditions of (23.16)?

Chapter 25

Collocation on a piecewise-uniform grid

This chapter introduces our first practical method for solving differential equations. The grids are piecewise uniform, and the resulting collocation solutions are piecewise polynomial.

25.1 ▪ First-order problems

Consider again the first-order linear differential equation

$$\alpha(t)u(t) + \beta(t)u'(t) = g(t). \tag{25.1}$$

Definition 25.1. *Let $[a, b]$ be an interval of the real line partitioned into subintervals $[a_j, b_j]$, $j = 1, \ldots, n$, and suppose that collocation nodes $s_{0j}, \ldots, s_{mj} \in [a_j, b_j]$ are provided for each subinterval. A piecewise-polynomial collocation solution to (25.1) is a function $p(t)$ that satisfies the following:*

1. *$p(t)$ is a polynomial of degree at most $m + 1$ on each subinterval $[a_j, b_j]$;*

2. *the collocation equations*

$$\alpha(s_{ij})p(s_{ij}) + \beta(s_{ij})p'(s_{ij}) = g(s_{ij}), \quad i = 0, \ldots, m, \tag{25.2}$$

   are satisfied on each subinterval $[a_j, b_j]$; and

3. *$p(t)$ is continuous on $[a, b]$.*

In (25.2), $p'(s_{ij})$ should be interpreted as the one-sided derivative from the right if $s_{ij} = a_j$ or as the one-sided derivative from the left if $s_{ij} = b_j$.

Note that $p(t)$ is explicitly required to be continuous to avoid jump discontinuities between subintervals.

For an IVP

$$\alpha(t)u(t) + \beta(t)u'(t) = g(t), \quad c_0 u(a) + c_1 u'(a) = d, \tag{25.3}$$

a piecewise-polynomial collocation solution $p(t)$ must also satisfy the initial condition $c_0 p(a) + c_1 p'(a) = d$.

Piecewise-polynomial collocation can be accomplished by solving n simple collocation problems, as the next theorem shows.

Theorem 25.2. *Define a partition $[a_j, b_j]$, $j = 1, \ldots, n$, and collocation nodes s_{ij}, $i = 0, \ldots, m$, $j = 1, \ldots, n$, as in Definition 25.1. Suppose*

1. *$p_1(t)$ is a collocation solution satisfying*

$$\alpha(s_{i1})p_1(s_{i1}) + \beta(s_{i1})p_1'(s_{i1}) = g(s_{i1}), \quad i = 0, \ldots, m,$$

and the initial condition $c_0 p_1(a) + c_1 p_1'(a) = d$, and

2. *for $j = 2, \ldots, n$, $p_j(t)$ is a collocation solution satisfying*

$$\alpha(s_{ij})p_j(s_{ij}) + \beta(s_{ij})p_j'(s_{ij}) = g(s_{ij}), \quad i = 0, \ldots, m,$$

and the initial condition $p_j(a_j) = p_{j-1}(b_{j-1})$.

Let $p(t)$ be the piecewise-polynomial function defined by $p(t) = p_j(t)$ for $t \in [a_j, b_j]$. Then $p(t)$ is a piecewise-polynomial collocation solution to (25.3).

Proof. Each $p_j(t)$ is a polynomial of degree at most $m + 1$ by Definition 23.1. Hence, $p(t)$ is a polynomial of degree at most $m + 1$ on each subinterval $[a_j, b_j]$.

Because $p_j(t)$ satisfies the collocation equations for $s_{0j}, \ldots, s_{mj}$, so too does $p(t)$ satisfy the same equations.

Because $p(t)$ is a piecewise-polynomial function, the only possible locations for discontinuities are at the joints between polynomial pieces. At each of these locations, the initial condition $p_j(a_j) = p_{j-1}(b_{j-1})$ enforces continuity.   □

Let's specialize to piecewise-uniform grids. Each subinterval is of width $l = (b - a)/n$, and on each subinterval are placed uniform grids $\mathbf{s}_j$, $\mathbf{t}_j$, and $\mathbf{t}_j'$ of degrees m, $m+1$, and m, respectively. The nodes of $\mathbf{s}_j$ are the collocation nodes, and the functions p, p', and g are sampled to produce $\mathbf{p}_j = p|_{\mathbf{t}_j}$, $\mathbf{q}_j = p'|_{\mathbf{t}_j'}$, and $\mathbf{g}_j = g|_{\mathbf{s}_j}$. The collocation solution is determined by n finite linear systems of the form (23.8). To specialize the general theory to uniform grids, we do the following:

- define $\mathbf{J}_m$ and $\hat{\mathbf{K}}_m$ as in Theorem 15.8;

- let $\mathbf{E}_{m,m+1}$ denote the resampling matrix from a uniform grid of degree $m + 1$ to a uniform grid of degree m (see Theorem 12.4 and the text just below it);

- define

$$\mathbf{A}_j = \begin{bmatrix} \alpha(s_{0j}) & & \\ & \ddots & \\ & & \alpha(s_{mj}) \end{bmatrix} \mathbf{E}_{m,m+1}, \quad \mathbf{B}_j = \begin{bmatrix} \beta(s_{0j}) & & \\ & \ddots & \\ & & \beta(s_{mj}) \end{bmatrix};$$

- let $\mathbf{e}_1^T$ denote a row vector consisting of a 1 followed by as many zeros as necessary; and

- let $p_{m+1,j-1}$ denote the last entry of $\mathbf{p}_{j-1}$.

(Note that no resampling matrix is required in the definition of $\mathbf{B}_j$ because $\mathbf{s}_j$ and $\mathbf{t}'_j$ are identical grids.) Then the collocation solution on the first subinterval is determined by

$$
\begin{bmatrix} \mathbf{J}_m & -l\hat{\mathbf{K}}_m \\ \mathbf{A}_1 & \mathbf{B}_1 \\ c_0\mathbf{e}_1^T & c_1\mathbf{e}_1^T \end{bmatrix} \begin{bmatrix} \mathbf{p}_1 \\ \mathbf{q}_1 \end{bmatrix} = \begin{bmatrix} \mathbf{0} \\ \mathbf{g}_1 \\ d \end{bmatrix} \tag{25.4}
$$

and on the jth subinterval, $j = 2, \ldots, n$, by

$$
\begin{bmatrix} \mathbf{J}_m & -l\hat{\mathbf{K}}_m \\ \mathbf{A}_j & \mathbf{B}_j \\ \mathbf{e}_1^T & \mathbf{0} \end{bmatrix} \begin{bmatrix} \mathbf{p}_j \\ \mathbf{q}_j \end{bmatrix} = \begin{bmatrix} \mathbf{0} \\ \mathbf{g}_j \\ p_{m+1,j-1} \end{bmatrix}. \tag{25.5}
$$

The n linear systems can be solved sequentially in the order $j = 1, 2, \ldots, n$. This is implemented by ivpl1uni in Listing 25.1.

Listing 25.1. Solve a first-order linear IVP by collocation on a piecewise-uniform grid.

```
function [ps,qs] = ivpl1uni(al,be,g,ab,c,d,m,n)

a = ab(1); b = ab(2);
% sample coefficient functions and right-hand side of DE
als = sampleuni(al,[a b],m,n);
bes = sampleuni(be,[a b],m,n);
gs = sampleuni(g,[a b],m,n);
% construct integration, resampling, and evaluation matrices
[J,K,E1,E2,Ea1,Ea2] = ivp1matuni(m,(b-a)/n);
% solve one subinterval at a time
ps = nan(m+2,n); qs = nan(m+1,n);
[ps(:,1),qs(:,1)] = ivpl1_(als(:,1),bes(:,1) ...
    ,gs(:,1),c,d,J,K,E1,E2,Ea1,Ea2);
for j = 2:n
  [ps(:,j),qs(:,j)] = ivpl1_(als(:,j),bes(:,j),gs(:,j),[1 0] ...
      ,ps(end,j-1),J,K,E1,E2,Ea1,Ea2);
end
% enforce continuity
ps(1,2:end) = ps(end,1:end-1);
if m>0, qs(1,2:end) = qs(end,1:end-1); end

function [J,K,E1,E2,Ea1,Ea2] = ivp1matuni(m,l)

[J,K] = indefiniteuni(m); K = l*K;
E1 = resamplematrixuni(m,m+1);
E2 = eye(m+1);
Ea1 = [ 1 zeros(1,m+1) ];
Ea2 = [ 1 zeros(1,m) ];
```

Example 25.3. Compute a collocation solution to

$$
tu(t) + u'(t) = 0, \quad u(0) = 1 \tag{25.6}
$$

on $0 \le t \le 4$. Use piecewise-uniform collocation with $m = 1$ and $n = 6$. Plot the solution components and the residual. Verify that the residual behaves as required by collocation.

Solution. The problem is defined.

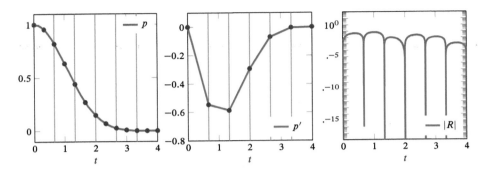

Figure 25.1. *A piecewise-polynomial collocation solution (left) and its derivative (middle) for a first-order IVP; the absolute residual (right).*

```
>> al = @(t) t; be = @(t) 1; g = @(t) 0;
>> a = 0; b = 4;
>> ua = 1;
```

The collocation solution is computed. Note that the initial condition can be expressed as $c_0 u(a) + c_1 u'(a) = d$, with $c_0 = 1$, $c_1 = 0$, and $d = 1$. The coefficients $c_0 = 1$ and $c_1 = 0$ are provided to ivpl1uni in the argument [1 0].

```
>> m = 1; n = 6;
>> [ps,qs] = ivpl1uni(al,be,g,[a b],[1 0],ua,m,n);
>> p = interpuni(ps,[a b]);
>> q = interpuni(qs,[a b]);
```

The collocation solution is plotted in the left and middle graphs of Figure 25.1. Note that p has quadratic pieces and p' has linear pieces.

```
>> ax = newfig(1,3); plotsample(griduni([a b],m+1,n),ps);
>> plotpartition([a b],n); plotfun(p,[a b]);
>> subplot(ax(2)); plotsample(griduni([a b],m,n),qs);
>> plotpartition([a b],n); plotfun(q,[a b]);
```

Also note that $p(t)$ resembles the exact solution $u(t) = e^{-t^2/2}$.

The residual is plotted in the graph on the right.

```
>> R = @(t) g(t)-al(t)*p(t)-be(t)*q(t);
>> subplot(ax(3)); ylog; ylim([1e-18 1e2]);
>> plotfun(@(x) abs(R(x)),[a b]);
```

Note that the absolute residual falls to zero at each collocation node. This is the defining feature of collocation. ∎

Example 25.4. Solve (25.6) again by piecewise-uniform collocation, this time using $m = 3$ and a sufficiently high n to produce $\|R\|_\infty \leq 10^{-6}$. Compare with the exact solution.

Solution. The problem is set up as before.

```
>> al = @(t) t; be = @(t) 1; g = @(t) 0;
>> a = 0; b = 4;
>> ua = 1;
```

By trial and error, we've found that $m = 3$ and $n = 30$ are sufficient to produce the requested residual.

```
>> m = 3; n = 30;
>> [ps,qs] = ivpl1uni(al,be,g,[a b],[1 0],ua,m,n);
```

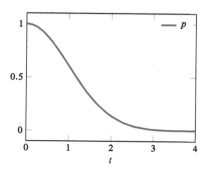

Figure 25.2. *A collocation solution on a finer grid.*

```
>> p = interpuni(ps,[a b]);
>> q = interpuni(qs,[a b]);
```
The solution is plotted in Figure 25.2.
```
>> newfig; plotfun(p,[a b]);
```
The absolute residual stays below 10^{-6} as specified.
```
>> R = @(t) g(t)-al(t)*p(t)-be(t)*q(t);
>> infnorm(R,[a b],100)
ans =
      9.013740212004961e-07
```
The errors are comparably small.
```
>> u = @(t) exp(-t^2/2);
>> v = @(t) -t*exp(-t^2/2);
>> infnorm(@(t) u(t)-p(t),[a b],100)
ans =
      1.577806653108027e-07
>> infnorm(@(t) v(t)-q(t),[a b],100)
ans =
      8.781895338705326e-07
```

■

25.2 ▪ Second-order problems

Next consider the second-order linear differential equation

$$\alpha(t)u(t) + \beta(t)u'(t) + \gamma(t)u''(t) = g(t). \tag{25.7}$$

Definition 25.5. *Let $[a,b]$ be an interval of the real line partitioned into subintervals $[a_j, b_j]$, $j = 1, \ldots, n$, and suppose that collocation nodes $s_{0j}, \ldots, s_{mj} \in [a_j, b_j]$ are provided for each subinterval. A* piecewise-polynomial collocation solution *to (25.7) is a piecewise-polynomial function $p(t)$ that satisfies the following:*

1. *$p(t)$ is a polynomial of degree at most $m + 2$ on each subinterval $[a_j, b_j]$;*

2. *on each subinterval $[a_j, b_j]$, the collocation equations*

$$\alpha(s_{ij})p(s_{ij}) + \beta(s_{ij})p'(s_{ij}) + \gamma(s_{ij})p''(s_{ij}) = g(s_{ij}), \quad i = 0, \ldots, m,$$

   are satisfied; and

3. *$p \in C^1[a,b]$.*

In the collocation equations, $p''(s_{ij})$ should be interpreted as the one-sided derivative from the right if $s_{ij} = a_j$ or as the one-sided derivative from the left if $s_{ij} = b_j$.

The condition $p \in C^1[a, b]$ guarantees that p transitions smoothly from one subinterval to the next.

For an IVP

$$\alpha(t)u(t) + \beta(t)u'(t) + \gamma(t)u''(t) = g(t), \quad c_{10}u(a) + c_{11}u'(a) + c_{12}u''(a) = d_1,$$
$$c_{20}u(a) + c_{21}u'(a) + c_{22}u''(a) = d_2,$$
$$(25.8)$$

a piecewise-polynomial collocation solution p must also satisfy the initial conditions $c_{10}p(a) + c_{11}p'(a) + c_{12}p''(a) = d_1$ and $c_{20}p(a) + c_{21}p'(a) + c_{22}p''(a) = d_2$.

Theorem 25.6. *Define $[a_j, b_j]$ and s_{ij} as in Definition 25.5. Suppose*

1. *$p_1(t)$ is a collocation solution that satisfies*

$$\alpha(s_{i1})p_1(s_{i1}) + \beta(s_{i1})p_1'(s_{i1}) + \gamma(s_{i1})p_1''(s_{i1}) = g(s_{i1}), \quad i = 0, \ldots, m,$$

and the initial conditions $c_{10}p_1(a) + c_{11}p_1'(a) + c_{12}p_1''(a) = d_1$ and $c_{20}p_1(a) + c_{21}p_1'(a) + c_{22}p_1''(a) = d_2$; and

2. *for $j = 2, \ldots, n$, $p_j(t)$ is a collocation solution that satisfies*

$$\alpha(s_{ij})p_j(s_{ij}) + \beta(s_{ij})p_j'(s_{ij}) + \gamma(s_{ij})p_j''(s_{ij}) = g(s_{ij}), \quad i = 0, \ldots, m,$$

and the initial conditions $p_j(a_j) = p_{j-1}(b_{j-1})$ and $p_j'(a_j) = p_{j-1}'(b_{j-1})$.

Let $p(t)$ be the piecewise-polynomial function defined by $p(t) = p_j(t)$ for $t \in [a_j, b_j]$. Then $p(t)$ is a piecewise-polynomial collocation solution to (25.8).

The proof is very similar to the proof of Theorem 25.2.

We concentrate on piecewise-uniform grids. Specifically, the subintervals $[a_j, b_j]$ are constructed to have equal width $l = (b - a)/n$, $\mathbf{s}_j$ is a uniform grid of degree m on $[a_j, b_j]$, and $\mathbf{t}_j$, $\mathbf{t}_j'$, and $\mathbf{t}_j''$ are uniform grids of degrees $m+2$, $m+1$, and m, respectively, on $[a_j, b_j]$. Let $\mathbf{p}_j = p|_{\mathbf{t}_j}$, $\mathbf{q}_j = p'|_{\mathbf{t}_j'}$, $\mathbf{r}_j = p''|_{\mathbf{t}_j''}$, and $\mathbf{g} = g|_{\mathbf{s}_j}$. Also, let

$$\mathbf{A}_j = \begin{bmatrix} \alpha(s_{0j}) & & \\ & \ddots & \\ & & \alpha(s_{mj}) \end{bmatrix} \mathbf{E}_{m,m+2},$$

$$\mathbf{B}_j = \begin{bmatrix} \beta(s_{0j}) & & \\ & \ddots & \\ & & \beta(s_{mj}) \end{bmatrix} \mathbf{E}_{m,m+1},$$

$$\mathbf{C}_j = \begin{bmatrix} \gamma(s_{0j}) & & \\ & \ddots & \\ & & \gamma(s_{mj}) \end{bmatrix}.$$

Then the piecewise-polynomial collocation solution is determined by the linear systems

$$
\begin{bmatrix}
\mathbf{J}_{m+1} & -l\hat{\mathbf{K}}_{m+1} & \mathbf{0} \\
\mathbf{0} & \mathbf{J}_m & -l\hat{\mathbf{K}}_m \\
\mathbf{A}_1 & \mathbf{B}_1 & \mathbf{C}_1 \\
c_{10}\mathbf{e}_1^T & c_{11}\mathbf{e}_1^T & c_{12}\mathbf{e}_1^T \\
c_{20}\mathbf{e}_1^T & c_{21}\mathbf{e}_1^T & c_{22}\mathbf{e}_1^T
\end{bmatrix}
\begin{bmatrix}
\mathbf{p}_1 \\
\mathbf{q}_1 \\
\mathbf{r}_1
\end{bmatrix}
=
\begin{bmatrix}
\mathbf{0} \\
\mathbf{0} \\
\mathbf{g}_1 \\
d_1 \\
d_2
\end{bmatrix}
$$

and, for $j = 2, \ldots, n$,

$$
\begin{bmatrix}
\mathbf{J}_{m+1} & -l\hat{\mathbf{K}}_{m+1} & \mathbf{0} \\
\mathbf{0} & \mathbf{J}_m & -l\hat{\mathbf{K}}_m \\
\mathbf{A}_j & \mathbf{B}_j & \mathbf{C}_j \\
\mathbf{e}_1^T & \mathbf{0} & \mathbf{0} \\
\mathbf{0} & \mathbf{e}_1^T & \mathbf{0}
\end{bmatrix}
\begin{bmatrix}
\mathbf{p}_j \\
\mathbf{q}_j \\
\mathbf{r}_j
\end{bmatrix}
=
\begin{bmatrix}
\mathbf{0} \\
\mathbf{0} \\
\mathbf{g}_j \\
p_{m+2,j-1} \\
q_{m+1,j-1}
\end{bmatrix}.
$$

Routine `ivpl2uni`, whose implementation is omitted here, solves a second-order linear IVP on a piecewise-uniform grid, starting at the left end of the interval and working to the right one subinterval at a time. The next two examples illustrate its usage.

Example 25.7. As seen in Example 22.5, the IVP

$$
Ku(t) + Bu'(t) + Mu''(t) = 0, \quad u(0) = u_0, \quad u'(0) = v_0
$$

describes the position of a mass attached to a damped spring. Set $K = 1$, $B = 0.1$, $M = 1$, $u_0 = 10$, and $v_0 = 0$, and solve the IVP over the first 30 seconds. Use collocation on a piecewise-uniform grid with $m = 3$ and n sufficiently large to make $\|R\|_\infty < 10^{-6}$. Demonstrate that the residual is as small as required and measure the absolute error $\|u - p\|_\infty$.

Solution. The problem is specified.

```
>> K = 1; B = 0.1; M = 1; g = @(t) 0;
>> a = 0; b = 30;
>> ua = 10; va = 0;
```

The solution to the associated collocation problem is found with `ivpl2uni`. Through trial and error, $n = 260$ has been found to satisfy the residual requirement.

```
>> m = 3; n = 260;
>> [ps,qs,rs] = ivpl2uni(@(t) K,@(t) B,@(t) M,g,[a b] ...
                      ,[ 1 0 0; 0 1 0 ],[ua; va],m,n);
>> p = interpuni(ps,[a b]);
>> q = interpuni(qs,[a b]);
>> r = interpuni(rs,[a b]);
```

The solution p is plotted in Figure 25.3. As expected, the mass oscillates back and forth about its equilibrium position, gradually losing energy.

```
>> newfig; plotfun(p,[a b]);
```

The residual is computed as follows.

```
>> R = @(t) g(t)-K*p(t)-B*q(t)-M*r(t);
>> infnorm(R,[a b],100)
ans =
     8.692765955942150e-07
```

It is uniformly below 10^{-6} as required.

The exact solution is known from Example 22.5.

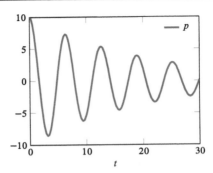

Figure 25.3. *A mass-spring system solved by collocation on a piecewise-uniform grid.*

```
>> la = sqrt(4*K*M-B^2)/(2*M);
>> u = @(t) ua*exp(-(B/(2*M))*t)*(cos(la*t)+(B/(2*la*M))*sin(la*t));
```
The absolute error is also small.
```
>> infnorm(@(t) u(t)-p(t),[a b],100)
ans =
      9.816548360852551e-07
```
The residual and error are small and of comparable magnitude. ∎

Example 25.8. Compute the Bessel function $J_0(x)$ on $0 \le x \le 20$ by solving

$$xu(x) + u'(x) + xu''(x) = 0, \quad u(0) = 1, \quad u''(0) = -1/2 \qquad (25.9)$$

by collocation. Use a piecewise-uniform grid with $m = 3$ and n sufficiently large to produce an absolute residual uniformly below 10^{-6}. Plot the solution $p(x)$. Also measure the absolute error on $p(x)$ by comparing with the MATLAB function besselj.

Solution. The problem is defined.
```
>> al = @(x) x; be = @(x) 1; ga = @(x) x; g = @(x) 0;
>> a = 0; b = 20;
>> ua = 1; wa = -1/2;
```
The grid parameters are set. We have found that $n = 140$ achieves the accuracy requirement.
```
>> m = 3; n = 140;
```
The IVP can now be solved. Note that the initial conditions constrain the zeroth and *second* derivatives. This is indicated by the matrix [1 0 0; 0 0 1] in the invocation below.
```
>> [ps,qs,rs] = ivpl2uni(al,be,ga,g,[a b] ...
                        ,[ 1 0 0; 0 0 1 ],[ua; wa],m,n);
>> p = interpuni(ps,[a b]);
>> q = interpuni(qs,[a b]);
>> r = interpuni(rs,[a b]);
```
The solution function is plotted in Figure 25.4.
```
>> newfig; plotfun(p,[a b]);
```
   The residual is measured.
```
>> R = @(x) g(x)-al(x)*p(x)-be(x)*q(x)-ga(x)*r(x);
>> infnorm(R,[a b],100)
```

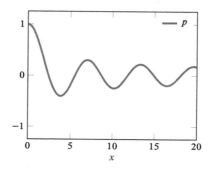

Figure 25.4. *The Bessel function J_0 computed by collocation on a piecewise-uniform grid.*

```
ans =
    7.403694120000637e-07
```
This satisfies the requirement.

Finally, the absolute error on p is found to be small as well.

```
>> u = @(x) besselj(0,x);
>> infnorm(@(x) u(x)-p(x),[a b],100)
ans =
    8.821825897564750e-08
```
■

25.3 ▪ An alternative formulation

The efficient way to compute a collocation solution on a piecewise grid is one subinterval at a time. For the first-order problem, this means sequentially solving the linear systems (25.4)–(25.5), each consisting of $2m + 3$ equations in $2m + 3$ unknowns. This is how routine ivp1luni works. The accuracy of this process seems to be difficult to analyze because the error committed on one subinterval perturbs the initial condition on the next subinterval. An error analysis would seem to have to track a cascade of errors across subintervals.

An easier approach to error analysis is to re-express the solution in terms of a single large linear system rather than a sequence of n little systems. In this last section of the chapter, we develop this alternative formulation for piecewise-polynomial collocation, and then in the next chapter we use this reformulation to analyze accuracy.

The alternative formulation we desire is driven by Theorem 15.10, which considers indefinite integration on piecewise-uniform grids. As in the lead-up to that theorem, the first step is to stack the polynomial samples into tall column vectors

$$\mathbf{p} = \begin{bmatrix} \mathbf{p}_1 \\ \mathbf{p}_2 \\ \vdots \\ \mathbf{p}_n \end{bmatrix}, \quad \mathbf{q} = \begin{bmatrix} \mathbf{q}_1 \\ \mathbf{q}_2 \\ \vdots \\ \mathbf{q}_n \end{bmatrix}, \tag{25.10}$$

with $(m+2)n$ and $(m+1)n$ entries, respectively, and the samples of g in the same way:

$$\mathbf{g} = \begin{bmatrix} \mathbf{g}_1 \\ \mathbf{g}_2 \\ \vdots \\ \mathbf{g}_n \end{bmatrix}. \tag{25.11}$$

Then the sequence of n linear systems (25.4)–(25.5) is replaced by the single linear system

$$\begin{bmatrix} \mathbf{J}_{mn} & -l\hat{\mathbf{K}}_{mn} \\ \mathbf{A} & \mathbf{B} \\ c_0\mathbf{e}_1^T & c_1\mathbf{e}_1^T \end{bmatrix} \begin{bmatrix} \mathbf{p} \\ \mathbf{q} \end{bmatrix} = \begin{bmatrix} \mathbf{0} \\ \mathbf{g} \\ d \end{bmatrix}, \tag{25.12}$$

consisting of $(2m+3)n$ equations in $(2m+3)n$ unknowns, in which $\mathbf{J}_{mn}$ and $\hat{\mathbf{K}}_{mn}$ are defined as in Theorem 15.10, $\mathbf{A} = \text{diag}(\mathbf{A}_1, \ldots, \mathbf{A}_n)$, $\mathbf{B} = \text{diag}(\mathbf{B}_1, \ldots, \mathbf{B}_n)$, and $\mathbf{e}_1^T$ denotes a row vector containing a 1 followed by as many zeros as necessary.

Theorem 25.9. *Let $p \in P_{m+1,n}[a,b]$. Then p is a piecewise-polynomial collocation solution to the first-order IVP (25.3) if and only if (25.12) holds.*

The proof is similar to the proof of Theorem 23.2 but relies on Theorem 15.10 in place of Theorem 14.8. The details are omitted.

A second-order problem can be solved by a similar system. Form the stacked vectors

$$\mathbf{p} = \begin{bmatrix} \mathbf{p}_1 \\ \mathbf{p}_2 \\ \vdots \\ \mathbf{p}_n \end{bmatrix}, \quad \mathbf{q} = \begin{bmatrix} \mathbf{q}_1 \\ \mathbf{q}_2 \\ \vdots \\ \mathbf{q}_n \end{bmatrix}, \quad \mathbf{r} = \begin{bmatrix} \mathbf{r}_1 \\ \mathbf{r}_2 \\ \vdots \\ \mathbf{r}_n \end{bmatrix}, \quad \mathbf{g} = \begin{bmatrix} \mathbf{g}_1 \\ \mathbf{g}_2 \\ \vdots \\ \mathbf{g}_n \end{bmatrix},$$

and let $\mathbf{A} = \text{diag}(\mathbf{A}_1, \ldots, \mathbf{A}_n)$, $\mathbf{B} = \text{diag}(\mathbf{B}_1, \ldots, \mathbf{B}_n)$, and $\mathbf{C} = \text{diag}(\mathbf{C}_1, \ldots, \mathbf{C}_n)$. The collocation solution is determined by the following system:

$$\begin{bmatrix} \mathbf{J}_{m+1,n} & -l\hat{\mathbf{K}}_{m+1,n} & \mathbf{0} \\ \mathbf{0} & \mathbf{J}_{mn} & -l\hat{\mathbf{K}}_{mn} \\ \mathbf{A} & \mathbf{B} & \mathbf{C} \\ c_{10}\mathbf{e}_1^T & c_{11}\mathbf{e}_1^T & c_{12}\mathbf{e}_1^T \\ c_{20}\mathbf{e}_1^T & c_{21}\mathbf{e}_1^T & c_{22}\mathbf{e}_1^T \end{bmatrix} \begin{bmatrix} \mathbf{p} \\ \mathbf{q} \\ \mathbf{r} \end{bmatrix} = \begin{bmatrix} \mathbf{0} \\ \mathbf{0} \\ \mathbf{g} \\ d_1 \\ d_2 \end{bmatrix}. \tag{25.13}$$

Theorem 25.10. *Let $p \in P_{m+2,n}[a,b]$. Then p is a piecewise-polynomial collocation solution to the second-order IVP (25.8) if and only if (25.13) is true.*

Notes

The collocation method on a piecewise-uniform grid of degree $m = 0$ is also called the *implicit midpoint method*. The solution satisfies $p(b_j) = p(a_j) + hf\big((a_j+b_j)/2, (p(a_j)+p(b_j))/2\big)$, in which $h = b_j - a_j$ and $f(t,y) = -\alpha(t)y + g(t)$. Notationally, this may appear to be a small twist on the classical Euler method, defined by $p(b_j) = p(a_j) + hf(a_j, p(a_j))$, but the differences are significant. Euler's method is an *explicit* method, the formula $p(a_j) + hf(a_j, p(a_j))$ consisting of just arithmetic and a function evaluation. The midpoint method, by contrast, is an *implicit* method because the unknown $p(b_j)$ occurs on both sides of the equation $p(b_j) = p(a_j) + hf\big((a_j + b_j)/2, (p(a_j) + p(b_j))/2\big)$. For the linear problems in this chapter, the implicit collocation equations are linear and can be solved directly, but when nonlinear problems are introduced later, an iterative method such as Newton's method is required. Implicit methods can be more stable than explicit ones, but they can also be slower for some problems. We emphasize stability over speed.

Exercises

Exercises 1–2: Find the collocation solution by hand, using nothing more than a hand-held calculator. Report the values to four decimal places of precision.

1. $2u + u' = t, u(0) = 5; 0 \le t \le 0.2; m = 0, n = 2$
2. $tu + u' = -4, u(1) = 0; 1 \le t \le 1.4; m = 0, n = 2$

3. Consider adding plots of $p'(t)$ and $R(t)$ to Example 25.4. What should they look like? Create the plots and see whether they match your expectation.

Exercises 4–7: Solve the IVP on the given interval using ivpl1uni with $m = 3$. Adjust n to achieve an absolute residual below 10^{-6}. Plot the solution function p and the absolute residual.

4. $-(\sin t)u + u' = -1, u(0) = 0; 0 \le t \le 10$
5. $(\log t)u + u' = 1, u(1) = -2; 1 \le t \le 10$
6. $u + u' = 1/(t^2 + 1), u(0) = 0; 0 \le t \le 10$
7. $(1 + t)u - tu' = 1 + t, u'(0) = -1; 0 \le t \le 2$

Exercises 8–11: Solve the IVP on the given interval using ivpl2uni with $m = 3$. Adjust n to achieve an absolute residual below 10^{-6}. Plot the solution function p and the absolute residual.

8. $e^t u + u' + u'' = t, u(0) = 0, u'(0) = 1; 0 \le t \le 4$
9. $u - t^2 u' + u'' = 0, u(0) = 1, u'(0) = 0; 0 \le t \le 2$
10. $u + t^2 u' + u'' = 0, u(0) = 1, u'(0) = 0; 0 \le t \le 10$
11. $(\sin t)u + (\cos t)u' + u'' = 1/(t + 1), u(0) = 0, u'(0) = 1; 0 \le t \le 15$

12. An IVP can be solved backward in time by choosing an initial time a and a final time b for which $b < a$. Solve the IVP $tu + u' = t^2 + 1, u(0) = 2$ on the interval $-5 \le t \le 0$ using ivpl1uni with $m = 2, n = 100$. Compare with the exact solution $u(t) = 2e^{-t^2/2} + t$.

13. Consider the IVP $u + \sqrt{t}\, u' = 0, u(0) = 2$ on $0 \le t \le 5$.

    (a) Try solving the IVP with ivpl1uni. What happens? Why?
    (b) Perform the change of variables $t = x^2$. Solve the resulting IVP in the independent variable x with ivpl1uni, obtaining an absolute residual below 10^{-6}. Then reverse the change of variables to find an approximation for $u(t)$. Plot the result.

14. The Bessel differential equation is usually stated as $(x^2 - v^2)u + xu' + x^2 u'' = 0$. The special case $v = 0$ is considered in Example 25.8, but the differential equation looks different. Explain why it is necessary to rewrite the equation in the form (25.9).

15. The Bessel function $J_0(x)$ satisfies the following properties at $x = 0$: $J_0(0) = 1$, $J'(0) = 0$, and $J''(0) = -1/2$. In Example 25.8, the Bessel differential equation is solved with initial conditions $u(0) = 1, u''(0) = -1/2$. What happens if the pair of initial conditions $u(0) = 1, u'(0) = 0$ is used instead? Explain why.

16. The Bessel function $J_{1/3}(x)$ satisfies the differential equation

$$\left(x^2 - \frac{1}{9}\right) y + xy' + x^2 y'' = 0.$$

However, the problem of computing $J_{1/3}(x)$ is an awkward fit for ivpl2uni because $\lim_{x \to 0+} J'_{1/3}(x) = +\infty$. Apply the change of variables $y = x^{1/3}u$ to show that the differential equation is equivalent to

$$xu + \frac{5}{3}u' + xu'' = 0.$$

17. Define $u(x)$ by $J_{1/3}(x) = x^{1/3}u(x)$ as in the previous exercise. Known asymptotics for the Bessel function imply that $u(x)$ satisfies the IVP

$$xu + \frac{5}{3}u' + xu'' = 0, \quad u(0) = \frac{3}{2^{1/3}\Gamma(1/3)}, \quad u''(0) = -\frac{9}{2^{10/3}\Gamma(1/3)}.$$

Solve this IVP on $0 \le x \le 20$ using ivpl2uni with $m = 3$. Adjust n to achieve an absolute residual below 10^{-6}. Finally, reverse the change of variables to find an approximation for $J_{1/3}(x)$ and plot this final approximation.

18. Develop a routine ivpl3uni for solving the following third-order IVP by collocation on a piecewise-uniform grid:

$$\alpha(t)u + \beta(t)u' + \gamma(t)u'' + \delta(t)u''' = g(t),$$
$$c_{10}u(a) + c_{11}u'(a) + c_{12}u''(a) + c_{13}u'''(a) = d_1,$$
$$c_{20}u(a) + c_{21}u'(a) + c_{22}u''(a) + c_{23}u'''(a) = d_2,$$
$$c_{30}u(a) + c_{31}u'(a) + c_{32}u''(a) + c_{33}u'''(a) = d_3.$$

Test your routine on the problem

$$u + 2u' + u'' + 2u''' = 0, \quad u(0) = 5, \quad u'(0) = 0, \quad u''(0) = 0$$

for $0 \le t \le 4\pi$, attaining an absolute residual below 10^{-6}. *Hint.* First solve Exercise 15 in Chapter 23.

19. Let $A(t)$ and $B(t)$ be r-by-r matrix-valued functions of t, and let $g(t)$ be an r-by-1 vector-valued function of t. Consider the vector-valued IVP

$$A(t)u(t) + B(t)u'(t) = g(t), \quad c_0 u(a) + c_1 u'(a) = d. \tag{25.14}$$

The solution is a vector-valued function $u(t) = (u_1(t), \ldots, u_r(t))$. Develop a MATLAB routine to solve this IVP by collocation on a piecewise-uniform grid. It should return an r-by-$(m+2)$-by-n array ps in which ps(i,j,k) equals $p_i(t_{j-1,k})$ and an r-by-$(m+1)$-by-n array qs in which qs(i,j,k) equals $p'_i(t'_{j-1,k})$. Demonstrate your implementation by solving

$$\begin{bmatrix} 1/2 & -1 \\ 1 & 1/2 \end{bmatrix} \begin{bmatrix} u_1 \\ u_2 \end{bmatrix} + \begin{bmatrix} 1 & 0 \\ 0 & 1 \end{bmatrix} \begin{bmatrix} u'_1 \\ u'_2 \end{bmatrix} = \begin{bmatrix} 0 \\ 0 \end{bmatrix}, \quad \begin{bmatrix} u_1(0) \\ u_2(0) \end{bmatrix} = \begin{bmatrix} 1 \\ 0 \end{bmatrix}$$

on $0 \le t \le 10$. *Hint.* Solve Exercises 16–17 in Chapter 23 first.

Chapter 26

Accuracy of collocation on a piecewise-uniform grid

How efficient is collocation on a piecewise-uniform grid? The answer depends in part on the smoothness of the unknown solution: from earlier results, e.g., Theorem 24.3, we can bound collocation error in terms of interpolation error, and interpolation works best on smooth functions. We start this chapter by analyzing the smoothness of solutions to differential equations. Then we analyze the implications of using piecewise-uniform grids.

26.1 ▪ Smoothness of solutions to differential equations

Linear differential equations with smooth coefficient functions have smooth solutions.

Theorem 26.1. *Let $\alpha_0, \ldots, \alpha_k \in C^r[a, b]$ and $g \in C^r[a, b]$, and suppose $\alpha_k(t) \neq 0$ for $t \in [a, b]$. If $u(t)$ is a solution to the differential equation*

$$\alpha_0(t)u(t) + \alpha_1(t)u'(t) + \cdots + \alpha_k(t)u^{(k)}(t) = g(t), \tag{26.1}$$

then $u \in C^{k+r}[a, b]$.

Proof. The proof is by induction on r.

For the base case, consider $r = 0$. By assumption, $u(t)$ is a solution to the differential equation and is therefore k-times differentiable. Thus, $u(t)$ and its first $k - 1$ derivatives must be continuous. In addition,

$$u^{(k)}(t) = \frac{g(t) - (\alpha_0(t)u(t) + \cdots + \alpha_{k-1}(t)u^{(k-1)}(t))}{\alpha_k(t)}, \tag{26.2}$$

which is continuous. Therefore, $u \in C^k[a, b]$.

For the induction step, suppose that the result has been proved for some value of r, and assume that $\alpha_0, \ldots, \alpha_k \in C^{r+1}[a, b]$ and $g \in C^{r+1}[a, b]$. Just as in the base case, $u^{(k)}(t)$ satisfies (26.2). But now the right-hand side of (26.2) is differentiable and therefore $u^{(k)}(t)$ is differentiable. Differentiating both sides of (26.1), we find

$$\tilde{\alpha}_0(t)u(t) + \tilde{\alpha}_1(t)u'(t) + \cdots + \tilde{\alpha}_{k+1}(t)u^{(k+1)}(t) = g'(t), \tag{26.3}$$

in which $\tilde{\alpha}_0(t) = \alpha_0'(t)$, $\tilde{\alpha}_{k+1}(t) = \alpha_k(t)$, and $\tilde{\alpha}_i(t) = \alpha_{i-1}(t) + \alpha_i'(t)$ for $i = 1, \ldots, k$. Note that $\tilde{\alpha}_0, \ldots, \tilde{\alpha}_{k+1} \in C^r[a, b]$, $g' \in C^r[a, b]$, and $\tilde{\alpha}_{k+1}(t) \neq 0$ for any $t \in [a, b]$. Because $u(t)$ is a solution to (26.3), the induction hypothesis implies $u \in C^{(k+1)+r}[a, b]$, i.e., $u \in C^{k+(r+1)}[a, b]$. $\qquad \square$

We write $f \in C^\infty[a, b]$ if f has derivatives of all orders on $[a, b]$. It follows from the theorem that if $\alpha_0, \ldots, \alpha_k, g \in C^\infty[a, b]$ and $\alpha_k(t) \neq 0$ for $t \in [a, b]$, then a solution u to (26.1) belongs to the class $C^\infty[a, b]$ as well.

In this book, we are primarily concerned with first- and second-order equations.

Corollary 26.2. *If $\alpha, \beta, g \in C^r[a, b]$ for some integer r, and $\beta(t) \neq 0$ for $t \in [a, b]$, then any solution to*

$$\alpha(t)u(t) + \beta(t)u'(t) = g(t)$$

satisfies $u \in C^{r+1}[a, b]$.

Corollary 26.3. *If $\alpha, \beta, \gamma, g \in C^r[a, b]$ for some integer r, and $\gamma(t) \neq 0$ for $t \in [a, b]$, then any solution to*

$$\alpha(t)u(t) + \beta(t)u'(t) + \gamma(t)u''(t) = g(t)$$

satisfies $u \in C^{r+2}[a, b]$.

These results, together with Theorems 22.8 and 22.10, can guarantee the existence of a well-behaved solution to an IVP, for which our numerical methods are well suited.

26.2 ▪ Rate of convergence for first-order problems

Consider the first-order IVP

$$\alpha(t)u(t) + \beta(t)u'(t) = g(t), \quad c_0 u(a) + c_1 u'(a) = d \tag{26.4}$$

on an interval $[a, b]$. We assume that $\|\alpha\|_\infty, \|\beta\|_\infty < \infty$ and that the problem has a unique solution over the whole interval.

We would like to bound the error from collocation on a piecewise-uniform grid. The notation from the previous chapter for first-order problems is adopted. In particular, the stacked samples

$$\mathbf{p} = \begin{bmatrix} \mathbf{p}_1 \\ \mathbf{p}_2 \\ \vdots \\ \mathbf{p}_n \end{bmatrix}, \quad \mathbf{q} = \begin{bmatrix} \mathbf{q}_1 \\ \mathbf{q}_2 \\ \vdots \\ \mathbf{q}_n \end{bmatrix}$$

represent the collocation solution and its derivative. These are to be compared with samples

$$\mathbf{u} = \begin{bmatrix} \mathbf{u}_1 \\ \mathbf{u}_2 \\ \vdots \\ \mathbf{u}_n \end{bmatrix}, \quad \mathbf{v} = \begin{bmatrix} \mathbf{v}_1 \\ \mathbf{v}_2 \\ \vdots \\ \mathbf{v}_n \end{bmatrix}$$

of the true solution, in which $\mathbf{u}_j = u|_{t_j}$ and $\mathbf{v}_j = u'|_{t'_j}$.

An example sets expectations.

Example 26.4. Solve

$$tu(t) + u'(t) = 0, \quad u(0) = 1 \tag{26.5}$$

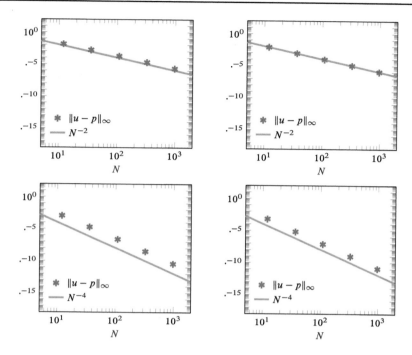

Figure 26.1. *Rate of convergence for first-order collocation on a piecewise-uniform grid of degree m: m = 0 (top-left), m = 1 (top-right), m = 2 (bottom-left), and m = 3 (bottom-right).*

on $0 \le t \le 4$ by piecewise-uniform collocation, and measure the absolute error $\|u - p\|_\infty$. For each degree $m = 0, 1, 2, 3$, measure the rate of convergence as $N = (m \vee 1)n$ approaches infinity.

Solution. The problem is specified in code, and the grid parameters are set.

```
>> al = @(t) t; be = @(t) 1; g = @(t) 0;
>> a = 0; b = 4;
>> ua = 1;
>> m = 0:3; N = 4*3.^(1:5);
```

The exact solution is known for this simple problem.

```
>> u = @(t) exp(-t^2/2);
```

The collocation equations are solved, and the absolute errors are measured.

```
>> err = nan(length(m),length(N));
>> for k = 1:length(m)
     for l = 1:length(N)
       n = N(l)/max(m(k),1);
       [ps,qs] = ivpl1uni(al,be,g,[a b],[1 0],ua,m(k),n);
       p = interpuni(ps,[a b]);
       err(k,l) = infnorm(@(t) u(t)-p(t),[a b],100);
     end
   end
```

The errors are plotted in Figure 26.1 along with functions that appear to converge at the same rates.

```
>> s = [-2 -2 -4 -4];
>> ax = newfig(2,2); ax = ax';
```

```
>> for k = 1:4
     subplot(ax(k)); xlog; ylog; ylim([1e-18 1e2]);
     plot(N,err(k,:),'*'); plotfun(@(N) N^s(k),[5 2000]);
   end
```

The convergence rates appear to be N^{-2}, N^{-2}, N^{-4}, and N^{-4}. ∎

The example suggests that $\|u - p\|_\infty$ may converge at the rate $N^{-(m+2)}$ when m is even or at the rate $N^{-(m+1)}$ when m is odd.

According to Theorem 25.9, the collocation solution is determined by the $(2m + 3)n$-by-$(2m + 3)n$ linear system

$$\mathbf{Lx} = \mathbf{b},$$

in which

$$\mathbf{L} = \begin{bmatrix} \mathbf{J}_{mn} & -l\hat{\mathbf{K}}_{mn} \\ \mathbf{A} & \mathbf{B} \\ c_0\mathbf{e}_1^T & c_1\mathbf{e}_1^T \end{bmatrix}, \quad \mathbf{x} = \begin{bmatrix} \mathbf{p} \\ \mathbf{q} \end{bmatrix}, \quad \mathbf{b} = \begin{bmatrix} \mathbf{0} \\ \mathbf{g} \\ d \end{bmatrix}.$$

Let

$$\mathbf{y} = \begin{bmatrix} \mathbf{u} \\ \mathbf{v} \end{bmatrix},$$

and set

$$\boldsymbol{\varepsilon} = \mathbf{b} - \mathbf{Ly}.$$

The accuracy analysis follows the outline of Chapter 24, bounding $\|\boldsymbol{\varepsilon}\|_\infty$ and then considering the implications.

Lemma 26.5. *Suppose the first-order IVP (26.4) is solved by piecewise-uniform collocation. If $u \in C^{m+2}[a, b]$, then*

$$\|\boldsymbol{\varepsilon}\|_\infty \leq CN^{-(m+1)} \tag{26.6}$$

for a coefficient C that is constant with respect to n. If m is even and $u \in C^{m+3}[a, b]$ and either $m \geq 2$ or $c_1 = 0$, then the bound can be improved to

$$\|\boldsymbol{\varepsilon}\|_\infty \leq CN^{-(m+2)}. \tag{26.7}$$

Proof. Let $s = m + 2$ if m is even and $u \in C^{m+3}[a, b]$ and either $m \geq 2$ or $c_1 = 0$. Otherwise, let $s = m + 1$.

The first several entries of $\boldsymbol{\varepsilon}$ are given by $-\mathbf{J}_{mn}\mathbf{u} + \frac{b-a}{n}\hat{\mathbf{K}}_{mn}\mathbf{v}$. By Theorem 16.6, these entries are bounded in magnitude by $C_1 N^{-s}$ for a scalar C_1 that is constant with respect to n.

The next entries of $\boldsymbol{\varepsilon}$ are given by $\mathbf{g} - \mathbf{Au} - \mathbf{Bv}$. Specifically, they are

$$g(s_{ij}) - \alpha(s_{ij})\tilde{p}(s_{ij}) - \beta(s_{ij})u'(s_{ij})$$

for $i = 0, \ldots, m$ and $j = 1, \ldots, n$, in which $\tilde{p}$ is the piecewise-polynomial interpolant for u on the $(m + 1)$-by-n piecewise-uniform grid. (Note that there is no need to introduce an interpolating polynomial $\tilde{q}$ in place of u' because there is no resampling involved in the definition of $\mathbf{B}$.) Because u solves the differential equation at all t, in particular it satisfies

$$g(s_{ij}) - \alpha(s_{ij})u(s_{ij}) - \beta(s_{ij})u'(s_{ij}) = 0$$

at every collocation node s_{ij}. Adding $\alpha(s_{ij})(u(s_{ij}) - \tilde{p}(s_{ij}))$ to both sides gives

$$g(s_{ij}) - \alpha(s_{ij})\tilde{p}(s_{ij}) - \beta(s_{ij})u'(s_{ij}) = \alpha(s_{ij})(u(s_{ij}) - \tilde{p}(s_{ij})).$$

Thus, each entry of $\mathbf{g} - \mathbf{Au} - \mathbf{Bv}$ is bounded in magnitude by $\|\alpha\|_\infty \|u - \tilde{p}\|_\infty$. By Theorem 8.1, $\|u - \tilde{p}\|_\infty \le C_2((m+1)n)^{-(m+2)} \le C_2 N^{-(m+2)} \le C_2 N^{-s}$ for a coefficient C_2 that is constant with respect to n.

The final entry of $\boldsymbol{\varepsilon}$ equals

$$d - c_0 u(t_{01}) - c_1 u'(t'_{01}), \tag{26.8}$$

in which t_{01} is the first interpolation node for p and t'_{01} is the first interpolation node for p'. Because u satisfies the initial condition, we have

$$d - c_0 u(a) - c_1 u'(a) = 0. \tag{26.9}$$

If $m \ge 1$, then $t_{01} = t'_{01} = a$, so (26.9) implies that (26.8) equals zero. Suppose for the rest of the proof that $m = 0$. Then $t_{01} = a$ but $t'_{01} = (a_1 + b_1)/2$. Let $\tilde{q}$ be the piecewise-constant interpolant for u' on the 0-by-n piecewise-uniform grid. Adding $c_1(u'(a) - u'(t'_{01}))$ to both sides of (26.9) gives

$$d - c_0 u(t_{01}) - c_1 u'(t'_{01}) = c_1(u'(a) - u'(t'_{01})) = c_1(u'(a) - \tilde{q}(a)).$$

If $c_1 = 0$, then this equals zero. If $c_1 \ne 0$, then Theorem 8.1 bounds the magnitude of this quantity by $C_3 N^{-1}$ for a constant C_3. In summary, the last entry of $\boldsymbol{\varepsilon}$ is nonzero only if $m = 0$ and $c_1 \ne 0$, in which case the entry is bounded in magnitude by $C_3 N^{-(m+1)}$. $\square$

Now we can bound the errors on $\mathbf{p}$ and $\mathbf{q}$. Below, $\kappa_{\text{abs}} = \|\mathbf{L}^{-1}\|_\infty$ is the absolute condition number of $\mathbf{L}$.

Theorem 26.6. *Suppose the first-order IVP (26.4) is solved by piecewise-uniform collocation. If $u \in C^{m+2}[a, b]$, then*

$$\|\mathbf{u} - \mathbf{p}\|_\infty \le C \kappa_{\text{abs}} N^{-(m+1)}, \tag{26.10}$$

$$\|\mathbf{v} - \mathbf{q}\|_\infty \le C \kappa_{\text{abs}} N^{-(m+1)}, \tag{26.11}$$

in which C is a coefficient that is constant with respect to n. If m is even and $u \in C^{m+3}[a, b]$ and either $m \ge 2$ or $c_1 = 0$, then these bounds can be improved to

$$\|\mathbf{u} - \mathbf{p}\|_\infty \le C \kappa_{\text{abs}} N^{-(m+2)}, \tag{26.12}$$

$$\|\mathbf{v} - \mathbf{q}\|_\infty \le C \kappa_{\text{abs}} N^{-(m+2)}. \tag{26.13}$$

Proof. The bounds are only meaningful if $\kappa_{\text{abs}} < \infty$, so assume that $\mathbf{L}$ is invertible. We have

$$\mathbf{L}(\mathbf{y} - \mathbf{x}) = \mathbf{Ly} - \mathbf{Lx} = (\mathbf{b} - \boldsymbol{\varepsilon}) - \mathbf{b} = -\boldsymbol{\varepsilon}.$$

Therefore,

$$\mathbf{y} - \mathbf{x} = -\mathbf{L}^{-1}\boldsymbol{\varepsilon},$$

and thus

$$\left\| \begin{bmatrix} \mathbf{u} - \mathbf{p} \\ \mathbf{v} - \mathbf{q} \end{bmatrix} \right\|_\infty \le \|\mathbf{L}^{-1}\|_\infty \|\boldsymbol{\varepsilon}\|_\infty = \kappa_{\text{abs}} \|\boldsymbol{\varepsilon}\|_\infty.$$

With C and s defined as in the preceding lemma, $\|\varepsilon\|_\infty \le CN^{-s}$. $\square$

The next two theorems extend the analysis to the function approximations $p(t)$ and $p'(t)$ and the continuous residual $R(t) = g(t) - \alpha(t)p(t) - \beta(t)p'(t)$.

Theorem 26.7. *Suppose the first-order IVP (26.4) is solved by piecewise-uniform collocation. If $u \in C^{m+2}[a,b]$, then*

$$\|u - p\|_\infty \le C_1(\kappa_{abs} \vee 1)N^{-(m+1)}, \tag{26.14}$$

$$\|u' - p'\|_\infty \le C_2(\kappa_{abs} \vee 1)N^{-(m+1)}, \tag{26.15}$$

in which C_1 and C_2 are constant with respect to n. If m is even and $u \in C^{m+3}[a,b]$ and either $m \ge 2$ or $c_1 = 0$, then the first bound can be improved to

$$\|u - p\|_\infty \le C_1(\kappa_{abs} \vee 1)N^{-(m+2)}. \tag{26.16}$$

Proof. Define the exponent s as in the proof of Lemma 26.5. From the previous theorem, we know $\|\mathbf{u} - \mathbf{p}\|_\infty \le C\kappa_{abs}N^{-s}$ and $\|\mathbf{v} - \mathbf{q}\|_\infty \le C\kappa_{abs}N^{-s}$. The desired bounds follow from Theorem 8.5. In the first case, the desired function is u, the approximating piecewise-polynomial function is p, the degree is $m + 1$, and the error at the interpolation nodes is bounded by $\|\mathbf{u} - \mathbf{p}\|_\infty \le C\kappa_{abs}N^{-s}$. Thus, the bound is

$$\|u - p\|_\infty \le C_3((m+1)n)^{-(m+2)} + \Lambda_{m+1}C\kappa_{abs}N^{-s}$$
$$\le C_3N^{-(m+2)} + \Lambda_{m+1}C\kappa_{abs}N^{-s} \le (C_3 + \Lambda_{m+1}C\kappa_{abs})N^{-s}$$

for a constant C_3. For the second bound, the desired function is u', the approximating function is p', the degree is m, and the error at the interpolation nodes is bounded by $\|\mathbf{v} - \mathbf{q}\|_\infty \le C\kappa_{abs}N^{-s}$. Thus, the bound is

$$\|u' - p'\|_\infty \le C_4((m \vee 1)n)^{-(m+1)} + \Lambda_m C\kappa_{abs}N^{-s}$$
$$= C_4N^{-(m+1)} + \Lambda_m C\kappa_{abs}N^{-s} \le (C_4 + \Lambda_m C\kappa_{abs})N^{-(m+1)}$$

for a constant C_4. Let $C_1 = C_3 + \Lambda_{m+1}C$ and $C_2 = C_4 + \Lambda_m C$. $\square$

Theorem 26.8. *Suppose the first-order IVP (26.4) is solved by piecewise-uniform collocation. If $u \in C^{m+2}[a,b]$, then the residual $R(t) = g(t) - \alpha(t)p(t) - \beta(t)p'(t)$ satisfies*

$$\|R\|_\infty \le C(\kappa_{abs} \vee 1)N^{-(m+1)} \tag{26.17}$$

for a coefficient C that is constant with respect to n.

Proof. Using Theorems 24.4 and 26.7, we find

$$\|R\|_\infty \le \|\alpha\|_\infty\|u - p\|_\infty + \|\beta\|_\infty\|u' - p'\|_\infty$$
$$\le \|\alpha\|_\infty C_1(\kappa_{abs} \vee 1)N^{-(m+1)} + \|\beta\|_\infty C_2(\kappa_{abs} \vee 1)N^{-(m+1)}.$$

Let $C = \|\alpha\|_\infty C_1 + \|\beta\|_\infty C_2$. $\square$

The previous theorem is perhaps misleading in one respect. It seems to suggest that the residual R may be harder to control when the condition number κ_{abs} is large, but the common phenomenon for ill-conditioned problems is for the errors $u - p$ and $u' - p'$ to be large even though the residual R is small. Although the bound (26.17) is correct, it can be loose when the condition number is large.

Theorems 26.6, 26.7, and 26.8 show that $\|\mathbf{u}-\mathbf{p}\|_\infty$, $\|\mathbf{v}-\mathbf{q}\|_\infty$, $\|u-p\|_\infty$, $\|u'-p'\|_\infty$, and $\|R\|_\infty$ are all bounded by expressions involving $N^{-(m+2)}$ or $N^{-(m+1)}$. Piecewise-polynomial collocation converges algebraically on well-conditioned first-order linear IVPs.

To simplify the discussion, the next few chapters focus on well-conditioned problems. Chapter 29 considers the effects of ill conditioning.

Example 26.9. Compare the bounds of Theorem 26.7 with the results of Example 26.4.

Solution. By Theorem 22.8, a unique solution to the IVP (26.5) exists. By Corollary 26.2, the solution is infinitely differentiable. (Note that $\alpha(t) = t$, $\beta(t) = 1$, and $g(t) = 0$ are infinitely differentiable and that $\beta(t)$ never equals zero.) Therefore, Theorem 26.7 applies.

For piecewise-uniform grids with degrees $m = 0, 1, 2, 3$, Theorem 26.7 shows that rates of convergence of N^{-2}, N^{-2}, N^{-4}, and N^{-4}, respectively, are possible when the problem is well conditioned.

The experimental data as seen in Example 26.4 shows that the predicted rates are achieved. ∎

26.3 ▪ Rate of convergence for second-order problems

For the second-order problem

$$\alpha(t)u(t) + \beta(t)u'(t) + \gamma(t)u''(t) = g(t), \quad c_{10}u(a) + c_{11}u'(a) + c_{12}u''(a) = d_1,$$
$$c_{20}u(a) + c_{21}u'(a) + c_{22}u''(a) = d_2,$$
$$(26.18)$$

analogous results hold. We assume that $\alpha(t)$, $\beta(t)$, and $\gamma(t)$ are bounded on an interval $[a, b]$ and that a unique solution $u(t)$ exists.

By Theorem 25.10, the collocation solution is determined by the solution

$$\mathbf{x} = \begin{bmatrix} \mathbf{p} \\ \mathbf{q} \\ \mathbf{r} \end{bmatrix}$$

to a matrix-vector equation

$$\mathbf{L}\mathbf{x} = \mathbf{b}.$$

Below, the collocation solution $\mathbf{x}$ is compared to

$$\mathbf{y} = \begin{bmatrix} \mathbf{u} \\ \mathbf{v} \\ \mathbf{w} \end{bmatrix},$$

in which $\mathbf{u}$, $\mathbf{v}$, and $\mathbf{w}$ are samples of the exact solution functions u, u', and u'' of (26.18), taken on the same grids as $\mathbf{p}$, $\mathbf{q}$, and $\mathbf{r}$, respectively. The discrete residual is $\boldsymbol{\varepsilon} = \mathbf{b} - \mathbf{L}\mathbf{y}$, and the absolute condition number is $\kappa_{abs} = \|\mathbf{L}^{-1}\|_\infty$.

Lemma 26.10. *Suppose the second-order IVP (26.18) is solved by piecewise-uniform collocation. If $u \in C^{m+3}[a,b]$, then*

$$\|\varepsilon\|_\infty \le CN^{-(m+1)}$$

for a coefficient C that is constant with respect to n. If m is even and $u \in C^{m+4}[a,b]$ and either $m \ge 2$ or $c_{12} = c_{22} = 0$, then the bound can be improved to

$$\|\varepsilon\|_\infty \le CN^{-(m+2)}.$$

The proof is omitted; it is very similar to the proof of Lemma 26.5. The proofs of the following three theorems, which are analogous to the proofs of Theorems 26.6–26.8, are also omitted.

Theorem 26.11. *Suppose the second-order IVP (26.18) is solved by piecewise-uniform collocation. If $u \in C^{m+3}[a,b]$, then*

$$\|\mathbf{u} - \mathbf{p}\|_\infty \le C\kappa_{\text{abs}} N^{-(m+1)}, \tag{26.19}$$

$$\|\mathbf{v} - \mathbf{q}\|_\infty \le C\kappa_{\text{abs}} N^{-(m+1)}, \tag{26.20}$$

$$\|\mathbf{w} - \mathbf{r}\|_\infty \le C\kappa_{\text{abs}} N^{-(m+1)}, \tag{26.21}$$

in which the coefficient C is constant with respect to n. If m is even and $u \in C^{m+4}[a,b]$ and either $m \ge 2$ or $c_{12} = c_{22} = 0$, then the bounds can be improved to the following:

$$\|\mathbf{u} - \mathbf{p}\|_\infty \le C\kappa_{\text{abs}} N^{-(m+2)}, \tag{26.22}$$

$$\|\mathbf{v} - \mathbf{q}\|_\infty \le C\kappa_{\text{abs}} N^{-(m+2)}, \tag{26.23}$$

$$\|\mathbf{w} - \mathbf{r}\|_\infty \le C\kappa_{\text{abs}} N^{-(m+2)}. \tag{26.24}$$

Theorem 26.12. *Suppose the second-order IVP (26.18) is solved by piecewise-uniform collocation. If $u \in C^{m+3}[a,b]$, then*

$$\|u - p\|_\infty \le C_1 (\kappa_{\text{abs}} \vee 1) N^{-(m+1)}, \tag{26.25}$$

$$\|u' - p'\|_\infty \le C_2 (\kappa_{\text{abs}} \vee 1) N^{-(m+1)}, \tag{26.26}$$

$$\|u'' - p''\|_\infty \le C_3 (\kappa_{\text{abs}} \vee 1) N^{-(m+1)}, \tag{26.27}$$

in which the coefficients C_1, C_2, and C_3 are constant with respect to n. If m is even and $u \in C^{m+4}[a,b]$ and either $m \ge 2$ or $c_{12} = c_{22} = 0$, then the first two bounds can be improved to

$$\|u - p\|_\infty \le C_1 (\kappa_{\text{abs}} \vee 1) N^{-(m+2)}, \tag{26.28}$$

$$\|u' - p'\|_\infty \le C_2 (\kappa_{\text{abs}} \vee 1) N^{-(m+2)}. \tag{26.29}$$

Theorem 26.13. *Suppose the second-order IVP (26.18) is solved by piecewise-uniform collocation. If $u \in C^{m+3}[a,b]$, then the residual $R(t) = g(t) - \alpha(t)p(t) - \beta(t)p'(t) - \gamma(t)p''(t)$ satisfies*

$$\|R\|_\infty \le C(\kappa_{\text{abs}} \vee 1) N^{-(m+1)}$$

for a coefficient C that is constant with respect to n.

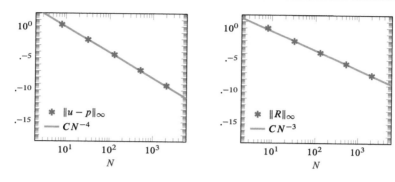

Figure 26.2. *Rates of convergence for second-order collocation on a piecewise-uniform grid of degree m = 2: absolute error (left) and absolute residual (right).*

Example 26.14. Predict the rates of convergence for $\|u - p\|_\infty$ and $\|R\|_\infty$ when the damped harmonic oscillator problem

$$u(t) + 0.1u'(t) + u''(t) = 0, \quad u(0) = 0.2, \quad u'(0) = 0$$

over $0 \le t \le 30$ is solved by piecewise-uniform collocation with $m = 2$. Then test your prediction.

Solution. Because the coefficient functions and the right-hand side of the differential equation are infinitely differentiable, the solution is also infinitely differentiable. With $m = 2$, we expect the error $\|u - p\|_\infty$ to converge at the rate $N^{-(m+2)} = N^{-4}$ and the residual $\|R\|_\infty$ to converge at the rate $N^{-(m+1)} = N^{-3}$, as long as the problem is reasonably well conditioned.

To test this, the problem is solved numerically.

```
>> K = 1; B = 0.1; M = 1; g = @(t) 0;
>> a = 0; b = 30;
>> ua = 0.2; va = 0;
```

The grids are uniform grids with $m = 2$ and increasing values of n.

```
>> m = 2; n = 4.^(1:5); N = m*n;
```

Recall that the exact solution is known for this problem.

```
>> la = sqrt(4*K*M-B^2)/(2*M);
>> u = @(t) ua*exp(-(B/(2*M))*t)*(cos(la*t)+(B/(2*la*M))*sin(la*t));
```

Now we can compute and compare.

```
>> err = nan(size(n)); res = nan(size(n));
>> for k = 1:length(n)
     [ps,qs,rs] = ivpl2uni(@(t) K,@(t) B,@(t) M,g,[a b] ...
                    ,[ 1 0 0; 0 1 0 ],[ua; va],m,n(k));
     p = interpuni(ps,[a b]);
     q = interpuni(qs,[a b]);
     r = interpuni(rs,[a b]);
     err(k) = infnorm(@(t) u(t)-p(t),[a b],100);
     R = @(t) g(t)-K*p(t)-B*q(t)-M*r(t);
     res(k) = infnorm(R,[a b],100);
   end
```

The results are plotted in Figure 26.2 with power functions that roughly fit the data.

```
>> ax = newfig(1,2); xlog; ylog; ylim([1e-18 1e2]);
>> plot(N,err,'*'); plotfun(@(N) 8000*N^(-4),[2 6000]);
>> subplot(ax(2)); xlog; ylog; ylim([1e-18 1e2]);
```

```
>> plot(N,res,'*'); plotfun(@(N) 300*N^(-3),[2 6000]);
```
The observed convergence rates match the predictions. ∎

Notes

Our error analysis involves the condition number of a single linear system that defines the solution over the entire domain. This allows us to avoid tracking errors across subintervals. There are other approaches [39, 40].

Exercises

Exercises 1–4: Suppose u is a solution to the given linear differential equation on the given interval. Does Corollary 26.2 or 26.3 imply that $u \in C^l[a, b]$ for some l? If so, what is the greatest value of l that can be justified by the theorem?

1. $t^{3/2}u + u' = t, 0 \le t \le 5$
2. $u + e^t u' = 1, 0 \le t \le 3$
3. $\sqrt{t}u + u'' = t^{-1}, 1 \le t \le 10$
4. $t^3(\log t)u + u' + u'' = 0, 0 \le t \le 5$. (Take $\alpha(0) = \lim_{t \to 0+} t^3(\log t)$.)

Exercises 5–8: Let p be the m-by-n piecewise-uniform collocation solution to the given IVP on the given interval. Predict the rates of convergence for $\|u - p\|_\infty$ and $\|R\|_\infty$ as $N \to \infty$ using the theory of this chapter. Then measure the rates experimentally. The exact solution u is provided so that you can measure $\|u - p\|_\infty$ directly.

5. $u + u' = 2, u(0) = 4; 0 \le t \le 5; m = 1; u(t) = 2(1 + e^{-t})$
6. $(\sin t)u + u' = 0, u(0) = 1; 0 \le t \le 5; m = 2; u(t) = e^{-1+\cos t}$
7. $-tu + u' = 1, u(0) = 1; 0 \le t \le 2; m = 2; u(t) = \frac{1}{2}e^{t^2/2}(2 + \sqrt{2\pi}\,\text{erf}(t/\sqrt{2}))$
8. $-u + tu' = t, u(1) = 0; 1 \le t \le 4; m = 3; u(t) = t \log t$

Exercises 9–12: Let p be the m-by-n piecewise-uniform collocation solution to the given IVP on the given interval. Predict the rate of convergence for $\|R\|_\infty$ as $N \to \infty$ using the theory of this chapter. Then measure the rate experimentally.

9. $2u + 2u' + u'' = 0, u(0) = 1, u'(0) = 0; 0 \le t \le 4\pi; m = 1$
10. $t^2 u + u' + u'' = 0, u(0) = 1, u'(0) = 0; 0 \le t \le 5; m = 0$
11. $(t^2 - 1)u + tu' + (t^2 + 1)u'' = 0, u(0) = 1/2, u'(0) = 0; 0 \le t \le 20; m = 2$
12. $u + u'' = \exp(-t^2), u(0) = 1, u'(0) = -1; 0 \le t \le 10; m = 3$

13. Consider the IVP $u + u' = |t|, u(-1) = 2 - e$ on $-1 \le t \le \log 2$.

   (a) What, if anything, do the theorems of this chapter imply about the rate of convergence of piecewise-uniform collocation on this problem? Explain.

   (b) Solve the IVP by collocation on a 0-by-n piecewise-uniform grid and measure the rates of convergence for $\|u - p\|_\infty$, $\|u' - p'\|_\infty$, and $\|R\|_\infty$ as $N \to \infty$. Measure the absolute errors by comparing with the exact solution $u(t) = \text{sign}(t)(-1 + t + e^{-t})$. (You may need to use a large number of sample points in infnorm.)

(c) Repeat part (b) with 1-by-n piecewise-uniform grids.

(d) Repeat part (b) with 2-by-n piecewise-uniform grids.

14. Consider solving the IVP $u + tu' = e^t$, $u'(0) = 1/2$ by collocation on a 2-by-n piecewise-uniform grid over $0 \leq t \leq 2$. What do the theorems of this chapter say about rate of convergence on this problem? Explain. Then measure the rate of convergence for $\|R\|_\infty$ experimentally.

15. Consider solving the IVP $\sqrt{t}\, u + u' = 0$, $u(0) = 1$ by collocation on a 2-by-n piecewise-uniform grid over $0 \leq t \leq 5$. What do the theorems of this chapter say about rate of convergence on this problem? Explain. Then measure the rate of convergence for $\|R\|_\infty$ experimentally.

16. The *elliptic integral* $E(x)$ satisfies the IVP

$$u + 4(1-x)\frac{du}{dx} + 4x(1-x)\frac{d^2u}{dx^2} = 0, \quad u|_{x=0} = \frac{\pi}{2}, \quad \frac{d^2u}{dx^2}\bigg|_{x=0} = -\frac{3\pi}{64}.$$
(26.30)

Deduce the value of $\frac{du}{dx}\big|_{x=0}$. Then apply the change of variables $x = 1 - e^{1-t}$ to show that the above IVP is equivalent to

$$\frac{1}{4}e^{1-t}u + \frac{du}{dt} + (1-e^{1-t})\frac{d^2u}{dt^2} = 0, \quad u|_{t=1} = \frac{\pi}{2}, \quad \frac{d^2u}{dt^2}\bigg|_{t=1} = \frac{5\pi}{64}. \quad (26.31)$$

17. Compute the elliptic integral $E(x)$ on $0 \leq x \leq 1 - 10^{-15}$ in two ways.

(a) Solve (26.30) by collocation on a 3-by-n piecewise-uniform grid over $0 \leq x \leq 1 - 10^{-15}$. Does the residual converge toward zero as $N \to \infty$? If so, at what rate? If not, why might we anticipate a problem based on the theory of this chapter?

(b) Solve (26.31) by collocation on a 3-by-n piecewise-uniform grid. (Over what interval?) Does the residual converge toward zero as $N \to \infty$? If so, at what rate? If not, why not?

(c) Compute $E(x)$ by solving (26.31) and then reversing the earlier change of variables to express E as a function of x rather than t. Plot the resulting approximation for $E(x)$ over $0 \leq x \leq 1 - 10^{-15}$.

18. Theorem 26.7 suggests that piecewise-uniform collocation may converge particularly slowly, at the rate N^{-1}, when $m = 0$ and $c_1 \neq 0$. Demonstrate this behavior.

19. Let $p(t)$ be a collocation solution for the first-order IVP (26.4) on an m-by-n piecewise-uniform grid. The *iterated collocation* solution for $u'(t)$ is $\hat{q}(t) = (g(t) - \alpha(t)p(t))/\beta(t)$. Repeat Example 26.4, but this time measure the rates of convergence for $\|u' - p'\|_\infty$ and $\|u' - \hat{q}\|_\infty$. For which degrees m does iterated collocation appear to converge at a strictly faster rate than ordinary collocation?

20. Let u be a solution to the first-order IVP (26.4), and define the iterated collocation solution $\hat{q}$ as in the previous exercise. Suppose $\|\alpha/\beta\|_\infty < \infty$, m is even, $u \in C^{m+3}[a,b]$, and either $m \geq 2$ or $c_1 = 0$. Prove that $\|u' - \hat{q}\|_\infty \leq C(\kappa_{abs} \vee 1)N^{-(m+2)}$ for a scalar C that is constant with respect to n.

21. Let $p(t)$ be a collocation solution for the second-order IVP (26.18) on an m-by-n piecewise-uniform grid, and define the iterated collocation solution $\hat{r}(t) = (g(t) - \alpha(t)p(t) - \beta(t)p'(t))/\gamma(t)$. Suppose $\|\alpha/\gamma\|_\infty < \infty$, $\|\beta/\gamma\|_\infty < \infty$, m is even, $u \in C^{m+4}[a,b]$, and either $m \geq 2$ or $c_{12} = c_{22} = 0$. Prove that $\|u'' - \hat{r}\|_\infty \leq C(\kappa_{abs} \vee 1)N^{-(m+2)}$ for a scalar C that is constant with respect to n.

Chapter 27

Collocation on a Chebyshev grid

In this chapter, we specialize collocation to Chebyshev grids. The process is remarkably simple, and the resulting methods are capable of remarkably high accuracy.

27.1 ▪ Collocation equations

To implement Chebyshev collocation, the methods of Chapter 23 are specialized to Chebyshev grids.

For the first-order problem

$$\alpha(t)u(t) + \beta(t)u'(t) = g(t), \quad c_0 u(a) + c_1 u'(a) = d \qquad (27.1)$$

on $[a, b]$, the grids $\mathbf{t}$, $\mathbf{t}'$, and $\mathbf{s}$ are Chebyshev grids of degrees $m + 1$, m, and m, respectively, on $[a, b]$. The linear system (23.8) becomes

$$\begin{bmatrix} \mathbf{J}_m & -l\hat{\mathbf{K}}_m \\ \mathbf{A} & \mathbf{B} \\ c_0 \mathbf{e}_1^T & c_1 \mathbf{e}_1^T \end{bmatrix} \begin{bmatrix} \mathbf{p} \\ \mathbf{q} \end{bmatrix} = \begin{bmatrix} \mathbf{0} \\ \mathbf{g} \\ d \end{bmatrix}, \qquad (27.2)$$

whose components are defined as follows:

- $l = (b - a)/2$, and $\mathbf{J}_m$ and $\hat{\mathbf{K}}_m$ are defined as in (17.4);

- $\mathbf{A} = \operatorname{diag}(\alpha(s_0), \ldots, \alpha(s_m))\mathbf{E}_{m,m+1}$ and $\mathbf{B} = \operatorname{diag}(\beta(s_0), \ldots, \beta(s_m))$, with $\mathbf{E}_{m,m+1}$ the resampling matrix from a Chebyshev grid of degree $m + 1$ to another of degree m (see Theorem 12.7);

- $\mathbf{e}_1$ denotes a vector with a single 1 followed by as many zeros as necessary; and

- $\mathbf{g} = g|_{\mathbf{s}}$.

The vector solutions $\mathbf{p}$ and $\mathbf{q}$ determine the collocation solution p and its derivative.

For the second-order problem

$$\alpha(t)u(t) + \beta(t)u'(t) + \gamma(t)u''(t) = g(t), \quad c_{10} u(a) + c_{11} u'(a) + c_{12} u''(a) = d_1,$$
$$c_{20} u(a) + c_{21} u'(a) + c_{22} u''(a) = d_2,$$

305

$\mathbf{t}$, $\mathbf{t'}$, $\mathbf{t''}$, and $\mathbf{s}$ are Chebyshev grids of degrees $m + 2$, $m + 1$, m, and m, respectively. The collocation solution is determined by

$$
\begin{bmatrix}
\mathbf{J}_{m+1} & -l\hat{\mathbf{K}}_{m+1} & \mathbf{0} \\
\mathbf{0} & \mathbf{J}_m & -l\hat{\mathbf{K}}_m \\
\mathbf{A} & \mathbf{B} & \mathbf{C} \\
c_{10}\mathbf{e}_1^T & c_{11}\mathbf{e}_1^T & c_{12}\mathbf{e}_1^T \\
c_{20}\mathbf{e}_1^T & c_{21}\mathbf{e}_1^T & c_{22}\mathbf{e}_1^T
\end{bmatrix}
\begin{bmatrix}
\mathbf{p} \\
\mathbf{q} \\
\mathbf{r}
\end{bmatrix}
=
\begin{bmatrix}
\mathbf{0} \\
\mathbf{0} \\
\mathbf{g} \\
d_1 \\
d_2
\end{bmatrix},
$$

whose components are defined analogously to those in the first-order problem.

Routines `ivpl1cheb` and `ivpl2cheb` compute collocation solutions for first-order and second-order linear IVPs, respectively, on Chebyshev grids. The implementation of `ivpl1cheb` is shown in Listing 27.1, along with its helper routine `ivp1matcheb`. The similar implementation of `ivpl2cheb` is omitted here.

Listing 27.1. Solve a first-order linear IVP by collocation on a Chebyshev grid.

```
function [ps,qs] = ivpl1cheb(al,be,g,ab,c,d,m)

a = ab(1);  b = ab(2);
als = samplecheb(al,[a b],m);
bes = samplecheb(be,[a b],m);
gs = samplecheb(g,[a b],m);
[J,K,E1,E2,Ea1,Ea2] = ivp1matcheb(m,b-a);
[ps,qs] = ivpl1_(als,bes,gs,c,d,J,K,E1,E2,Ea1,Ea2);
```

```
function [J,K,E1,E2,Ea1,Ea2] = ivp1matcheb(m,l)

[J,K] = indefinitecheb(m);  K = 1/2*K;
E1 = resamplematrixcheb(m,m+1);
E2 = eye(m+1);
Ea1 = [ 1 zeros(1,m+1) ];
Ea2 = [ 1 zeros(1,m) ];
```

27.2 ▪ Examples

Recall that a collocation solution $p(t)$ to (27.1) makes the residual $R(t) = g(t) - \alpha(t)p(t) - \beta(t)p'(t)$ equal zero at the collocation nodes, which are now Chebyshev nodes. The next example illustrates this.

Example 27.1. Compute the collocation solution to

$$
tu(t) + u'(t) = 0, \quad u(0) = 1
$$

on a Chebyshev grid of degree $m = 15$ over $0 \le t \le 4$. Plot the solution components and the absolute residual.

Solution. The problem is defined.
```
>> al = @(t) t;  be = @(t) 1;  g = @(t) 0;
>> a = 0;  b = 4;
>> ua = 1;
```
The collocation solution is computed.

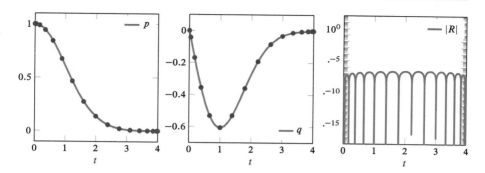

Figure 27.1. *A collocation solution to a first-order linear IVP on a Chebyshev grid (left), the derivative of the solution (middle), and the absolute residual (right).*

```
>> m = 15;
>> [ps,qs] = ivpl1cheb(al,be,g,[a b],[1 0],ua,m);
>> p = interpcheb(ps,[a b]);
>> q = interpcheb(qs,[a b]);
```

The solution components are graphed in the left and middle plots of Figure 27.1.

```
>> ax = newfig(1,3);
>> plotsample(gridcheb([a b],m+1),ps); plotfun(p,[a b]);
>> subplot(ax(2));
>> plotsample(gridcheb([a b],m),qs); plotfun(q,[a b]);
```

The absolute residual is graphed in the plot on the right.

```
>> R = @(t) g(t)-al(t)*p(t)-be(t)*q(t);
>> subplot(ax(3)); ylog; ylim([1e-18 1e2]);
>> plotfun(@(t) abs(R(t)),[a b]);
```

As expected, the absolute residual equals zero at the Chebyshev nodes. ∎

As evident from the three examples below, collocation on a Chebyshev grid is capable of very high accuracy.

Example 27.2. Compute

$$tu(t) + u'(t) = 0, \quad u(0) = 1$$

on $0 \leq t \leq 4$ by Chebyshev collocation. Find a degree m for which the absolute residual is below 10^{-14}, and measure the absolute error $\|u - p\|_\infty$ for this choice of m as well.

Solution. The problem is defined as in the previous example.

```
>> al = @(t) t; be = @(t) 1; g = @(t) 0;
>> a = 0; b = 4;
>> ua = 1;
```

Through trial and error, a Chebyshev grid of degree $m = 26$ has been found to satisfy the residual requirement.

```
>> m = 26;
>> [ps,qs] = ivpl1cheb(al,be,g,[a b],[1 0],ua,m);
>> p = interpcheb(ps,[a b]);
>> q = interpcheb(qs,[a b]);
>> infnorm(@(t) g(t)-al(t)*p(t)-be(t)*q(t),[a b],100)
ans =
      7.938094626069869e-15
```

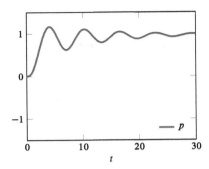

Figure 27.2. *Chebyshev collocation solution for a forced mass-spring system.*

Finally, we compare the computed solution with the known exact solution.

```
>> u = @(t) exp(-t^2/2);
>> infnorm(@(t) u(t)-p(t),[a b],100)
ans =
      1.554312234475219e-15
```

The absolute error is also tiny. ∎

In the previous example, a grid of degree just 26 nearly exhausts the precision of the computer hardware.

Example 27.3. The following IVP models a damped mass-spring system that is also subject to an external force:

$$u(t) + 0.2u'(t) + u''(t) = 1 - \frac{1}{t+1}, \quad u(0) = 0, \quad u'(0) = 0.$$

The right-hand side of the differential equation, known as a "forcing function," represents an external force that is applied slowly and approaches 1 N in the $t \to \infty$ limit. Note also that the mass is initially at rest in its natural position, so any movement is caused by the external force. Solve the IVP over $0 \le t \le 30$, finding a Chebyshev grid for which the absolute residual is below 10^{-12}.

Solution. The problem is defined.

```
>> K = 1; B = 0.2; M = 1; g = @(t) 1-1/(t+1);
>> a = 0; b = 30;
>> ua = 0; va = 0;
```

By trial and error, $m = 80$ has been found to satisfy the residual requirement.

```
>> m = 80;
>> [ps,qs,rs] = ivpl2cheb(@(t) K,@(t) B,@(t) M,g,[a b] ...
                         ,[ 1 0 0; 0 1 0 ],[ua; va],m);
>> p = interpcheb(ps,[a b]);
>> q = interpcheb(qs,[a b]);
>> r = interpcheb(rs,[a b]);
>> infnorm(@(t) g(t)-K*p(t)-B*q(t)-M*r(t),[a b],100)
ans =
      2.328692794151266e-13
```

The collocation solution is plotted in Figure 27.2.

```
>> newfig; plotfun(p,[a b]);
```

In the long term, the position of the mass approaches $u = 1$; at this position, the restoring force of the spring perfectly opposes the external force. ∎

In Chapter 25, the Bessel function $J_0(x)$ is computed by collocation on a piecewise-uniform grid. To keep things interesting, the next two examples consider the Bessel function $J_{-2/3}(x)$, which is a bit more exotic and a bit less well behaved than $J_0(x)$.

Example 27.4. The Bessel function $J_{-2/3}(x)$ satisfies the differential equation

$$\left(x^2 - \frac{4}{9}\right) J_{-2/3}(x) + x J'_{-2/3}(x) + x^2 J''_{-2/3}(x) = 0 \tag{27.3}$$

and the asymptotic relationship

$$J_{-2/3}(x) \sim \frac{2^{2/3}}{\Gamma(1/3)} x^{-2/3} - \frac{3}{2^{4/3}\Gamma(1/3)} x^{4/3} \quad (x \to 0^+). \tag{27.4}$$

(The gamma function $\Gamma(x)$ is an extension of the factorial function to real- and complex-valued arguments and is implemented in the MATLAB library as gamma.) Roughly, the asymptotic relationship means that the right-hand side of (27.4) becomes a better and better approximation for $J_{-2/3}(x)$ as x approaches zero. In particular, it implies that the graph of $J_{-2/3}(x)$ has a vertical asymptote at $x = 0$.

To remove the vertical asymptote, the function

$$u(x) = x^{2/3} J_{-2/3}(x)$$

is introduced. Find an IVP that uniquely determines $u(x)$.

Solution. We have the following relationships between $J_{-2/3}(x)$ and $u(x)$:

$$J_{-2/3}(x) = x^{-2/3} u(x),$$

$$J'_{-2/3}(x) = -\frac{2}{3} x^{-5/3} u(x) + x^{-2/3} u'(x),$$

$$J''_{-2/3}(x) = \frac{10}{9} x^{-8/3} u(x) - \frac{4}{3} x^{-5/3} u'(x) + x^{-2/3} u''(x).$$

Substituting into (27.3) gives

$$x^{4/3} u(x) - \frac{1}{3} x^{1/3} u'(x) + x^{4/3} u''(x) = 0,$$

which is equivalent to

$$x u(x) - \frac{1}{3} u'(x) + x u''(x) = 0 \tag{27.5}$$

for $x \neq 0$. The asymptotic behavior for $J_{-2/3}(x)$ as $x \to 0^+$ implies

$$u(x) \sim \frac{2^{2/3}}{\Gamma(1/3)} - \frac{3}{2^{4/3}\Gamma(1/3)} x^2 \quad (x \to 0^+),$$

so the solution can be extended in a continuous way to $x = 0$ with

$$u(0) = \frac{2^{2/3}}{\Gamma(1/3)}, \quad u'(0) = 0, \quad u''(0) = -\frac{3}{2^{1/3}\Gamma(1/3)}. \tag{27.6}$$

At $x = 0$, equation (27.5) reduces to $-(1/3)u'(0) = 0$. Hence, the restriction $u'(0) = 0$ in (27.6) is redundant. Discarding this condition gives the IVP

$$x u(x) - \frac{1}{3} u'(x) + x u''(x) = 0, \quad u(0) = \frac{2^{2/3}}{\Gamma(1/3)}, \quad u''(0) = -\frac{3}{2^{1/3}\Gamma(1/3)}. \quad \blacksquare$$

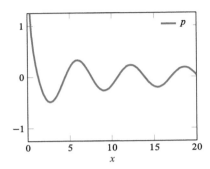

Figure 27.3. *The Bessel function* $J_{-2/3}$ *computed by collocation on a Chebyshev grid.*

Example 27.5. Compute the Bessel function $J_{-2/3}(x)$ as follows. First, solve

$$xu(x) - \frac{1}{3}u'(x) + xu''(x) = 0, \quad u(0) = \frac{2^{2/3}}{\Gamma(1/3)}, \quad u''(0) = -\frac{3}{2^{1/3}\Gamma(1/3)},$$

the IVP derived in the previous example, by Chebyshev collocation on $0 \le x \le 20$ and obtain an absolute residual below 10^{-13}. Then reverse the change of variables $u(x) = x^{2/3}J_{-2/3}(x)$ to find and plot the Bessel function. Finally, compare with the built-in MATLAB function `besselj(-2/3,x)`.

Solution. The IVP is specified.
```
>> al = @(x) x; be = @(x) -1/3; ga = @(x) x; g = @(x) 0;
>> a = 0; b = 20;
>> ua = 2^(2/3)/gamma(1/3); wa = -3/(2^(1/3)*gamma(1/3));
```
Next, it is solved. A Chebyshev grid of degree 34 has been found to satisfy the residual requirement.
```
>> m = 34;
>> [ps,qs,rs] = ivpl2cheb(al,be,ga,g,[a b] ...
                        ,[ 1 0 0; 0 0 1 ],[ua; wa],m);
>> p = interpcheb(ps,[a b]);
>> q = interpcheb(qs,[a b]);
>> r = interpcheb(rs,[a b]);
>> R = @(x) g(x)-al(x)*p(x)-be(x)*q(x)-ga(x)*r(x);
>> infnorm(R,[a b],100)
ans =
      9.059419880941277e-14
```
The Bessel function is reconstructed by reversing the earlier change of variables. It is plotted in Figure 27.3.
```
>> newfig; plotfun(@(x) x^(-2/3)*p(x),[a b]);
```
Finally, we find that the absolute error is similar in size to the absolute residual.
```
>> infnorm(@(x) x^(2/3)*besselj(-2/3,x)-p(x),[a b],100)
ans =
      3.039235529911366e-14
```
Note that the error $u(x) - p(x)$, rather than $J_{-2/3}(x) - x^{-2/3}p(x)$, is measured. We cannot hope to achieve small absolute error for a quantity that tends to infinity. ∎

Notes

The Chebfun system, which was earlier mentioned in the preface of this book, has made Chebyshev collocation available to a wider audience and has spurred technical advancements in the research community. A few references that are relevant in the present context are [23, 24, 25, 73].

Exercises

Exercises 1–8: Solve the IVP over the domain using Chebyshev collocation. Adjust the degree to achieve an absolute residual below 10^{-12}. Plot the solution.

1. $2u + tu' = 0, u(1) = 1; 1 \leq t \leq 5$
2. $(1 - t)u + tu' = 0, u(1) = e; 1 \leq t \leq 4$
3. $(\cos t)u + (\sin t)u' = 1, u'(0) = 0; 0 \leq t \leq 3\pi/4$
4. $(\cos t)u - (\sin t)u' = t, u(\pi/2) = 0; \pi/2 \leq t \leq 0.99\pi$
5. $u + (-3 + t)u' + u'' = 0, u(0) = 5, u'(0) = -1; 0 \leq t \leq 10$
6. $0.1e^t u + 3u' + u'' = 3t, u(0) = 0, u'(0) = 0.1; 0 \leq t \leq 8$
7. $t^{-2}u + u'' = 0, u(1) = 3, u'(1) = 3/2; 1 \leq t \leq 100$
8. $xu + u' + xu'' = 2/\pi, u(0) = 0, u''(0) = 0; 0 \leq x \leq 20$ (Struve function)

9. Solve the forced harmonic oscillator problem

$$u + \frac{1}{2}u' + u'' = \sin(5t), \quad u(0) = 0, \quad u'(0) = 0$$

over $0 \leq t \leq 20$ by Chebyshev collocation. Seek an absolute residual below 10^{-12}. Plot the position function.

10. Consider the forced harmonic oscillator problem

$$u + \frac{1}{2}u' + u'' = \sin(\omega t), \quad u(0) = 0, \quad u'(0) = 0.$$

Solve this IVP by Chebyshev collocation. By trial and error, find, to two significant digits, the value for ω that produces the largest long-term amplitude.

11. Solve the following IVP for the *Airy* Ai *function* on $-20 \leq x \leq 0$:

$$-xu + u'' = 0, \quad u(0) = \frac{1}{3^{2/3}\Gamma(2/3)}, \quad u'(0) = -\frac{1}{3^{1/3}\Gamma(1/3)}.$$

Note that the "initial" condition is located at the right end of the interval. Seek numerical convergence.

12. Let $J_0(x)$ be the Bessel function of order zero, which satisfies the IVP (22.8), and let $J_1(x) = -J_0'(x)$.

    (a) Prove that $xJ_0'''(x) = -J_0(x) - xJ_0'(x) - 2J_0''(x)$.
    (b) Show that $J_1(x)$ satisfies the IVP

$$(x^2 - 1)u + xu' + x^2 u'' = 0, \quad u'(0) = \frac{1}{2}, \quad u''(0) = 0. \quad (27.7)$$

    Hint. See Exercise 13 in Chapter 22.

  (c) Compute $J_1(x)$ on $0 \leq x \leq 20$ by solving (27.7) by Chebyshev collocation. Obtain an absolute residual below 10^{-10}. Plot the approximation.

13. Recall that the Bessel function $J_0(x)$ satisfies the differential equation

$$xu + u' + xu'' = 0. \tag{27.8}$$

Because this equation is second order, we can find a second, linearly independent solution. A common approach is to seek a solution of the form

$$Y_0(x) = \frac{2}{\pi} \left(J_0(x) \left(\log \frac{x}{2} + \gamma \right) + y(x) \right),$$

in which $\gamma = 0.5772156649015328\ldots$ is a constant known as the Euler–Mascheroni constant.

  (a) Prove that $Y_0(x)$ satisfies (27.8) if and only if $y(x)$ satisfies

$$xy + y' + xy'' = 2J_1(x).$$

  (The function $J_1(x) = -J_0'(x)$ is introduced in Exercise 12.)

  (b) Solve the IVP

$$xy + y' + xy'' = 2J_1(x), \quad y(0) = 0, \quad y''(0) = 1/2$$

on $0 \leq x \leq 20$ by Chebyshev collocation, achieving an absolute residual below 10^{-10}. From this solution $y(x)$, construct and plot $Y_0(x)$, a *Bessel function of the second kind*.

14. The following is a second-order linear BVP:

$$\alpha(x)u + \beta(x)u' + \gamma(x)u'' = g(x), \quad c_{10}u(a) + c_{11}u'(a) + c_{12}u''(a) = d_1,$$
$$c_{20}u(b) + c_{21}u'(b) + c_{22}u''(b) = d_2.$$

(See also Exercise 14 in Chapter 23.) Develop a routine for solving the BVP above by Chebyshev collocation. (*Hint.* Duplicate and modify ivpl2_, ivp2matcheb, and ivpl2cheb.) Demonstrate your routine on the hanging-cable problem $u - u'' = 0$, $u(-1) = 1, u(1) = 1$, whose solution is $u(x) = (e^x + e^{-x})/(e + e^{-1})$.

15. Let $A(t)$ and $B(t)$ be r-by-r matrix-valued functions of t, and let $g(t)$ be an r-by-1 vector-valued function of t. Consider the vector-valued IVP

$$A(t)u(t) + B(t)u'(t) = g(t), \quad c_0u(a) + c_1u'(a) = d.$$

The solution is a vector-valued function $u(t) = (u_1(t), \ldots, u_r(t))$. Develop a MATLAB routine to solve this IVP by collocation on a Chebyshev grid. It should return an r-by-$(m+2)$ matrix $\mathbf{p} = [p_{ij}]$ in which $p_{ij} = p_i(x_j)$ and an r-by-$(m+1)$ matrix $\mathbf{q} = [q_{ij}]$ in which $q_{ij} = p_i'(x_j')$. Demonstrate your implementation by solving

$$\begin{bmatrix} 1/2 & -1 \\ 1 & 1/2 \end{bmatrix} \begin{bmatrix} u_1 \\ u_2 \end{bmatrix} + \begin{bmatrix} 1 & 0 \\ 0 & 1 \end{bmatrix} \begin{bmatrix} u_1' \\ u_2' \end{bmatrix} = \begin{bmatrix} 0 \\ 0 \end{bmatrix}, \quad \begin{bmatrix} u_1(0) \\ u_2(0) \end{bmatrix} = \begin{bmatrix} 1 \\ 0 \end{bmatrix}$$

on $0 \leq t \leq 10$. *Hint.* Solve Exercises 16–17 in Chapter 23 first.

Chapter 28

Accuracy of collocation on a Chebyshev grid

Chebyshev collocation performs best on linear differential equations that have very smooth coefficient functions. Below, we see that a class of linear differential equations with analytic coefficients has analytic solutions and that Chebyshev collocation converges geometrically on such problems.

28.1 ▪ Analytic equations

In this chapter, we focus on linear differential equations with analytic coefficients. Recall that a function is analytic in a domain of the complex plane if it is complex differentiable everywhere in the domain. Such a function is necessarily infinitely differentiable and therefore very smooth.

Theorem 28.1. *Let D be the interior of a Bernstein ellipse, and let $a \in D$. Consider the IVP*

$$\alpha(t)u(t) + \beta(t)u'(t) = g(t), \quad u(a) = u_a. \tag{28.1}$$

If $\alpha(z)$, $\beta(z)$, and $g(z)$ are analytic in D and $\beta(z) \neq 0$ for all $z \in D$, then the IVP has a unique analytic solution in D. The statement is also true with the entire complex plane in place of D.

A point z where $\alpha(z)$, $\beta(z)$, or $g(z)$ fails to be analytic or where $\beta(z) = 0$ is called a *singular point* of the differential equation in (28.1). An analytic solution may or may not exist in a neighborhood of such a point. Although the theory of singular points is beyond the scope of this text, we can experiment with such problems using our numerical methods.

An analogous result holds for second-order equations.

Theorem 28.2. *Let D be the interior of a Bernstein ellipse, and let $a \in D$. Consider the IVP*

$$\alpha(t)u(t) + \beta(t)u'(t) + \gamma(t)u''(t) = g(t), \quad u(a) = u_a, \quad u'(a) = v_a. \tag{28.2}$$

If $\alpha(z)$, $\beta(z)$, $\gamma(z)$, and $g(z)$ are analytic in D and $\gamma(z) \neq 0$ for all $z \in D$, then the IVP has a unique analytic solution in D. The statement is also true with the entire complex plane in place of D.

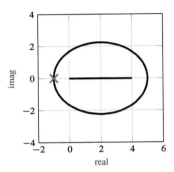

Figure 28.1. *The singular point of* $(t + 1)^{-1}u(t) + u'(t) = 0$ *and a Bernstein ellipse with foci at* $t = 0$ *and* $t = 4$.

Example 28.3. What does Theorem 28.1 imply about the smoothness of the solution to

$$\frac{1}{t+1}u(t) + u'(t) = 0, \quad u(0) = 1 \tag{28.3}$$

on $0 \le t \le 4$?

Solution. The problem has the form (28.1) with $\alpha(z) = 1/(z + 1)$, $\beta(z) = 1$, and $g(z) = 0$. The only singularity among this trio is at $z = -1$, which sits on the Bernstein ellipse $E_\rho(a, b)$ with $a = 0$, $b = 4$, and the following value of ρ:

```
>> a = 0; b = 4;
>> w = 2/(b-a)*(-1-(a+b)/2);
>> rho = abs(w-sqrt(w^2-1))
rho =
    2.618033988749895
```

(Compare with (11.5)–(11.6).) See Figure 28.1.

```
>> newfig; plot(-1,0,'x'); bernstein([a b],rho);
```

In addition, $\beta(z)$ never vanishes. By Theorem 28.1, a unique analytic solution exists in the interior of the Bernstein ellipse in Figure 28.1, which contains the real interval $0 \le t \le 4$. ∎

Once again, a question about the real line is answered by appealing to the complex plane.

28.2 ▪ Rate of convergence for first-order problems

If the exact solution u to

$$\alpha(t)u(t) + \beta(t)u'(t) = g(t), \quad c_0 u(a) + c_1 u'(a) = d \tag{28.4}$$

is analytic in a sufficiently large region, then Chebyshev collocation converges geometrically. The next few results make this statement precise. The requirement in each of the theorems below is that, when the problem is stated on the interval $[a, b]$, the solution u is analytic inside a Bernstein ellipse with foci at a and b. To avoid excessive repetition, the foci are not mentioned again explicitly.

The Chebyshev collocation solution p is determined by the linear system

$$\mathbf{Lx} = \mathbf{b}$$

of the previous chapter, in which $\mathbf{p} = p|_\mathbf{t}$ and $\mathbf{q} = p'|_{\mathbf{t'}}$ and

$$\mathbf{L} = \begin{bmatrix} \mathbf{J}_m & -l\hat{\mathbf{K}}_m \\ \mathbf{A} & \mathbf{B} \\ c_0\mathbf{e}_1^T & c_1\mathbf{e}_1^T \end{bmatrix}, \quad \mathbf{x} = \begin{bmatrix} \mathbf{p} \\ \mathbf{q} \end{bmatrix}, \quad \mathbf{b} = \begin{bmatrix} \mathbf{0} \\ \mathbf{g} \\ d \end{bmatrix}.$$

The first result concerns the residual $\boldsymbol{\varepsilon} = \mathbf{b} - \mathbf{L}\mathbf{y}$, in which

$$\mathbf{y} = \begin{bmatrix} \mathbf{u} \\ \mathbf{v} \end{bmatrix}$$

contains samples $\mathbf{u} = u|_\mathbf{t}$ and $\mathbf{v} = u'|_{\mathbf{t'}}$ of the exact solution to the IVP.

Lemma 28.4. *If u is analytic in the interior of a Bernstein ellipse of parameter ρ and its derivative is bounded in the ellipse, then*

$$\|\boldsymbol{\varepsilon}\|_\infty \le C\rho^{-m}$$

for a coefficient C that is constant with respect to m.

Proof. Let $\tilde{p}$ and $\tilde{q}$ be the Chebyshev interpolating polynomials of degrees $m + 1$ and m for u and u', respectively. By Lemma 24.1,

$$\|\boldsymbol{\varepsilon}\|_\infty \le C_1\|u - \tilde{p}\|_\infty + C_2\|u' - \tilde{q}\|_\infty.$$

The derivative u' must be analytic in the same Bernstein ellipse as u (Appendix B). Because u' is bounded by some constant M_2 in the ellipse, u must also be bounded by some constant M_1 in the ellipse. By Theorem 11.3,

$$\|u - \tilde{p}\|_\infty \le \frac{4M_1}{\rho - 1}\rho^{-(m+1)} = \frac{4M_1}{\rho(\rho - 1)}\rho^{-m}$$

and

$$\|u' - \tilde{q}\|_\infty \le \frac{4M_2}{\rho - 1}\rho^{-m}.$$

Let $C = (4C_1M_1)/(\rho(\rho - 1)) + (4C_2M_2)/(\rho - 1)$. $\quad\square$

The next result concerns the discrete errors $\mathbf{u} - \mathbf{p}$ and $\mathbf{v} - \mathbf{q}$. As usual, κ_abs denotes the absolute condition number of $\mathbf{L}$.

Theorem 28.5. *Suppose (28.4) is solved by Chebyshev collocation. If the solution u is analytic in the interior of a Bernstein ellipse of parameter ρ and its derivative is bounded in the ellipse, then*

$$\|\mathbf{u} - \mathbf{p}\|_\infty \le C\kappa_\text{abs}\rho^{-m}, \tag{28.5}$$
$$\|\mathbf{v} - \mathbf{q}\|_\infty \le C\kappa_\text{abs}\rho^{-m}, \tag{28.6}$$

in which C is constant with respect to m.

Proof. This is an immediate consequence of Theorem 24.2 and Lemma 28.4. $\quad\square$

The continuous errors $\|u - p\|_\infty$ and $\|u' - p'\|_\infty$ are considered next.

Theorem 28.6. *Suppose* (28.4) *is solved by collocation on a Chebyshev grid of degree* $m > 1$. *If the solution u is analytic in the interior of a Bernstein ellipse of parameter ρ and its derivative is bounded in the ellipse, then*

$$\|u - p\|_\infty \le C_1(\kappa_{\text{abs}} \vee 1)(\log m)\rho^{-m}, \tag{28.7}$$
$$\|u' - p'\|_\infty \le C_2(\kappa_{\text{abs}} \vee 1)(\log m)\rho^{-m} \tag{28.8}$$

for coefficients C_1 and C_2 that are constant with respect to m.

Proof. By Theorem 11.14, the Lebesgue constant Λ_m satisfies $\lim_{m\to\infty} \Lambda_m/\log m = 2/\pi$. Thus, $\Lambda_m/\log m$ is bounded by some constant C_3 for all $m \ge 2$. Theorems 24.3 and 28.5 imply

$$\|u - p\|_\infty \le \|u - \tilde{p}\|_\infty + \Lambda_{m+1}\|\mathbf{u} - \mathbf{p}\|_\infty \le \frac{4M_1}{\rho - 1}\rho^{-(m+1)} + (C_3 \log(m+1))(C\kappa_{\text{abs}}\rho^{-m})$$

and

$$\|u' - p'\|_\infty \le \|u' - \tilde{q}\|_\infty + \Lambda_m\|\mathbf{v} - \mathbf{q}\|_\infty \le \frac{4M_2}{\rho - 1}\rho^{-m} + (C_3 \log m)(C\kappa_{\text{abs}}\rho^{-m}).$$

Because $\log m > 1/2$ and $\log(m + 1) < 2\log m$ for all $m \ge 2$,

$$\|u - p\|_\infty \le \left(\frac{8M_1}{\rho(\rho - 1)} + 2C_3 C\right)(\kappa_{\text{abs}} \vee 1)(\log m)\rho^{-m}$$

and

$$\|u' - p'\|_\infty \le \left(\frac{8M_2}{\rho - 1} + C_3 C\right)(\kappa_{\text{abs}} \vee 1)(\log m)\rho^{-m}.$$

Let $C_1 = (8M_1)/(\rho(\rho - 1)) + 2C_3 C$ and $C_2 = (8M_2)/(\rho - 1) + C_3 C$.   □

A factor of $\log m$ appears in the error bounds (28.7)–(28.8). Although $\log m$ grows without bound, its very slow growth is insignificant compared to the rapid decay of ρ^{-m} as $m \to \infty$, so that the rate of convergence for $(\log m)\rho^{-m}$ is still geometric (Exercise 13).

The final accuracy theorem for first-order problems concerns the continuous residual.

Theorem 28.7. *Suppose* (28.4) *is solved by collocation on a Chebyshev grid of degree* $m > 1$. *If the solution u is analytic in the interior of a Bernstein ellipse of parameter ρ and its derivative is bounded in the ellipse, then the continuous residual $R(t) = g(t) - \alpha(t)p(t) - \beta(t)p'(t)$ satisfies*

$$\|R\|_\infty \le C(\kappa_{\text{abs}} \vee 1)(\log m)\rho^{-m} \tag{28.9}$$

for a coefficient C that is constant with respect to m.

Proof. By Theorems 24.4 and 28.6,

$$\|R\|_\infty \le (C_1\|\alpha\|_\infty + C_2\|\beta\|_\infty)(\kappa_{\text{abs}} \vee 1)(\log m)\rho^{-m}.$$

Let $C = C_1\|\alpha\|_\infty + C_2\|\beta\|_\infty$.   □

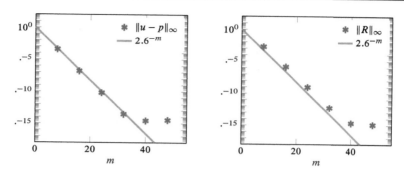

Figure 28.2. *Rate of convergence for Chebyshev collocation on a problem with an analytic solution: absolute error (left) and absolute residual (right).*

Remark 28.8. The proofs of the preceding theorems involve bounds M_1 and M_2 on $|u(z)|$ and $|u'(z)|$ in the Bernstein ellipse. What happens if u is analytic but unbounded? A similar concern was raised earlier in Corollary 11.4, and the same resolution is possible here. Although u may be unbounded in an ellipse of parameter ρ, it and its derivative must be bounded in any smaller ellipse of parameter $\rho_* < \rho$. Hence, the lemma and theorems apply with ρ_* in place of ρ.

Example 28.9. Predict the rate of convergence of Chebyshev collocation on the problem (28.3) over $0 \leq t \leq 4$. Then test your prediction experimentally.

Solution. As discussed in Example 28.3, the IVP has a solution u that is analytic in a Bernstein ellipse of parameter $\rho = 2.6\ldots$ with foci at $a = 0$ and $b = 4$. Thus, if the problem is well conditioned, we expect Chebyshev collocation to converge at a rate within a logarithmic factor of ρ^{-m}, or a slightly slower rate if Remark 28.8 must be invoked.

To test the prediction, we measure the absolute error $\|u - p\|_\infty$ and the residual $\|R\|_\infty$. Note that the exact solution to the IVP is $u(t) = 1/(t + 1)$.

```
>> al = @(t) 1/(t+1); be = @(t) 1; g = @(t) 0;
>> a = 0; b = 4;
>> ua = 1;
>> m = 8:8:48;
>> u = @(t) 1/(t+1);
>> err = nan(size(m)); res = nan(size(m));
>> for k = 1:length(m)
       [ps,qs] = ivpllcheb(al,be,g,[a b],[1 0],ua,m(k));
       p = interpcheb(ps,[a b]);
       q = interpcheb(qs,[a b]);
       err(k) = infnorm(@(t) u(t)-p(t),[a b],100);
       res(k) = infnorm(@(t) g(t)-al(t)*p(t)-be(t)*q(t),[a b],100);
   end
```

The results in Figure 28.2 indicate geometric convergence, at a rate nearly equal to 2.6^{-m}, as predicted.

```
>> ax = newfig(1,2); ylog; ylim([1e-18 1e2]);
>> plot(m,err,'*'); plotfun(@(m) 2.6^(-m),[0 55]);
>> subplot(ax(2)); ylog; ylim([1e-18 1e2]);
>> plot(m,res,'*'); plotfun(@(m) 2.6^(-m),[0 55]);
```                                                                  ∎

In review, Chebyshev collocation converges geometrically or even supergeometrically on problems with analytic solutions.

For problems with solutions that are differentiable but not analytic, Chebyshev collocation can converge algebraically. Some of the exercises at the end of this chapter consider these sorts of problems.

28.3 ∎ Rate of convergence for second-order problems

The story for second-order problems is similar. Chebyshev collocation works best on the second-order IVP

$$\alpha(t)u(t) + \beta(t)u'(t) + \gamma(t)u''(t) = g(t), \quad c_{10}u(a) + c_{11}u'(a) + c_{12}u''(a) = d_1,$$
$$c_{20}u(a) + c_{21}u'(a) + c_{22}u''(a) = d_2 \tag{28.10}$$

when the solution is analytic. We adopt the notation used in the discussion of second-order problems in Chapter 24, but now the grids are Chebyshev grids. The following lemma and theorems are presented without proof. The Bernstein ellipses should be understood to have foci at the endpoints a and b of the interval on which a solution is computed.

Lemma 28.10. *Suppose* (28.10) *is solved by Chebyshev collocation. If the solution u is analytic in the interior of a Bernstein ellipse of parameter* ρ *and its second derivative is bounded in the ellipse, then*

$$\|\varepsilon\|_\infty \leq C\rho^{-m}$$

for a coefficient C that is constant with respect to m.

Theorem 28.11. *Suppose* (28.10) *is solved by Chebyshev collocation. If the solution u is analytic in a Bernstein ellipse of parameter* ρ *and its second derivative is bounded in the ellipse, then*

$$\|\mathbf{u} - \mathbf{p}\|_\infty \leq C\kappa_{\text{abs}}\rho^{-m}, \tag{28.11}$$
$$\|\mathbf{v} - \mathbf{q}\|_\infty \leq C\kappa_{\text{abs}}\rho^{-m}, \tag{28.12}$$
$$\|\mathbf{w} - \mathbf{r}\|_\infty \leq C\kappa_{\text{abs}}\rho^{-m}, \tag{28.13}$$

in which C is constant with respect to m.

Theorem 28.12. *Suppose* (28.10) *is solved by collocation on a Chebyshev grid of degree* $m > 1$. *If the solution u is analytic in the interior of a Bernstein ellipse of parameter* ρ *and its second derivative is bounded in the ellipse, then*

$$\|u - p\|_\infty \leq C_1(\kappa_{\text{abs}} \vee 1)(\log m)\rho^{-m}, \tag{28.14}$$
$$\|u' - p'\|_\infty \leq C_2(\kappa_{\text{abs}} \vee 1)(\log m)\rho^{-m}, \tag{28.15}$$
$$\|u'' - p''\|_\infty \leq C_3(\kappa_{\text{abs}} \vee 1)(\log m)\rho^{-m} \tag{28.16}$$

for coefficients C_1, C_2, *and* C_3 *that are constant with respect to m.*

Theorem 28.13. *Suppose* (28.10) *is solved by collocation on a Chebyshev grid of degree* $m > 1$. *If the solution u is analytic in the interior of a Bernstein ellipse of parameter* ρ *and its second derivative is bounded in the ellipse, then the continuous residual* $R(t) = g(t) - \alpha(t)p(t) - \beta(t)p'(t) - \gamma(t)p''(t)$ *satisfies*

$$\|R\|_\infty \leq C(\kappa_{\text{abs}} \vee 1)(\log m)\rho^{-m} \tag{28.17}$$

for a coefficient C that is constant with respect to m.

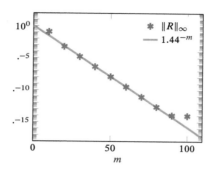

Figure 28.3. *Rate of convergence for Chebyshev collocation applied to a forced harmonic oscillator.*

Example 28.14. Predict the rate of convergence of Chebyshev collocation on the forced harmonic oscillator problem

$$u(t) + 0.2u'(t) + u''(t) = 1 - \frac{1}{t+1}, \quad u(0) = 0, \quad u'(0) = 0$$

over $0 \le t \le 30$. Then test the prediction experimentally by measuring residuals.

Solution. The forcing function $g(z) = 1 - 1/(z+1)$ has a singularity at $z = -1$. This point lies on the Bernstein ellipse $E_\rho(0, 30)$ with the following value of ρ:

```
>> a = 0; b = 30;
>> w = 2/(b-a)*(-1-(a+b)/2);
>> rho = abs(w-sqrt(w^2-1))
rho =
    1.437850957522001
```

The problem has the form (28.2); the functions $\alpha(z) = 1$, $\beta(z) = 0.2$, $\gamma(z) = 1$, and $g(z) = 1 - 1/(z+1)$ are analytic inside the Bernstein ellipse; and $\gamma(z)$ never equals zero. Therefore, the IVP has a solution that is analytic in the Bernstein ellipse as well. As long as the problem does not suffer from ill conditioning, we expect the collocation solution to converge as fast as 1.44^{-m}.

To test this, the problem is entered.

```
>> K = 1; B = 0.2; M = 1; g = @(t) 1-1/(t+1);
>> ua = 0; va = 0;
```

Then it is solved on a sequence of finer and finer grids, and the residuals are measured.

```
>> m = 10:10:100;
>> res = nan(size(m));
>> for k = 1:length(m)
      [ps,qs,rs] = ivpl2cheb(@(t) K,@(t) B,@(t) M,g,[a b] ...
                          ,[ 1 0 0; 0 1 0 ],[ua; va],m(k));
      p = interpcheb(ps,[a b]);
      q = interpcheb(qs,[a b]);
      r = interpcheb(rs,[a b]);
      R = @(t) g(t)-K*p(t)-B*q(t)-M*r(t);
      res(k) = infnorm(R,[a b],100);
   end
```

The absolute residuals are graphed in Figure 28.3, along with the predicted rate of convergence 1.44^{-m}.

```
>> ax = newfig; ylog; ylim([1e-18 1e2]);
>> plot(m,res,'*'); plotfun(@(m) 1.44^(-m),[0 110]);
```

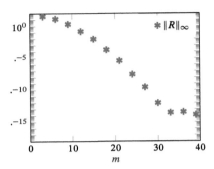

Figure 28.4. *Rate of convergence for Chebyshev collocation on a Bessel equation.*

The prediction appears to be correct. ∎

Example 28.15. How fast does Chebyshev collocation converge on the Bessel IVP

$$xu(x) + u'(x) + xu''(x) = 0, \quad u(0) = 1, \quad u''(0) = -1/2,$$

computed over $0 \le x \le 20$?

Solution. In this case, Theorem 28.2 provides no help. The coefficient function $\gamma(x) = x$ vanishes at $x = 0$, which lies on the interval $0 \le x \le 20$ where the solution is to be computed. Therefore, the theorem does not apply.

The inapplicability of Theorem 28.2 does *not* imply that the solution to the IVP necessarily has a singularity or that collocation will necessarily perform poorly. So, let's solve the collocation equations and see what happens.

The problem is defined.

```
>> al = @(x) x; be = @(x) 1; ga = @(x) x; g = @(x) 0;
>> a = 0; b = 20;
>> ua = 1; wa = -1/2;
```

The collocation solutions are computed on a sequence of Chebyshev grids, and the absolute residuals are measured.

```
>> m = 3:3:39;
>> res = nan(size(m));
>> for k = 1:length(m)
      [ps,qs,rs] = ivpl2cheb(al,be,ga,g,[a b] ...
                            ,[ 1 0 0; 0 0 1 ],[ua; wa],m(k));
      p = interpcheb(ps,[a b]);
      q = interpcheb(qs,[a b]);
      r = interpcheb(rs,[a b]);
      R = @(x) g(x)-al(x)*p(x)-be(x)*q(x)-ga(x)*r(x);
      res(k) = infnorm(R,[a b],100);
   end
```

The residuals are plotted in Figure 28.4.

```
>> newfig; ylog; ylim([1e-18 1e3]); plot(m,res,'*');
```

The residuals appear to converge supergeometrically. We are happy to see this, even if we haven't established enough theory to show that the solution is analytic on the entire complex plane. ∎

Notes

A proof of Theorem 28.1 can be found, for example, in the book by Coddington and Levinson [18, p. 91].

The BVP in Exercise 19 is solved in a 1991 paper by Greengard [37]. Greengard credits the textbook by Stoer and Bulirsch [68] as the source of the problem.

Exercises

Exercises 1–4: A linear differential equation and an interval of the t-axis are given. Is a solution to the differential equation guaranteed by Theorem 28.1 or 28.2 to be analytic in an open region containing the interval? Explain. If the answer is yes, specify the largest Bernstein ellipse within which the solution is guaranteed to be analytic, or state that the solution is guaranteed to be analytic in the entire complex plane.

1. $u - \sqrt{1 - t^2}u' = 0, 0 \le t \le 0.9$
2. $|t - 1|u + u' = 0, 0 \le t \le 2$
3. $(t - 2)u - (t - 10)u'' = t - 3, 0 \le t \le 8$
4. $(\cos t)u + (\sin t)u' + u'' = 0, 0 \le t \le 8\pi$

Exercises 5–12: Predict, using the theory of this chapter, the class of convergence for Chebyshev collocation: algebraic, geometric, or supergeometric. If geometric, find a value for ρ in the bound $C\rho^{-m}$. Test your prediction experimentally by measuring the absolute residual.

5. $-u + t(\log t)u' = 0, u(e) = 1; e \le t \le 10$
6. $-tu + u' = 0, u(0) = 1; 0 \le t \le 2$
7. $(\sin t)u - u' = 1, u(0) = 1; 0 \le t \le 4\pi$
8. $|t - 2|^{5/2}u + u' = t, u(0) = 1/2; 0 \le t \le 5$
9. $u + t^2u' + u'' = 0, u(0) = 1, u'(0) = 0; 0 \le t \le 6$
10. $u + tu' + (t + 1)u'' = 0, u(0) = 1, u'(0) = 0; 0 \le t \le 15$
11. $\sqrt{t}u + u'' = t, u(0) = 1, u'(0) = 1; 0 \le t \le 1$
12. $(1/(1 + t^2))u - u'' = \cos t, u(0) = 0, u'(0) = 0; 0 \le t \le 10$

13. Let $1 < \tau < \rho$. Prove that $(\log m)\rho^{-m}$ converges to zero as fast as τ^{-m} as $m \to \infty$. (See the definition of *converges as fast as* on p. 33.) Hence, $(\log m)\rho^{-m}$ converges to zero geometrically.

14. Prove the following statement, which is analogous to Lemma 28.4: If the solution u to (28.4) is $(r + 1)$-times weakly differentiable on $[a, b]$ for some $r \ge 1$ and its $(r + 1)$th derivative has bounded variation, then, for any $m > r + 1$, the degree-m Chebyshev collocation solution satisfies $\|\varepsilon\|_\infty \le Cm^{-r}$ for a coefficient C that is constant with respect to m.

15. *Struve's equation* is

$$(x^2 - v^2)u + xu' + x^2u'' = \frac{x^{v+1}}{2^{v-1}\sqrt{\pi}\,\Gamma(v + 1/2)}.$$

We'll consider $v = 4/3$ and $0 \le x \le 30$.

(a) Can Theorem 28.12 be applied in this situation? Discuss.

(b) Solve the differential equation by Chebyshev collocation using initial conditions $u'(0) = 0$, $u''(0) = 0$. Measure the absolute residual and classify the rate of convergence as algebraic, geometric, or supergeometric.

16. Struve's equation is introduced in the previous exericse.

(a) By applying a change of the dependent variable of the form $u = x^r y$, show that Struve's equation with $v = 4/3$ is equivalent to

$$\left(x^2 + \frac{11}{3}\right) y + \frac{17}{3} xy' + x^2 y'' = \frac{1}{2^{1/3}\sqrt{\pi}\,\Gamma(11/6)}.$$

(One nice feature of this differential equation is that the right-hand side is analytic everywhere—in fact, constant.)

(b) Solve the differential equation of (a) on $0 \le x \le 30$ by Chebyshev collocation using initial conditions $y'(0) = 0$, $y''(0) = -1/(2^{4/3} \times 3\sqrt{\pi}\,\Gamma(23/6))$. Measure the absolute residual and classify the rate of convergence as algebraic, geometric, or supergeometric.

(c) Plot an approximate $u(x) = x^r y(x)$ to see the Struve function.

17. Develop an adaptive routine for solving a first-order linear IVP on a piecewise-Chebyshev grid. The routine should first attempt to solve the equation over the entire domain $[a, b]$ using ivp1cheb with $m = 32$. If the absolute residual is estimated to be greater than a provided tolerance, then the domain should be split in half, and the problem should be solved recursively first on $[a, (a+b)/2]$ and then on $[(a+b)/2, b]$. To keep the cost manageable, the recursion should be limited to a user-supplied depth. Demonstrate your routine on $|\sin t|\, u + u' = \cos t$, $u(0) = 1$, $0 \le t \le 20$, seeking an absolute residual below 10^{-12}. How many subintervals are used?

18. Develop an adaptive piecewise-Chebyshev solver analogous to the one described in Exercise 17 but for second-order problems. Demonstrate your routine on

$$u + 0.1u' + u'' = H(t - 5), \quad u(0) = 0, \quad u'(0) = 0$$

over $0 \le t \le 30$, in which

$$H(t) = \begin{cases} 0 & \text{if } t \le 0, \\ 1 & \text{if } t > 0 \end{cases}$$

is the *Heaviside unit step function*. (This IVP models a damped mass-spring system subject to a discontinuous external force.) Seek an absolute residual below 10^{-12}. Plot the position, velocity, and acceleration (p, p', and p'', respectively) and verify that the residual is below 10^{-12}.

19. Solve the BVP

$$400u - u'' = -400\cos^2(\pi x) - 2\pi^2 \cos(2\pi x), \quad u(0) = 0, \quad u(1) = 0$$

using your solution to Exercise 14 of Chapter 27, and measure the absolute residual. Does the rate of convergence appear to be algebraic, geometric, or supergeometric? Is the behavior surprising or not surprising given what you've learned about Chebyshev collocation for IVPs?

Chapter 29

Conditioning of linear differential equations

The condition number has appeared repeatedly in error bounds for collocation methods. In this chapter, we look more closely at the role of the condition number and study examples of ill-conditioned problems.

29.1 ▪ Conditioning for continuous and discrete problems

Recall that when a linear system $\mathbf{Ax} = \mathbf{b}$ is ill conditioned, a small residual $\mathbf{b} - \mathbf{A}\hat{\mathbf{x}}$ does not imply a small error $\mathbf{x} - \hat{\mathbf{x}}$. (See Theorem 21.3 and the surrounding discussion.) A similar phenomenon is possible for differential equations.

Consider a solution $u(t)$ to an IVP

$$\alpha(t)u(t) + \beta(t)u'(t) = g(t), \quad c_0 u(a) + c_1 u'(a) = d,$$

and a modified function $\hat{u}(t) = u(t) + \Delta u(t)$, which satisfies

$$\alpha(t)\hat{u}(t) + \beta(t)\hat{u}'(t) = g(t) + \varepsilon_1(t), \quad c_0\hat{u}(a) + c_1\hat{u}'(a) = d + \varepsilon_2,$$

with $\varepsilon_1(t) = \alpha(t)\Delta u(t) + \beta(t)\Delta u'(t)$ and $\varepsilon_2 = c_0\Delta u(a) + c_1\Delta u'(a)$. For some problems, it is possible to find a change $\Delta u(t)$ to the solution function that is fairly large in magnitude but that nevertheless produces very small residual components $\varepsilon_1(t)$ and ε_2. In other words, the modified function $\hat{u}(t)$ nearly solves the original IVP even though it is significantly different from the exact solution $u(t)$. Such a problem is considered ill conditioned, and the danger is that a computation seeking $u(t)$ could easily locate $\hat{u}(t)$ or some other similarly inaccurate function instead.

A bit of good news is that we can detect ill conditioning. Suppose the IVP above is ill conditioned and $\Delta u(t)$ is large in magnitude while $\varepsilon_1(t)$ and ε_2 are very small. This can lead to sample vectors

$$\begin{bmatrix} \mathbf{u} \\ \mathbf{v} \end{bmatrix}, \quad \begin{bmatrix} \hat{\mathbf{u}} \\ \hat{\mathbf{v}} \end{bmatrix}$$

(of $u(t)$, $u'(t)$, $\hat{u}(t)$, and $\hat{u}'(t)$) that are quite different, even though the residuals associated with (24.3), specifically

$$\mathbf{b} - \mathbf{L}\begin{bmatrix} \mathbf{u} \\ \mathbf{v} \end{bmatrix}, \quad \mathbf{b} - \mathbf{L}\begin{bmatrix} \hat{\mathbf{u}} \\ \hat{\mathbf{v}} \end{bmatrix},$$

are quite close (both nearly zero). Subtracting the residuals leaves the product

$$\mathbf{L} \begin{bmatrix} \Delta\mathbf{u} \\ \Delta\mathbf{v} \end{bmatrix},$$

which is very small even though $\Delta\mathbf{u}$ and $\Delta\mathbf{v}$ are not considered small. This implies that the matrix $\mathbf{L}$ is ill conditioned (Exercise 1).

Hence, an ill-conditioned collocation matrix can be a clue that the IVP itself is ill conditioned, and therefore the problem at hand is delicate.

From the theory of linear systems, we know that the absolute condition number κ_{abs} for a matrix $\mathbf{A}$ provides an upper bound on the ratio of error to residual:

$$\|\mathbf{x} - \hat{\mathbf{x}}\|_\infty \le \kappa_{abs}\|\mathbf{b} - \mathbf{A}\hat{\mathbf{x}}\|_\infty.$$

For an ill-conditioned first-order IVP, the errors $u(t) - p(t)$, $u'(t) - p'(t)$ could be much larger than the residual $R(t) = g(t) - \alpha(t)p(t) - \beta(t)p'(t)$, and the absolute condition number for the collocation matrix is a rough estimate for the worst-case ratio. Keep in mind that what qualifies as a "large" absolute condition number depends on the units of measurement chosen for the model.

29.2 ▪ Measuring the condition number

Because condition numbers can be costly to measure, special-purpose routines are provided for estimating condition numbers of collocation systems fairly quickly. These are ivp1conduni, ivp2conduni, ivp1condcheb, and ivp2condcheb.

Example 29.1. The IVP

$$tu(t) + u'(t) = 0, \quad u(0) = 1 \tag{29.1}$$

is to be solved on $0 \le t \le 4$ using piecewise-uniform collocation with $m = 3$. Measure the absolute condition number for the collocation system for several values of N using ivp1conduni. Does ill conditioning appear to be a problem?

Solution. The IVP is defined.
```
>> al = @(t) t; be = @(t) 1; g = @(t) 0;
>> a = 0; b = 4;
>> ua = 1;
```
A sequence of grids is specified.
```
>> m = 3; N = 4*3.^(1:5);
```
The absolute condition number is measured for each grid using ivp1conduni.
```
>> kappa = nan(size(N));
>> for k = 1:length(N)
      n = N(k)/m;
      kappa(k) = ivp1conduni(al,be,[a b],[1 0],m,n);
   end
>> disptable('N',N,'%3d','kappa',kappa,'%5.1f');
   N    kappa
   12    8.5
   36    8.3
  108    9.0
  324    9.4
  972    9.4
```

The absolute condition number does not appear to grow significantly with N, so conditioning should not affect the convergence rate N^{-4} suggested by Theorems 26.6–26.8. (In fact, this convergence rate has already been observed in Example 26.4.)

The problem is "well scaled" in the sense that the coefficient functions $\alpha(t) = t$, $\beta(t) = 1$ and the solution $u(t) = e^{-t^2/2}$ and its derivative $u'(t) = -te^{-t^2/2}$ have magnitudes near 1. Therefore, whatever number, say, 10^{-15}, that might be considered a small error should also be considered a small residual. Because the condition numbers are not much larger than 1, the errors $\|u - p\|_\infty$ and $\|u' - p'\|_\infty$ should not be much bigger than the residual $\|R\|_\infty$. If the residual can be made small, then we expect the errors to be small as well; the problem is well conditioned. ∎

Example 29.2. Repeat the previous example with a sequence of Chebyshev grids, and use ivp11condcheb to measure the condition number.

Solution. The IVP is defined.

```
>> al = @(t) t; be = @(t) 1; g = @(t) 0;
>> a = 0; b = 4;
>> ua = 1;
```

We consider Chebyshev grids of degrees $m = 20, 40, 60, 80, 100$:

```
>> m = 20:20:100;
```

The absolute condition number is measured for each grid.

```
>> kappa = nan(size(m));
>> for k = 1:length(m)
      kappa(k) = ivp11condcheb(al,be,[a b],[1 0],m(k));
   end
>> disptable('m',m,'%3d','kappa',kappa,'%5.1f');
   m    kappa
   20     8.8
   40     9.8
   60    10.1
   80    11.5
  100    11.1
```

The absolute condition number does not appear to grow quickly with m, so conditioning should not affect the rate of convergence significantly. Because the coefficient functions $\alpha(z) = z$ and $\beta(z) = 1$ and the right-hand side $g(z) = 0$ are analytic in the entire complex plane and $\beta(z)$ never equals zero, the solution is analytic in the entire complex plane by Theorem 28.1. Therefore, the rate of convergence is expected to be supergeometric.

Because the problem is well scaled as discussed in the previous example and because the condition numbers are not much larger than 1, the problem is well conditioned. We expect that a small residual $\|R\|_\infty$ should be sufficient to achieve small errors $\|u - p\|_\infty$ and $\|u' - p'\|_\infty$. ∎

Example 29.3. Solve (29.1) by Chebyshev collocation on $0 \le t \le 4$. Compare with the results of the previous example.

Solution. The IVP is solved on a sequence of more and more refined Chebyshev grids.

```
>> al = @(t) t; be = @(t) 1; g = @(t) 0;
>> a = 0; b = 4;
>> ua = 1;
```

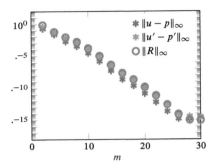

Figure 29.1. *Absolute errors and absolute residual on a well-conditioned problem.*

```
>> m = 2:2:30;
>> uerr = nan(size(m)); verr = nan(size(m));
>> res = nan(size(m));
>> u = @(t) exp(-t^2/2);
>> v = @(t) -t*exp(-t^2/2);
>> for k = 1:length(m)
      [ps,qs] = ivp1lcheb(al,be,g,[a b],[1 0],ua,m(k));
      p = interpcheb(ps,[a b]);
      q = interpcheb(qs,[a b]);
      uerr(k) = infnorm(@(t) u(t)-p(t),[a b],100);
      verr(k) = infnorm(@(t) v(t)-q(t),[a b],100);
      res(k) = infnorm(@(t) g(t)-al(t)*p(t)-be(t)*q(t),[a b],100);
   end
```

The absolute errors and residuals are plotted on the same axes in Figure 29.1.

```
>> newfig; ylog; ylim([1e-18 1e2]);
>> plot(m,uerr,'*'); plot(m,verr,'*'); plot(m,res,'o');
```

The rate of convergence appears to be supergeometric as expected. The absolute errors and absolute residuals are similar in magnitude. ∎

29.3 ▪ Sensitivity to perturbations

An IVP may be ill conditioned because it is sensitive to perturbations. In mathematical usage, a *perturbation* is a small change, often suggesting that a problem or solution has been slightly corrupted from its original true form. Perturbations are unavoidable in applications whenever real-world data are measured inexactly, and they are inevitable in numerical computations because of roundoff error. To say that a problem is sensitive to perturbations is to say that a small error in the problem statement can result in a much larger error in the computed answer. This is a dangerous situation.

Our first example concerns the IVP

$$-tu(t) + u'(t) = 1, \quad u(0) = -\sqrt{\pi/2}, \tag{29.2}$$

which is sensitive to perturbations of its initial condition. The differential equation has the infinite family of solutions

$$u(t) = e^{t^2/2} \left(\int_0^t e^{-\xi^2/2} \, d\xi + C \right), \tag{29.3}$$

and the solution satisfying the initial condition is the one with $C = -\sqrt{\pi/2}$. Several

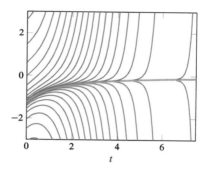

Figure 29.2. *An IVP that is sensitive to perturbations.*

solutions to the differential equation are graphed in Figure 29.2, with the particular solution satisfying the initial condition highlighted in orange. Note that some solutions very nearly satisfy the initial condition at time $t = 0$ but veer far away from the highlighted solution as t increases.

Example 29.4. Solve (29.2) by collocation on a 2-by-1000 piecewise-uniform grid over $0 \le t \le 7.5$. Measure the error and the residual and comment on conditioning.

Solution. The problem is defined.
```
>> al = @(t) -t; be = @(t) 1; g = @(t) 1;
>> a = 0; b = 7.5;
>> ua = -sqrt(pi/2);
```
The collocation grid is set.
```
>> m = 2; n = 1000;
```
The solution is computed numerically, and its residual is measured.
```
>> [ps,qs] = ivpl1uni(al,be,g,[a b],[1 0],ua,m,n);
>> p = interpuni(ps,[a b]);
>> q = interpuni(qs,[a b]);
>> R = @(t) g(t)-al(t)*p(t)-be(t)*q(t);
>> format short;
>> infnorm(R,[a b],100)
ans =
    1.2977e-04
```
The residual is reasonably small, on the order of 10^{-4}.

What can we expect for the error $\|u - p\|_\infty$? The condition number gives an indication.
```
>> kappa = ivpl1conduni(al,be,[a b],[1 0],m,n)
kappa =
    3.9987e+13
```
The absolute condition number, at over 10^{13}, indicates that the absolute error could be several orders of magnitude larger than the absolute residual. Let's see if this is actually the case.
```
>> u = @(t) -sqrt(pi/2)*exp(t^2/2)*erfc(t/sqrt(2));
>> infnorm(@(t) u(t)-p(t),[a b],100)
ans =
    11.2811
>> format long;
```
Indeed, the absolute error is much larger than the residual and in fact is visible to the

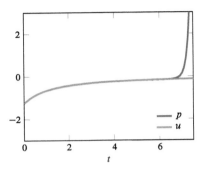

Figure 29.3. *The desired solution and a collocation solution to an ill-conditioned initial-value problem.*

naked eye, as seen in Figure 29.3.

```
>> newfig; plotfun(p,[a b]); plotfun(u,[a b]);
```

The discrepancy between the residual and the absolute error is simply described: the numerical routine approximately follows one of the trajectories suggested by Figure 29.2, producing a small residual, but it follows the wrong trajectory, producing a large absolute error. The problem is ill conditioned because of its sensitivity to perturbations. ∎

Ultimately, collocation fails on the last example, but any numerical scheme would be challenged by the sensitivity to perturbations. Solving this IVP is like walking a high wire—a single slip is catastrophic.

Note that the residual alone does not indicate the extent of the inaccuracy—it is important to check both the condition number and the residual when assessing the accuracy of a numerical solution.

Example 29.5. Repeat Example 29.4 using a Chebyshev grid of degree 50.

Solution. The collocation solution is computed.

```
>> al = @(t) -t; be = @(t) 1; g = @(t) 1;
>> a = 0; b = 7.5;
>> ua = -sqrt(pi/2);
>> m = 50;
>> [ps,qs] = ivpl1cheb(al,be,g,[a b],[1 0],ua,m);
>> p = interpcheb(ps,[a b]);
>> q = interpcheb(qs,[a b]);
```

The residual is as small as we could hope, on the order of 10^{-15}:

```
>> R = @(t) g(t)-al(t)*p(t)-be(t)*q(t);
>> format short;
>> infnorm(R,[a b],50)
ans =
    2.6749e-15
```

However, the condition number is again greater than 10^{13}:

```
>> kappa = ivpl1condcheb(al,be,[a b],[1 0],m)
kappa =
    3.9960e+13
```

Thus only a few decimal places of accuracy can be expected. For this simple problem, it is possible to compare with the known solution.

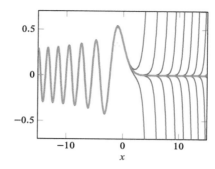

Figure 29.4. *Several solutions to the Airy differential equation.*

```
>> u = @(t) -sqrt(pi/2)*exp(t^2/2)*erfc(t/sqrt(2));
>> infnorm(@(t) u(t)-p(t),[a b],100)
ans =
    7.1932e-04
>> format long;
```

The absolute error $\|u - p\|_\infty$ is small but not tiny by our standards, on the order of 10^{-4}.

As in the previous example, the IVP is ill conditioned and the absolute error is much larger than the absolute residual. ∎

Notice the closeness between the condition numbers in Examples 29.4 and 29.5. These measurements largely reflect the sensitivity of the underlying IVP rather than the specifics of the solution method.

The next example considers the *Airy* Ai *function,* an important function in mathematical theory and scientific applications that satisfies the second-order IVP

$$-xu(x) + u''(x) = 0, \quad u(0) = \frac{1}{3^{2/3}\Gamma(2/3)}, \quad u'(0) = -\frac{1}{3^{1/3}\Gamma(1/3)}. \quad (29.4)$$

The solution to the IVP is plotted in orange in Figure 29.4, and several other solutions to the differential equation are plotted in blue. Where the blue curves disappear they are hidden behind the orange curve. All of the plotted solutions satisfy the initial condition $u(0) = 3^{-2/3}/\Gamma(2/3)$, but they differ slightly in their initial derivatives $u'(0)$. Notice how solutions that are nearly equal near $x = 0$ diverge widely as x increases.

Example 29.6. Compute the Airy Ai function by solving (29.4) by collocation on a degree-60 Chebyshev grid over $0 \le x \le 15$. Measure and discuss the error, the residual, and conditioning. The absolute error may be measured by comparing with the built-in airy function.

Solution. The problem is defined.
```
>> al = @(x) -x; be = @(x) 0; ga = @(x) 1; g = @(x) 0;
>> a = 0; b = 15;
>> ua = 1/(3^(2/3)*gamma(2/3)); va = -1/(3^(1/3)*gamma(1/3));
```
The collocation grid is set.
```
>> m = 60;
```
The collocation solution is computed.
```
>> [ps,qs,rs] = ivpl2cheb(al,be,ga,g,[a b] ...
```

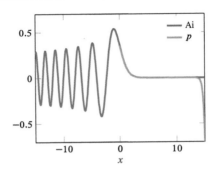

Figure 29.5. *Airy's* Ai *function and a collocation solution for* $0 \leq x \leq 15$.

```
                                    ,[ 1 0 0; 0 1 0 ],[ua; va],m);
>> p = interpcheb(ps,[a b]);
>> q = interpcheb(qs,[a b]);
>> r = interpcheb(rs,[a b]);
```

The residual is a good indication of whether the Chebyshev grid is sufficiently fine.

```
>> R = @(x) g(x)-al(x)*p(x)-be(x)*q(x)-ga(x)*r(x);
>> format short;
>> infnorm(R,[a b],100)
ans =
   9.9476e-14
```

This is a very small residual, and it is unlikely that a finer Chebyshev grid could produce a significantly better result.

The residual is small; what about the condition number?

```
>> kappa = ivp12condcheb(al,be,ga,[a b],[ 1 0 0; 0 1 0 ],m)
kappa =
   1.6786e+16
```

Because the condition number is above 10^{16}, the absolute error could be far larger than the absolute residual.

To assess the actual error, the function is first plotted.

```
>> u = @(x) airy(x);
>> newfig; plotfun(u,[-b b]); plotfun(p,[a b]); xlim([-b b]);
```

Looking at Figure 29.5, the collocation solution clearly deviates from the exact solution for larger values of x. (Note that we have not attempted to solve the IVP for $x < 0$, so the orange curve does not extend into the left half-plane.) The absolute error is indeed large:

```
>> infnorm(@(x) u(x)-p(x),[a b],100)
ans =
    0.7750
>> format long;
```

Once again, ill conditioning is responsible for making the error much larger than the residual. ∎

Although there is no universal way to ameliorate sensitivity to perturbations, sometimes problem-specific knowledge can be used to formulate a well-conditioned problem. For example, the Airy Ai function satisfies the *boundary condition* $\lim_{x \to \infty} \text{Ai}(x) = 0$, unlike the other solutions to the Airy differential equation in Figure 29.4, a fact that can be used to compute the function accurately (Exercise 11).

Notes

The error measure $\|u - p\|_\infty \vee \|u' - q\|_\infty$ is related to the residual measure $\|p(t) - p(a) - \int_a^t q(\xi)\,d\xi\|_\infty \vee \|g(t) - \alpha(t)p(t) - \beta(t)q(t)\|_\infty \vee |d - c_0 p(a) - c_1 q(a)|$ by the condition number of an *integral operator* that is discretized as the matrix $\mathbf{L}$ in (23.8). The condition number $\kappa_{\text{abs}} = \|\mathbf{L}^{-1}\|_\infty$ is a good but imperfect estimate for the condition number of the integral operator because it reflects both the conditioning of the integral operator and the conditioning of the interpolation problem associated with the choice of grids [2, pp. 90–93].

Routines ivpl1conduni and ivpl2conduni estimate condition numbers using a method of Hager from 1984 [38]. Hager's method is designed for general linear systems and is specialized here to the collocation systems (25.12) and (25.13).

Exercises

1. Let $\mathbf{A}$ be an n-by-n matrix, and let $\mathbf{x}$ be an n-by-1 vector. Suppose $\|\mathbf{Ax}\|_\infty < \varepsilon$. Prove that $\kappa_{\text{abs}} = \|\mathbf{A}^{-1}\|_\infty > \|\mathbf{x}\|_\infty / \varepsilon$.

2. Repeat Example 26.14 and measure the condition number of each collocation system using ivpl2conduni. What does the condition number indicate about the relationship between the residual $\|R\|_\infty$ and the error $\|u - p\|_\infty$? Is this consistent with the data from the example?

3. Repeat Example 28.9 and measure the condition number of each collocation system using ivpl1condcheb. What does the condition number indicate about the relationship between the residual $\|R\|_\infty$ and the error $\|u - p\|_\infty$? Is this consistent with the data from the example?

4. Repeat Example 28.14 and measure the condition number of each collocation system using ivpl2condcheb. What does the condition number indicate about the relationship between the residual $\|R\|_\infty$ and the error $\|u - p\|_\infty$?

5. Repeat Example 28.15 and measure the condition number of each collocation system using ivpl2condcheb. What does the condition number indicate about the relationship between the residual $\|R\|_\infty$ and the error $\|u - p\|_\infty$? Measure the error by comparing with the built-in besselj(0,x) function. Is the observed error consistent with your interpretation of the condition number?

Exercises 6–9: Solutions to the given differential equation on the given interval are known to be sensitive to perturbations of their initial condition(s). Demonstrate this by finding two solutions that satisfy nearly identical initial conditions but then diverge, one rising and the other falling. Plot the two solutions and compute and comment on the absolute condition number.

6. $tu - (t + 1)u' = t^3, 0 \le t \le 15$
7. $(t + 1)^{-1/2}u - 2u' = 1, 0 \le t \le 100$
8. $u - 2u' + u'' = t, 0 \le t \le 10$
9. $-4t^2 u - (t + 1)^{-1}u' + u'' = t, 0 \le t \le 3$

10. Verify that (29.3) is a solution to the differential equation of (29.2) for any value of C.

11. The Airy Ai function decays rapidly toward zero as x approaches $+\infty$. Solve the BVP

$$-xu + u'' = 0, \quad u(0) = \frac{1}{3^{2/3}\Gamma(2/3)}, \quad u(15) = 0$$

on $0 \le x \le 15$ using your solution to Exercise 14 in Chapter 27 and compare with the MATLAB `airy` function. Does the BVP appear to be well conditioned or ill conditioned?

12. The *parabolic cylinder function* D_ν is a particular solution to the differential equation

$$\left(\nu + \frac{1}{2} - \frac{1}{4}x^2\right)u + u'' = 0.$$

Let's consider $\nu = 8.5$.

(a) It is known that $D_{8.5}(x)$ satisfies the IVP

$$\left(9 - \frac{1}{4}x^2\right)u + u'' = 0, \quad u(0) = \frac{2^{17/4}\sqrt{\pi}}{\Gamma(-15/4)}, \quad u'(0) = -\frac{2^{19/4}\sqrt{\pi}}{\Gamma(-17/4)}.$$

Solve this IVP on $0 \le x \le 20$ by Chebyshev collocation with $m = 100$. Are you confident that the result is accurate in the sense of achieving low absolute error? Explain.

(b) It is also known that $D_{8.5}(x)$ and $D'_{8.5}(x)$ converge rapidly to zero in the $x \to +\infty$ limit, falling well below 10^{-16} by $x = 20$. Solve the BVP

$$\left(9 - \frac{1}{4}x^2\right)u + u'' = 0, \quad u(0) = \frac{2^{17/4}\sqrt{\pi}}{\Gamma(-15/4)}, \quad u(20) = 0$$

on $0 \le x \le 20$ by Chebyshev collocation with $m = 100$. (See Exercise 14 in Chapter 27 on BVPs.) Are you confident that the result is accurate in the sense of achieving low absolute error? Explain.

13. Uranium-235 is an isotope used in radiometric dating, with a half-life of about 700 million years. Let $u(t)$ be the proportion of uranium-235 remaining in a sample at time t, with t measured in years. The decay is modeled by

$$10^{-10}u + u' = 0, \quad u(0) = 1.$$

Solve this IVP over the time period $0 \le t \le 10^9$ by collocation on a Chebyshev grid of degree 30. Verify that the absolute condition number is over 10^8. Should the problem be considered ill conditioned? Measure $\|u - p\|_\infty$, $\|u' - p'\|_\infty$, and $\|R\|_\infty$, and comment on conditioning.

14. The previous exercise provides a model for the decay of uranium-235. Suppose that time is measured in billions of years rather than years. Use a change of variables to show that the decay obeys

$$0.1u + u' = 0, \quad u(0) = 1$$

under this new definition for t. Solve this IVP using degree-30 Chebyshev collocation to compute the proportion of uranium-235 over the first billion years. Verify that the absolute condition number is below 10. How does the accuracy compare with the solution to the previous exercise? Consider $\|u - p\|_\infty$, $\|u' - p'\|_\infty$, and $\|R\|_\infty$.

Part VI

Zero finding

Chapter 30

Zeros as eigenvalues

In this chapter, we seek solutions to a nonlinear equation $f(x) = 0$. In other words, we seek the *zeros* of the function $f(x)$. As with many of our earlier methods, the first step is to replace $f(x)$ by an interpolant $p(x)$. Two questions arise: First, how can the zeros of $p(x)$ be located? Second, how accurately do the zeros of $p(x)$ approximate the zeros of $f(x)$?

30.1 ▪ The companion matrix pair

A clever idea of Frobenius is to express a given polynomial as a determinant of a specially designed matrix function. The zeros of the polynomial are the *eigenvalues* of the matrix, which can be computed with an eigenvalue solver. The reader is not expected to have prior experience with eigenvalues, but he or she may want to review methods for computing the determinant of a matrix and the identity

$$\det \left(\begin{bmatrix} \mathbf{A} & \mathbf{B} \\ \mathbf{C} & \mathbf{D} \end{bmatrix} \right) = \det(\mathbf{D}) \det(\mathbf{A} - \mathbf{B}\mathbf{D}^{-1}\mathbf{C}),$$

which involves the *Schur complement* of $\mathbf{D}$.

Let $p(x)$ be a polynomial of degree at most m, determined by its values at $m + 1$ distinct nodes $x_0, \ldots, x_m$. Let $l_i(x) = \prod_{k \neq i}(x - x_k)$ be the ith Lagrange basis polynomial and let $w_i = C/l_i(x_i)$, $i = 0, \ldots, m$, be a sequence of barycentric weights for the grid. Define $(m + 2)$-by-$(m + 2)$ matrices $\mathbf{A}$ and $\mathbf{B}$ by

$$\mathbf{A} = \begin{bmatrix} 0 & -p(x_0) & -p(x_1) & \cdots & -p(x_m) \\ w_0 & x_0 & & & \\ w_1 & & x_1 & & \\ \vdots & & & \ddots & \\ w_m & & & & x_m \end{bmatrix} \tag{30.1}$$

and

$$\mathbf{B} = \begin{bmatrix} 0 & & & & \\ & 1 & & & \\ & & 1 & & \\ & & & \ddots & \\ & & & & 1 \end{bmatrix}. \tag{30.2}$$

As usual, the blank entries equal zero. With

$$
\mathbf{p} = \begin{bmatrix} p(x_0) \\ \vdots \\ p(x_m) \end{bmatrix}, \quad
\mathbf{w} = \begin{bmatrix} w_0 \\ \vdots \\ w_m \end{bmatrix}, \quad
\mathbf{X} = \begin{bmatrix} x_0 & & \\ & \ddots & \\ & & x_m \end{bmatrix},
$$

the matrices can be expressed more compactly as

$$
\mathbf{A} = \begin{bmatrix} 0 & -\mathbf{p}^T \\ \mathbf{w} & \mathbf{X} \end{bmatrix}, \quad
\mathbf{B} = \begin{bmatrix} 0 & \\ & \mathbf{I} \end{bmatrix}.
$$

Theorem 30.1. *If $p(x)$ is a polynomial of degree at most m, $x_0, \ldots, x_m$ are distinct nodes, and $\mathbf{A}$ and $\mathbf{B}$ are constructed as above, then*

$$
\det(x\mathbf{B} - \mathbf{A}) = C p(x). \tag{30.3}
$$

Proof. The matrix $x\mathbf{I} - \mathbf{X}$ is diagonal, and its determinant is the product of its diagonal entries:

$$
\det(x\mathbf{I} - \mathbf{X}) = \prod_{k=0}^{m} (x - x_k). \tag{30.4}
$$

Assume for now that x does not coincide with any of the nodes. Then $\det(x\mathbf{I} - \mathbf{X}) \neq 0$ and therefore $x\mathbf{I} - \mathbf{X}$ is invertible.

We have

$$
x\mathbf{B} - \mathbf{A} = \begin{bmatrix} 0 & \mathbf{p}^T \\ -\mathbf{w} & x\mathbf{I} - \mathbf{X} \end{bmatrix}.
$$

The Schur complement formula for the determinant gives

$$
\det(x\mathbf{B} - \mathbf{A}) = \det(x\mathbf{I} - \mathbf{X}) \det(0 + \mathbf{p}^T (x\mathbf{I} - \mathbf{X})^{-1} \mathbf{w}). \tag{30.5}
$$

The product $\mathbf{p}^T (x\mathbf{I} - \mathbf{X})^{-1} \mathbf{w}$ is a scalar, and its value is

$$
\mathbf{p}^T (x\mathbf{I} - \mathbf{X})^{-1} \mathbf{w} = \sum_{i=0}^{m} p(x_i) \frac{1}{x - x_i} w_i = \sum_{i=0}^{m} \frac{p(x_i) w_i}{x - x_i}. \tag{30.6}
$$

Substituting (30.4) and (30.6) into (30.5), we find

$$
\det(x\mathbf{B} - \mathbf{A}) = \prod_{k=0}^{m} (x - x_k) \sum_{i=0}^{m} \frac{p(x_i) w_i}{x - x_i} = \sum_{i=0}^{m} \left(\frac{p(x_i) w_i}{x - x_i} \prod_{k=0}^{m} (x - x_k) \right)
$$

$$
= \sum_{i=0}^{m} \left(p(x_i) w_i \prod_{k \neq i} (x - x_k) \right) = C \sum_{i=0}^{m} \frac{p(x_i)}{l_i(x_i)} l_i(x),
$$

in which the last step follows from the definition of barycentric weights $w_i = C / l_i(x_i)$. The final summation is the Lagrange form (6.2) for $p(x)$. Hence, $\det(x\mathbf{B} - \mathbf{A}) = C p(x)$.

Because the left- and right-hand sides of (30.3) are continuous functions of x, the equation must be true at each $x = x_i$ as well. $\quad\square$

In linear algebra, the zeros of $\det(x\mathbf{B} - \mathbf{A})$ are called the *generalized eigenvalues* for the matrix pair $\mathbf{A}, \mathbf{B}$. Fortunately, we do not need a detailed knowledge of generalized eigenvalues in this text. The important facts are that the solutions of

$$p(x) = 0$$

are identical to the solutions of

$$\det(x\mathbf{B} - \mathbf{A}) = 0 \tag{30.7}$$

and that there are stable, efficient, and widely available computer codes for solving (30.7), known as the *characteristic equation*. In the MATLAB standard library, the necessary routine is called eig. Our routines zerosgen and zeros_ construct $\mathbf{A}$ and $\mathbf{B}$ and then use eig to compute the zeros. See Listing 30.1.

Listing 30.1. Locate the zeros of a polynomial specified by a sample on a provided grid.

```
function z = zerosgen(xs,ps)

ws = baryweights(xs);
% compute generalized eigenvalues
z = zeros_(xs,ws,ps);
% discard infinite eigenvalues, leaving polynomial zeros
z = z(isfinite(z));
```

```
function z = zeros_(xs,ws,ps)

if max(abs(ps))==0, z = mean([ min(xs) max(xs) ]); return; end
m = length(ps)-1;
A = [ 0 -ps'; ws diag(xs) ];
B = blkdiag(0,eye(m+1));
z = eig(A,B);
```

Remark 30.2. The matrix pair $\mathbf{A}, \mathbf{B}$ also has a generalized eigenvalue at infinity. This is because $\det(x\mathbf{B} - \mathbf{A}) = 0$ is equivalent to $\det(\mathbf{B} - x^{-1}\mathbf{A}) = 0$ for nonzero x, and $\lim_{x \to \infty} \det(\mathbf{B} - x^{-1}\mathbf{A}) = \det(\mathbf{B}) = 0$. Our computer codes discard all infinite eigenvalues.

30.2 ▪ Uniform and Chebyshev grids

Routine zerosgen computes the zeros of a polynomial using any given grid of distinct nodes. More convenient are zerosuni and zeroscheb, which are designed specifically for piecewise-uniform and Chebyshev grids, respectively. These are displayed in Listings 30.2 and 30.3.

Note that when a function f is interpolated by a (piecewise-)polynomial p on an interval $[a, b]$, there is no expectation that p is a good fit for f outside of the interval. Thus, any zeros of p away from $[a, b]$ should be discarded from the final list of zeros z. In each computer code, a user-adjustable tolerance determines which zeros are close enough to the interval to keep.

When a function f has a zero ζ, the goal is to compute an accurate approximation z. Ideally, the absolute residual $|f(z)|$ and the absolute error $|\zeta - z|$ should both be small.

Listing 30.2. Locate the zeros of a piecewise polynomial specified on a piecewise-uniform grid.

```
function z = zerosuni(ps,ab,tol)

if nargin<3, tol = 1e-6;  end
m = size(ps,1)-1;
n = size(ps,2);
a = ab(1);
b = ab(2);
l = (b-a)/n;
[xs,ws] = griduni([0 1],m);
z = [];
for j = 1:n
  % find zeros in subinterval
  e = zeros_(xs,ws,ps(:,j));
  e = e(abs(imag(e))<=tol&real(e)>=-tol&real(e)<=1+tol);
  % append zeros to running list
  z = [ z; a+l*(j-1+e) ];
end
z = sort(z);
```

Listing 30.3. Locate the zeros of a polynomial specified on a Chebyshev grid.

```
function z = zeroscheb(ps,ab,tol)

if nargin<3, tol = 1e-6;  end
a = ab(1);
b = ab(2);
m = size(ps,1)-1;
[xs,ws] = gridcheb([-1 1],m);
% compute generalized eigenvalues
e = zeros_(xs,ws,ps);
% select zeros in domain and discard other eigenvalues
e = e(abs(imag(e))<=tol&abs(real(e))<=1+tol);
z = sort((a+b)/2+(b-a)/2*e);
```

Example 30.3. Compute the zeros of $f(x) = \cos x^2$, $0 \le x \le 5$, using zerosuni with $m = 4$. Adjust the number of subintervals in the grid to produce absolute residuals and absolute errors below 10^{-8}.

Solution. First, the function is defined and plotted in Figure 30.1.

```
>> f = @(x) cos(x^2); a = 0; b = 5;
>> newfig; plotfun(f,[a b]);
```

Computing the function's zeros using zerosuni is a two-step process. First, f is interpolated, and second, the zeros of the interpolant are computed.

```
>> m = 4; n = 250;
>> ps = sampleuni(f,[a b],m,n);
>> z = zerosuni(ps,[a b])
z =
   1.253314137314753
   2.170803763679563
   2.802495608281044
   3.315957521916778
   3.759942411925234
   4.156772737629296
```

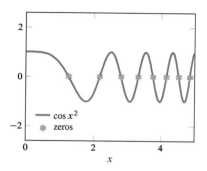

Figure 30.1. *A function and computed zeros.*

```
4.518888385814722
4.854064781598345
```

The computed zeros $z_1, \ldots, z_8$ are plotted in Figure 30.1.

```
>> plot(z,zeros(size(z)),'*');
```

The parameter $n = 250$ was found by trial and error. It is worth noting that this choice produces a close-fitting interpolant:

```
>> p = interpuni(ps,[a b]);
>> infnorm(@(x) f(x)-p(x),[a b],100)
ans =
     7.949743974222656e-09
```

More importantly, every approximate zero z produces a residual $|f(z)|$ below the specified 10^{-8}:

```
>> res = arrayfun(f,z);
>> disptable('z'      ,z          ,'%17.15f' ...
          ,'|f(z)|',abs(res),'%7.1e');
            z      |f(z)|
1.253314137314753  1.9e-12
2.170803763679563  2.1e-11
2.802495608281044  4.6e-10
3.315957521916778  4.1e-10
3.759942411925234  1.6e-10
4.156772737629296  2.4e-09
4.518888385814722  4.9e-09
4.854064781598345  2.0e-09
```

The differences between the exact zeros $\zeta_k = \sqrt{(2k-1)\pi/2}$, $k = 1, \ldots, 8$, and the computed zeros are below 10^{-8} as well.

```
>> zeta = sqrt((2*(1:8)'-1)*pi/2);
>> disptable('zeta'    ,zeta       ,'%17.15f' ...
          ,'z'      ,z          ,'%17.15f' ...
          ,'|zeta-z|',abs(zeta-z),'%8.1e');
          zeta               z     |zeta-z|
1.253314137315500  1.253314137314753   7.5e-13
2.170803763674803  2.170803763679563   4.8e-12
2.802495608198964  2.802495608281044   8.2e-11
3.315957521978271  3.315957521916778   6.1e-11
3.759942411946501  3.759942411925234   2.1e-11
4.156772737923480  4.156772737629296   2.9e-10
4.518888386354951  4.518888385814722   5.4e-10
4.854064781389248  4.854064781598345   2.1e-10
```

The previous example used a piecewise-uniform grid. With a Chebyshev grid, it should be easy to achieve higher accuracy.

Example 30.4. Compute the zeros of $f(x) = \cos x^2$, $0 \le x \le 5$, using zeroscheb. Adjust the grid to get as many accurate digits as possible.

Solution. The function is the same as before.

```
>> f = @(x) cos(x^2); a = 0; b = 5;
```

Through trial and error, a Chebyshev interpolant with $m = 55$ is found to fit f to full precision:

```
>> m = 55;
>> ps = samplecheb(f,[a b],m);
>> p = interpcheb(ps,[a b]);
>> infnorm(@(x) f(x)-p(x),[a b],100)
ans =
      2.942091015256665e-15
```

The zeros of the interpolant are computed with zeroscheb.

```
>> z = zeroscheb(ps,[a b])
z =
    1.253314137315496
    2.170803763674802
    2.802495608198965
    3.315957521978273
    3.759942411946504
    4.156772737923484
    4.518888386354956
    4.854064781389257
```

As in the previous example, we check accuracy in two ways. First, the residuals are measured:

```
>> res = arrayfun(f,z);
>> disptable('z'        ,z          ,'%17.15f' ...
             ,'|f(z)|',abs(res),'%7.1e');
            z         |f(z)|
    1.253314137315496    1.0e-14
    2.170803763674802    4.6e-15
    2.802495608198965    5.9e-15
    3.315957521978273    1.0e-14
    3.759942411946504    2.3e-14
    4.156772737923484    3.7e-14
    4.518888386354956    4.0e-14
    4.854064781389257    8.3e-14
```

Second, the absolute errors are measured:

```
>> zeta = sqrt((2*(1:8)'-1)*pi/2);
>> disptable('zeta'      ,zeta       ,'%17.15f' ...
             ,'z'        ,z          ,'%17.15f' ...
             ,'|zeta-z|',abs(zeta-z),'%8.1e');
             zeta                  z        |zeta-z|
    1.253314137315500    1.253314137315496     4.0e-15
    2.170803763674803    2.170803763674802     8.9e-16
    2.802495608198964    2.802495608198965     8.9e-16
    3.315957521978271    3.315957521978273     1.8e-15
    3.759942411946501    3.759942411946504     3.1e-15
    4.156772737923480    4.156772737923484     4.4e-15
    4.518888386354951    4.518888386354956     4.4e-15
```

```
4.854064781389248     4.854064781389257     8.9e-15
```
Both are close to the level of roundoff error. ∎

30.3 • Residual bound

There are two sources of error in the zero-finding method, as indicated by the following theorem. When reading the theorem, think of f as the original function of interest, p as an interpolant, and z as an approximate zero.

Theorem 30.5. *For any functions f and p and any number z,*

$$|f(z)| \leq \|f - p\|_\infty + |p(z)|. \tag{30.8}$$

Proof. Use the triangle inequality and the definition of the infinity norm:

$$|f(z)| = |f(z) - p(z) + p(z)| \leq |f(z) - p(z)| + |p(z)| \leq \|f - p\|_\infty + |p(z)|. \qquad \square$$

In practice, the absolute error $\|f - p\|_\infty$ is made small by interpolating on a fine grid, and the built-in `eig` function, as called by `zeros_`, is responsible for finding an approximate zero z that makes $|p(z)|$ small. Then the residual $|f(z)|$ must be small in turn.

30.4 • Conditioning

Theorem 30.5 concentrates on the absolute residual $|f(z)|$ for an approximate zero z. This section considers the absolute error $|\zeta - z|$, in which ζ is an exact zero of f.

Theorem 30.6. *Let $f \in C^1[a,b]$ have a zero $\zeta \in [a,b]$, and let z be another point in $[a,b]$. If there exists a constant M such that $1/|f'(x)| \leq M$ for all x between ζ and z, then*

$$|\zeta - z| \leq M|f(z)|. \tag{30.9}$$

Proof. By the mean-value theorem, there exists a location η between ζ and z where

$$f'(\eta) = \frac{f(\zeta) - f(z)}{\zeta - z} = -\frac{f(z)}{\zeta - z}.$$

Hence,

$$|\zeta - z| = \frac{1}{|f'(\eta)|}|f(z)| \leq M|f(z)|. \qquad \square$$

Our algorithms strive to make $|f(z)|$ small. The theorem suggests, however, that if M is large, this may not guarantee that z is especially close to a zero of f.

Remark 30.7. In most problems, finding M exactly is prohibitively difficult—the exact zero ζ is unknown, and therefore M is defined to be the maximum of $1/|f'(x)|$ over an unknown interval. In this case, $1/|f'(z)|$ serves as a reasonable estimate for M.

Example 30.8. Let $f(x) = \sin(x^5 - 10^{-10})$, $-1 \le x \le 1$. Consider computing the zero of the function using zeroscheb with $m = 40$. What should be expected for the residual and error based on Theorems 30.5 and 30.6? Test your predictions experimentally.

Solution. The Chebyshev interpolant produces a very good fit:

```
>> f = @(x) sin(x^5-1e-10); a = -1; b = 1;
>> m = 40;
>> ps = samplecheb(f,[a b],m);
>> p = interpcheb(ps,[a b]);
>> disptable('||f-p||',infnorm(@(x) f(x)-p(x),[a b],100),'%7.1e');
||f-p||
1.1e-15
```

We have $\|f - p\|_\infty$ close to 10^{-15}, and we expect zeroscheb to compute an approximate zero z for which $|p(z)|$ is similarly small. By Theorem 30.5, the absolute residual $|f(z)|$ should also be close to unit roundoff.

However, the absolute error could be substantially larger. We have $f'(x) = \cos(x^5 - 10^{-10}) \times 5x^4$. The zero is at $\zeta = 0.01$, and the derivative there is $f'(\zeta) = 5 \times 10^{-8}$. Therefore, an appropriate value for M in Theorem 30.6 must be at least as large as $1/(5 \times 10^{-8}) = 2 \times 10^7$. We would not be surprised if the absolute error were larger than the absolute residual by a factor of 10^7.

Let's see what happens experimentally.

```
>> zeta = 0.01;
>> z = zeroscheb(ps,[a b]);
>> res = abs(f(z));
>> err = abs(zeta-z);
>> disptable('zeta'    ,zeta,'%17.15f' ...
            ,'z'       ,z   ,'%17.15f' ...
            ,'|f(z)|'  ,res ,'%7.1e'   ...
            ,'|zeta-z|',err ,'%7.1e');
        zeta                   z        |f(z)|   |zeta-z|
0.010000000000000    0.009999992342684    3.8e-16   7.7e-09
```

The absolute residual is as small as predicted, and the absolute error is as large as we feared. ∎

The zero in the previous example is ill conditioned because a tiny change in the problem statement can result in a large change in the solution. The difference between the functions $\sin(x^5)$ and $\sin(x^5 - 10^{-10})$ is tiny, but the difference in their zeros— $\zeta = 0$ versus $\zeta = 0.01$—is much larger. Such sensitivity poses a challenge to any numerical method.

30.5 ▪ Multiple zeros

A zero ζ of a function f has *multiplicity* r if $f(\zeta) = f'(\zeta) = f''(\zeta) = \cdots = f^{(r-1)}(\zeta) = 0$, i.e., the function and its first $r - 1$ derivatives equal zero. When a zero has multiplicity greater than 1, it is called a multiple or repeated zero; otherwise it is a simple zero. In the case of a polynomial p, a zero ζ has multiplicity r if and only if $(x - \zeta)^r$ divides $p(x)$, perhaps a more familiar notion of multiplicity; see Theorem A.1 in Appendix A.

Multiple zeros are especially difficult to locate accurately. The following example illustrates common phenomena.

Example 30.9. Compute the zeros of $f(x) = (x - 1)^2 e^x$ using Chebyshev grids of degrees $m = 14$, $m = 15$, and $m = 50$ on $-2 \leq x \leq 2$. Discuss.

Solution. The only zero of $f(x) = (x - 1)^2 e^x$ is a zero of multiplicity two at $x = 1$.
First, we try degree $m = 14$.

```
>> f = @(x) (x-1)^2*exp(x); a = -2; b = 2;
>> ps14 = samplecheb(f,[a b],14);
>> zeroscheb(ps14,[a b])
ans =
   0x1 empty double column vector
```

Curiously, no zeros are found. (On your computer, a different result may appear; see Remark 30.10 below.) The following code reveals where the zeros are hiding:

```
>> zeroscheb(ps14,[a b],1e-5)
ans =
   0.999999999697083  -  0.0000133300757112i
   0.999999999697083  +  0.0000133300757112i
```

The double zero at $x = 1$ has been perturbed into a pair of complex-conjugate zeros. To include these zeros in the result, a fairly large value for tol must be supplied to zeroscheb.

When the degree of the interpolant is increased to $m = 15$, the zeros reveal themselves more easily.

```
>> ps15 = samplecheb(f,[a b],15);
>> zeroscheb(ps15,[a b])
ans =
   0.999999929684657
   1.000000070188467
```

Again, the double zero separates into a pair of simple zeros, but this time they are both on the real line.

With degree $m = 50$, the story changes little.

```
>> ps50 = samplecheb(f,[a b],50);
>> zeroscheb(ps50,[a b])
ans =
   0.999999977588666
   1.000000022411338
```

Again, there are two real zeros, and again they differ from the true zero of f by about 10^{-8}. This is despite the fact that the interpolant produces a very close fit.

```
>> p50 = interpcheb(ps50,[a b]);
>> infnorm(@(x) f(x)-p50(x),[a b],100)
ans =
       3.552713678800501e-15
```

As in Example 30.8, the problem is ill conditioned and the error on the zero computation is much larger than the interpolation error.

In conclusion, the numerical results leave us uncertain about both the location and the multiplicity of the zero. The location of the zero is accurate to about eight digits rather than 16, and it is impossible to tell from the output alone whether there is a single zero of multiplicity two or two separate zeros of multiplicity one. ∎

Measuring the multiplicity of a zero is a difficult problem because perturbations can produce sudden changes in the number of zeros and their multiplicities. In the most recent example, the original function has a single zero of multiplicity two, but the first computation perturbs the zero into a complex-conjugate pair and the second perturbs the zero into a pair of distinct real zeros. A polynomial interpolant may not have the same zero multiplicities as the function it replaces.

Remark 30.10. Because it is an ill-posed problem to count the zero multiplicities in the previous example, tiny changes to the eigenvalue algorithm or the computer environment can have outsized effects on the computation. Therefore, the output on your computer may not match the output shown in Example 30.9.

Remark 30.11. Suppose a function is interpolated on a piecewise-defined grid with subintervals $[a_j, b_j]$, $j = 1, \ldots, n$. If there is a zero very close to one of the subdivision points $b_j = a_{j+1}$, then that zero may be identified as a zero on both $[a_j, b_j]$ and $[a_{j+1}, b_{j+1}]$ and hence reported with double its true multiplicity. Because the measurement of a zero's multiplicity is so delicate, no attempt is made to refine the reported multiplicity.

Notes

The companion matrix pair in this chapter was introduced by Corless in 2004 [19, 20]. Lawrence and Corless [49] recommend some adjustments to the companion matrix pair to support numerically stable eigenvalue computation, which we omit for simplicity.

The general idea of a companion matrix has a longer history. Frobenius introduced the original companion matrix, sometimes called a *Frobenius matrix*, in 1879 [33]. Its eigenvalues are the zeros of a polynomial $c_0 + c_1 x + c_2 x^2 + \cdots + c_m x^m$ specified by coefficients $c_0, \ldots, c_m$. Loewy coined the term *companion matrix* in 1917 [50], as told by Hawkins [42]. A "colleague matrix" for finding the zeros of a polynomial sampled on a Chebyshev grid was discovered independently by Specht in 1957 [66, 67] and Good in 1961 [36]. For the matrices of Frobenius and Specht/Good, a simple eigenvalue solver is sufficient, rather than the generalized eigenvalue solver required by Corless's companion matrix pair. On the other hand, Corless's approach allows a unified attack on both uniform and Chebyshev grids.

A 2013 article of Boyd [9] reviews proxy rootfinding, in which a function is replaced by a polynomial whose zeros are then computed, and discusses potential difficulties and resolutions.

As Moler wrote in 1991, companion matrices "turned the tables" on polynomial zero computation [57]. In early computer codes, eigenvalues were found by computing polynomial zeros. The MATLAB routine `roots`, in contrast, uses an eigenvalue solver to find polynomial zeros. This approach was put on firmer footing with the stability work of Edelman and Murakami in 1995 [26].

Exercises

Exercises 1–4: Sample the given polynomial on the given grid; construct the companion matrix pair $\mathbf{A}$, $\mathbf{B}$; and verify that $\det(x\mathbf{B} - \mathbf{A}) = Cp(x)$ for some constant C.

1. $4 - 5x$; uniform grid on $[0, 1]$ with $m = 1$
2. $5 - 9x - 7x^2$; uniform grid on $[-2, 2]$ with $m = 2$
3. $8 - 8x + x^2$; Chebyshev grid on $[-1, 7]$ with $m = 2$
4. $-9 + 6x - 9x^2 - 5x^3$; Chebyshev grid on $[-3, 1]$ with $m = 3$

Exercises 5–8: A function, an interval, and a degree m are given. Sample the function on an m-by-n piecewise-uniform grid and use `zerosuni` to compute its zeros. Increase the number of subintervals n until all zeros are found and all computed zeros have residuals below 10^{-6}.

5. $e^x - 1/9; -4 \leq x \leq 0; m = 2$

6. $\sin x + e^x; -10 \leq x \leq 0; m = 1$

7. $x + \cot x; 1 \leq x \leq 3; m = 3$

8. $-11 + 7x + 9x^2 + 8x^3 - 8x^4; -1 \leq x \leq 2; m = 2$

Exercises 9–12: A function and an interval are given. Sample the function on a Chebyshev grid and use zeroscheb to compute its zeros. Increase the degree of the grid until all zeros are found and all computed zeros have residuals below 10^{-12}.

9. $\log x - \tan x, 11 \leq x \leq 14$

10. $4 + 9x - 8x^2 - 6x^3, -2 \leq x \leq 2$

11. $e^x - x^2, -2 \leq x \leq 2$

12. $\sin(x^3 + 1), -2 \leq x \leq 2$

13. Compute the first 10 zeros of the Bessel function $J_0(x)$. Seek numerical convergence.

14. Recall that the Airy Ai function satisfies the IVP (29.4). (See also Exercise 11 in Chapter 27.) All of the function's zeros lie on the negative real axis. Solve the IVP for the Ai function and then compute the 10 rightmost zeros to nine-decimal-place accuracy. Compare with the table of zeros in the *Digital Library of Mathematical Functions* [22, Table 9.9.1].

15. Suppose a degree-m polynomial $p(x)$ has a zero at $x = \lambda$. Find a *generalized eigenvector* $\mathbf{v}$ for the companion matrix pair that is nonzero and that satisfies $\mathbf{Av} = \lambda \mathbf{Bv}$. Prove that your solution is correct. *Hint.* Try setting the first entry of $\mathbf{v}$ equal to 1.

16. Let $p(x) = c_0 + c_1 x + c_2 x^2 + \cdots + c_m x^m$ be a degree-m monic polynomial expressed in the monomial basis. Let

$$\mathbf{A} = \begin{bmatrix} 0 & 0 & \cdots & 0 & -c_0 \\ 1 & 0 & \cdots & 0 & -c_1 \\ 0 & 1 & \cdots & 0 & -c_2 \\ \vdots & \vdots & \ddots & \ddots & \vdots \\ 0 & 0 & \cdots & 1 & -c_{m-1} \end{bmatrix}.$$

Prove that $\det(x\mathbf{I} - \mathbf{A}) = p(x)$.

17. Compute the stationary points (where the derivative equals zero) of the following *Hermite function*:

$$f(x) = (2\sqrt{15}\pi^{1/4})^{-1}(4x^5 - 20x^3 + 15x)e^{-x^2/2}.$$

Seek numerical convergence.

18. Compute the stationary points of $f(x) = \sin(\sqrt{5}x) + \sin(\sqrt{23}x) + \sin(\sqrt{43}x)$ in $0 \leq x \leq 20$. Seek numerical convergence. Then locate the global maximum and report its (x, y) coordinates.

19. Develop a MATLAB routine for computing the zeros of a piecewise-Chebyshev interpolant. (See Example 10.10 on piecewise-Chebyshev interpolants.) Demonstrate it by computing the zeros of $f(x) = \sin(x^3 + x + 1)$ in $0 \leq x \leq 10$. Partition the interval into subintervals $[(10(j - 1))^{1/3}, (10j)^{1/3}], j = 1, \ldots, 100$, and use a Chebyshev interpolant of degree $m = 32$ on each subinterval.

20. Compute the zeros of $f(x) = (1/x) + \log(x) - 2$, $0.1 \le x \le 8$, using zeroscheb, achieving absolute error below 10^{-6}. Then estimate the absolute error for each zero using Theorem 30.6 and Remark 30.7.

21. The function $f(x) = \sin(x) + \sin(7x/3) + \sin(10x/3)$ has two zeros in the interval $[7, 10]$. They are computed as follows:

```
>> f = @(x) sin(x)+sin(7*x/3)+sin(10*x/3); a = 7; b = 10;
>> m = 10:10:50; z = nan(2,length(m));
>> for k = 1:length(m)
       ps = samplecheb(f,[a b],m(k));
       z(:,k) = zeroscheb(ps,[a b]);
   end
>> disptable('m',m,'%2d' ...
             ,'z1',z(1,:),'%17.15f','z2',z(2,:),'%17.15f');
```

m	z1	z2
10	7.539969330426779	9.482725428841833
20	7.539822368617783	9.424898640827148
30	7.539822368615503	9.424782951175297
40	7.539822368615504	9.424785748695690
50	7.539822368615503	9.424766868788904

The approximations for the first zero quickly converge to 15 consistent digits, while the approximations for the second zero have converged in only about five digits. Why should we expect the first zero to be computed more accurately?

22. Theorem 30.6 is not useful when ζ is a multiple zero. Explain.

23. Suppose $f \in C^2[a, b]$ has a double zero $\zeta \in [a, b]$ and z is another point in $[a, b]$. Prove that if there is a constant M for which $1/|f''(x)| \le M$ for all x between ζ and z, then

$$|\zeta - z| \le \sqrt{2M|f(z)|}.$$

Hint. Use the Lagrange remainder of Exercise 14 in Chapter 4.

Chapter 31

Newton's method

An older method for finding zeros is *Newton's method*. This method is designed to locate a single zero rather than all of a function's zeros at once. Given a sufficiently accurate starting guess, a succession of more and more accurate approximations is produced, until the process converges numerically on the desired zero. Newton's method is known as an *iterative* method because, rather than directly computing the answer in a single step, a transformation is applied repeatedly to move closer and closer to the solution.

31.1 ▪ The method

Suppose $x^{(0)}$ is a starting guess for a zero ζ of a given function $f(x)$. Newton's method produces a sequence of values $x^{(1)}, x^{(2)}, x^{(3)}, \ldots$ that may, under the right circumstances, converge to the desired zero ζ. The sequence of estimates is defined recursively:

$$x^{(k+1)} = x^{(k)} - \frac{f(x^{(k)})}{f'(x^{(k)})}, \quad k = 0, 1, 2, 3, \ldots. \tag{31.1}$$

Here is the idea behind each Newton step: Rather than trying to solve $f(x) = 0$ directly, we replace $f(x)$ by its first-order Taylor polynomial about $x = x^{(0)}$, a process known as *linearization*. Equivalently, we replace the graph of $f(x)$ by its tangent line at the point $(x^{(0)}, f(x^{(0)}))$, as in the left plot of Figure 31.1. If $x^{(0)}$ is somewhat close to a zero ζ, then the tangent line should roughly follow the graph of $f(x)$ toward the zero. The tangent line

$$y = f(x^{(0)}) + f'(x^{(0)})(x - x^{(0)})$$

intersects the x-axis at

$$x = x^{(0)} - \frac{f(x^{(0)})}{f'(x^{(0)})},$$

so this value becomes the next iterate $x^{(1)}$. The process is repeated as in the middle plot of Figure 31.1 to produce $x^{(2)}$, and then again as in the right plot to produce $x^{(3)}$, and again and again to produce $x^{(4)}, x^{(5)}, \ldots$. If everything goes right, then the sequence converges to the zero ζ.

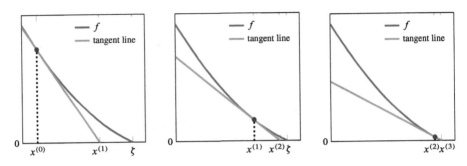

Figure 31.1. *Three steps of Newton's method.*

Example 31.1. Approximate $\sqrt{2}$ using Newton's method. Show steps.

Solution. By definition, $\sqrt{2}$ is the positive solution to $x^2 = 2$. In other words, it is the positive zero ζ of the function $f(x) = x^2 - 2$.

The Newton iteration formula for this particular function is

$$x^{(k+1)} = x^{(k)} - \frac{(x^{(k)})^2 - 2}{2x^{(k)}} = x^{(k)} - \left(\frac{x^{(k)}}{2} - \frac{1}{x^{(k)}} \right) = \frac{x^{(k)}}{2} + \frac{1}{x^{(k)}}.$$

(Note that this is equivalent to the Babylonian method of Chapter 1.)

To initialize the iteration, a starting guess is required. We'll use $x^{(0)} = 1.4$.

Newton's method then produces the following sequence of iterates:

```
>> x0 = 1.4
x0 =
    1.400000000000000
>> x1 = x0/2+1/x0
x1 =
    1.414285714285714
>> x2 = x1/2+1/x1
x2 =
    1.414213564213564
>> x3 = x2/2+1/x2
x3 =
    1.414213562373095
>> x4 = x3/2+1/x3
x4 =
    1.414213562373095
```

Numerical convergence occurs after just three steps.

Let's compare with the built-in value:

```
>> sqrt(2)
ans =
    1.414213562373095
```

The answers agree. ∎

In the previous example, we terminated Newton's method when numerical convergence was observed. A computer routine must detect convergence—or divergence—automatically so that it does not run forever. Our code uses four termination criteria:

1. If the change $\Delta x = -f(x^{(k)})/f'(x^{(k)})$ falls below a given tolerance, then numerical convergence has likely occurred, and the iteration terminates.

2. If $f(x^{(k+1)})$ is indistinguishable from 0 in the computer's representation, then a zero has evidently been found and the iteration terminates.

3. If $x^{(k+1)}$ is undefined, then the iteration has failed and must be terminated.

4. If the number of steps reaches a fixed cap, then the iteration terminates. This is a safety measure to force termination when the sequence $x^{(0)}, x^{(1)}, x^{(2)}, \ldots$ diverges.

Newton's method is implemented in routine newton in Listing 31.1.

Listing 31.1. Locate a zero using Newton's method.

```
function [z,x] = newton(f,fprime,x0,kmax,tol)

if nargin<4, kmax = 100; end
if nargin<5, tol = 1e-13; end
x = nan(kmax+1,1);
% initialize
x(1) = x0;
% iterate
for k = 1:kmax
  fx = f(x(k));
  % terminate if solution found
  if fx==0
    x = x(1:k); z = x(end); return;
  end
  % take a Newton step
  fpx = fprime(x(k));
  dx = -fx/fpx;
  x(k+1) = x(k)+dx;
  % signal failure on NaN or infinity
  if ~isfinite(x(k+1))
    x = x(1:k+1); z = NaN; return;
  end
  % terminate on numerical convergence
  if abs(dx)<=tol*max(abs(x(k+1)),1)
    x = x(1:k+1); z = x(end); return;
  end
end
% signal failure to converge
z = NaN;
```

Example 31.2. Solve $\log x = 1$ using newton.

Solution. Solving the equation is equivalent to finding a zero of $f(x) = \log(x) - 1$.
```
>> f = @(x) log(x)-1;
>> fprime = @(x) 1/x;
```
For the starting guess, we use $x^{(0)} = 1$—not a great guess, but hopefully good enough.
```
>> x0 = 1;
```
Now solve.
```
>> [z,x] = newton(f,fprime,x0);
```
The sequence of Newton iterates is

```
>> x
x =
    1.000000000000000
    2.000000000000000
    2.613705638880109
    2.716243926355791
    2.718281064358139
    2.718281828458938
    2.718281828459045
```

The final approximation is extremely close to the known value.

```
>> exp(1)
ans =
    2.718281828459046
```

■

31.2 ▪ Rate of convergence

Newton's method can be startlingly fast.

Theorem 31.3. *Suppose Newton's method is used to solve* $f(x) = 0$ *and that the sequence* $x^{(0)}, x^{(1)}, x^{(2)}, x^{(3)}, \ldots$ *of Newton iterates converges to a zero* ζ *for which* $f'(\zeta) \neq 0$. *In addition, suppose that* $f(x)$ *has a continuous second derivative in an interval containing the Newton iterates. Then*

$$f(x^{(k+1)}) = C_{k+1} f(x^{(k)})^2, \tag{31.2}$$

in which $C_1, C_2, C_3, \ldots$ *is a sequence of numbers converging to* $f''(\zeta)/(2f'(\zeta)^2)$.

The important point is that when Newton's method converges to a zero ζ (and $f'(\zeta) \neq 0$), the residual is more or less squared at each step. If, for example, $|f(x^{(k)})| \approx 10^{-5}$, then after only one additional step, $|f(x^{(k+1)})|$ should be on the order of 10^{-10}. The number of correct digits roughly doubles with each step. This is a form of supergeometric convergence that is faster than anything we've seen previously.

The proof requires a rigorous notion of Taylor polynomial approximation. The statement

$$f(x + \Delta x) \approx f(x) + f'(x)\Delta x$$

is an intuitive but imprecise statement that f can be approximated by a linear function in a neighborhood of x. A rigorous statement of this fact incorporates the *Lagrange remainder*. If $f''(x)$ exists and is continuous on an interval containing x and $x + \Delta x$, then

$$f(x + \Delta x) = f(x) + f'(x)\Delta x + \frac{f''(\eta)}{2}(\Delta x)^2$$

for some η between x and $x + \Delta x$. A proof of this fact is the subject of Exercises 13–14 in Chapter 4, and a second proof appears in Appendix A.

Proof of Theorem 31.3. Let $\Delta x = x^{(k+1)} - x^{(k)}$. Expanding $f(x)$ in a first-order Taylor polynomial and using the Lagrange remainder, we have

$$f(x^{(k+1)}) = f(x^{(k)}) + f'(x^{(k)})\Delta x + \frac{f''(\eta_k)}{2}(\Delta x)^2$$

for some η_k between $x^{(k)}$ and $x^{(k+1)}$. By (31.1), $\Delta x = -f(x^{(k)})/f'(x^{(k)})$. This reduces the previous equation as follows:

$$f(x^{(k+1)}) = \frac{f''(\eta_k)}{2}(\Delta x)^2 = \frac{f''(\eta_k)f(x^{(k)})^2}{2f'(x^{(k)})^2} = C_{k+1}f(x^{(k)})^2,$$

with

$$C_{k+1} = \frac{f''(\eta_k)}{2f'(x^{(k)})^2}.$$

As $k \to \infty$, the iterates $x^{(k)}$ converge to ζ and in turn squeeze the η_k to ζ. Hence, $\lim_{k \to \infty} C_k = f''(\zeta)/(2f'(\zeta)^2)$. $\square$

Example 31.4. Verify Theorem 31.3 on the function $f(x) = \log(x) - 1$ starting from $x^{(0)} = 1$.

Solution. The first several Newton iterates are the following.
```
>> f = @(x) log(x)-1;
>> fprime = @(x) 1/x;
>> [z,x] = newton(f,fprime,1);
>> x
x =
   1.000000000000000
   2.000000000000000
   2.613705638880109
   2.716243926355791
   2.718281064358139
   2.718281828458938
   2.718281828459045
```
Theorem 31.3 predicts that the absolute residual should be roughly squared with each step. Let's measure the residuals.
```
>> res = arrayfun(f,x)
res =
  -1.000000000000000
  -0.306852819440055
  -0.039231000595620
  -0.000749983454204
  -0.000000281097054
  -0.000000000000039
                   0
```
Once the method "kicks in," the number of zeros immediately to the right of the decimal point approximately doubles with each step. This confirms the prediction. The absolute residuals are plotted in Figure 31.2.
```
>> newfig; ylog; ylim([1e-18 1e2]);
>> plot(0:length(x)-1,abs(res),'*');
```
The convergence is supergeometric as expected.

The theorem also predicts that the ratio $C_{k+1} = f(x^{(k+1)})/f(x^{(k)})^2$ should converge to

$$\frac{f''(\zeta)}{2f'(\zeta)^2} = \frac{-1/\zeta^2}{2/\zeta^2} = -\frac{1}{2}.$$

Let's check.

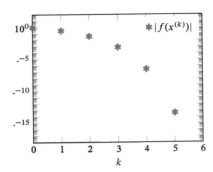

Figure 31.2. *Rate of convergence for Newton's method.*

```
>> C = arrayfun(@(k) f(x(k+1))/f(x(k))^2,1:length(x)-1)'
C =
  -0.306852819440055
  -0.416647853699156
  -0.487296073746148
  -0.499750145933081
  -0.498799482799810
                   0
```

We see a clear tendency toward -0.5, as expected. The final value is 0 because of roundoff error and should be ignored. ∎

The particular limiting value for C_k is not so important; what is important is that C_k does not diverge to infinity and therefore the residual $f(x^{(k+1)})$ on the left-hand side of (31.2) is proportional to the factor $f(x^{(k)})^2$ on the right-hand side of the equation.

31.3 ▪ Possibilities for failure

Newton's method should succeed if the starting guess $x^{(0)}$ is "sufficiently close" to a zero ζ. However, the method can fail in multiple ways.

1. If $f(x^{(k)})$ or $f'(x^{(k)})$ is undefined, then $x^{(k+1)} = x^{(k)} - f(x^{(k)})/f'(x^{(k)})$ is undefined.

2. If $f'(x^{(k)}) = 0$, then $x^{(k+1)} = x^{(k)} - f(x^{(k)})/f'(x^{(k)})$ is undefined.

3. The sequence of Newton iterates could diverge, for example, by getting stuck in a cycle or by tending toward infinity.

Exercises 13 and 14 ask you to explore these possibilities.

Note that Theorem 31.3 says nothing about these failure modes. The theorem describes the rate of convergence when Newton's method converges; it does not predict whether the method will succeed or fail in the first place.

The use of Newton's method requires care. If a starting guess is sufficiently close to a zero, then the method is expected to succeed. However, any user of Newton's method must be ready to compensate for failure.

31.4 ▪ Finding a starting guess

Newton's method solves a different problem from the companion matrix pair of the previous chapter. With the companion matrix pair, all zeros of a polynomial are computed without supervision. In contrast, Newton's method finds only one zero, and it requires a starting guess. The companion matrix pair is often easier to use, but Newton's method can be highly accurate and very fast.

The most obvious question regarding Newton's method is, How can we realistically find a starting guess? We explore two possibilities. First, we look at examples in which initial estimates are computed with the companion matrix pair and then refined using Newton's method. Second, we look at an example in which the physics of the problem provide a natural starting guess.

Example 31.5. Compute the zeros of $f(x) = x^2 + \cos(4x) + \sin(4x)$, $-4 \le x \le 4$, as follows. First, find rough estimates using the companion-matrix-pair method with a 3-by-10 piecewise-uniform grid. Then refine the estimates using Newton's method. Measure the residuals before and after the application of Newton's method.

Solution. The zeros are approximated with the companion-matrix-pair method.

```
>> f = @(x) x^2+cos(4*x)+sin(4*x); a = -4; b = 4;
>> m = 3; n = 10;
>> ps = sampleuni(f,[a b],m,n);
>> zbefore = zerosuni(ps,[a b]);
```

Then they are refined with Newton's method.

```
>> fprime = @(x) 2*x-4*sin(4*x)+4*cos(4*x);
>> zafter = nan(size(zbefore));
>> for i = 1:length(zbefore)
     zafter(i) = newton(f,fprime,zbefore(i));
   end
```

The residuals are measured, before and after Newton's method.

```
>> resbefore = arrayfun(f,zbefore);
>> resafter = arrayfun(f,zafter);
>> disptable('before Newton',zbefore         ,'%18.15f' ...
             ,'abs res'      ,abs(resbefore),'%7.1e'    ...
             ,'after Newton' ,zafter         ,'%18.15f' ...
             ,'abs res'      ,abs(resafter) ,'%7.1e');
```

before Newton	abs res	after Newton	abs res
-0.844207943426836	2.7e-02	-0.848290801630888	3.1e-16
-0.199473420450717	2.2e-02	-0.203684601667735	0.0e+00
0.678137448379387	3.3e-02	0.669734186036642	2.2e-16
1.107331742953797	1.3e-02	1.110012815631033	3.3e-16

The final residuals are tiny. In this experiment, the eigenvalues of the companion matrix pair provide good starting guesses rather than precise approximations for the function zeros. ∎

Example 31.6. Repeat the previous example, but use a Chebyshev grid of degree 50 in place of the piecewise-uniform grid.

Solution. The zeros are approximated by substituting the function by a polynomial on a Chebyshev grid.

```
>> f = @(x) x^2+cos(4*x)+sin(4*x); a = -4; b = 4;
>> m = 50;
>> ps = samplecheb(f,[a b],m);
>> zbefore = zeroscheb(ps,[a b]);
```

Then they are refined with Newton's method.

```
>> fprime = @(x) 2*x-4*sin(4*x)+4*cos(4*x);
>> zafter = nan(size(zbefore));
>> for i = 1:length(zbefore)
      zafter(i) = newton(f,fprime,zbefore(i));
   end
```

The residuals are measured.

```
>> resbefore = arrayfun(f,zbefore);
>> resafter = arrayfun(f,zafter);
>> disptable('before Newton',zbefore           ,'%18.15f' ...
            ,'abs res'      ,abs(resbefore),'%7.1e'   ...
            ,'after Newton' ,zafter            ,'%18.15f' ...
            ,'abs res'      ,abs(resafter) ,'%7.1e');
    before Newton    abs res        after Newton    abs res
 -0.848290801630890  1.8e-14     -0.848290801630888  3.1e-16
 -0.203684601667735  4.4e-16     -0.203684601667735  1.1e-16
  0.669734186036640  9.2e-15      0.669734186036642  2.2e-16
  1.110012815631037  2.1e-14      1.110012815631033  3.3e-16
```

In this example, a degree-50 Chebyshev interpolant is sufficient to produce very high accuracy. Still, a step or two of Newton's method lowers the residual to the level of roundoff error. ∎

Cleaning up approximate zeros with Newton's method, as in the previous example, is sometimes called "Newton polishing."

In the next example, a starting guess for Newton's method is evident from the physics of the problem itself. It illustrates a common scenario in which an equation changes gradually over time. With each short time step, its solution moves by a rather small amount and needs to be tracked. The solution at one moment in time serves as a reasonable starting guess for the solution at the next moment.

Example 31.7. A planet orbits a star. The orbit is an ellipse lying in a plane, and the planet's position (x, y) at time t obeys the following parametric equations:

$$t = \theta - \frac{1}{2}\sin\theta,$$
$$x = \cos\theta,$$
$$y = \frac{\sqrt{3}}{2}\sin\theta.$$

Curiously, the equations are most simply stated with time t as a function of angle θ rather than vice versa.

Compute the planet's position at times $t_i = (\pi/10)i$ for $i = 0, 1, 2, \ldots, 20$.

Solution. The planet's position is desired at linearly spaced times t_i.

```
>> t = (pi/10)*(0:20);
```

To find the position, we first compute the angles θ_i satisfying

$$t_i - \theta_i + \frac{1}{2}\sin\theta_i = 0, \quad i = 0, 1, 2, \ldots, 20. \tag{31.3}$$

The first equation $0 - \theta_0 + (1/2)\sin\theta_0 = 0$ is especially simple and has exactly one solution, specifically $\theta_0 = 0$.

```
>> th = nan(size(t));
>> th(1) = 0;
```

Newton's method is well suited to solving each of the remaining equations in (31.3). For each equation, just one solution is expected, and there is a reasonable starting guess for θ_{i+1}: the previous value θ_i.

```
>> for i = 1:length(t)-1
     f = @(th) t(i+1)-th+(1/2)*sin(th);
     fprime = @(th) -1+(1/2)*cos(th);
     th(i+1) = newton(f,fprime,th(i));
   end
```

The position is given explicitly in terms of θ:

```
>> x = cos(th);
>> y = sqrt(3)/2*sin(th);
```

The complete data table is formed.

```
>> disptable('t'    ,t ,'%6.4f' ...
             ,'theta',th,'%6.4f' ...
             ,'x'    ,x ,'%7.4f' ...
             ,'y'    ,y ,'%7.4f');
```

t	theta	x	y
0.0000	0.0000	1.0000	0.0000
0.3142	0.5940	0.8287	0.4847
0.6283	1.0659	0.4837	0.7580
0.9425	1.4381	0.1323	0.8584
1.2566	1.7487	-0.1770	0.8524
1.5708	2.0210	-0.4351	0.7797
1.8850	2.2682	-0.6422	0.6638
2.1991	2.4988	-0.8004	0.5191
2.5133	2.7185	-0.9118	0.3555
2.8274	2.9316	-0.9780	0.1805
3.1416	3.1416	-1.0000	0.0000
3.4558	3.3515	-0.9780	-0.1805
3.7699	3.5646	-0.9118	-0.3555
4.0841	3.7844	-0.8004	-0.5191
4.3982	4.0150	-0.6422	-0.6638
4.7124	4.2622	-0.4351	-0.7797
5.0265	4.5344	-0.1770	-0.8524
5.3407	4.8451	0.1323	-0.8584
5.6549	5.2172	0.4837	-0.7580
5.9690	5.6892	0.8287	-0.4847
6.2832	6.2832	1.0000	-0.0000

The planet's motion is shown in Figure 31.3. The planet starts at $(1, 0)$ and travels counterclockwise. It traces an ellipse with the star at one of its foci, and it moves fastest when it is closest to the star.

```
>> newfig; plot(x,y,'.'); plot(0.5,0,'o','markersize',10);
>> axis([-1.5 1.5 -1.5 1.5]); axis square;
```

The accuracy of the computation is indicated by the following absolute residuals:

```
>> max(abs(t-th+(1/2)*sin(th)))
ans =
     4.996003610813204e-16
>> max(abs(x-cos(th)))
ans =
     0
>> max(abs(y-sqrt(3)/2*sin(th)))
ans =
     0
```

The residuals are tiny. ∎

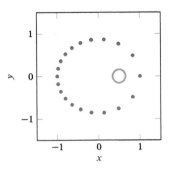

Figure 31.3. *Sample points from the trajectory of an orbiting planet.*

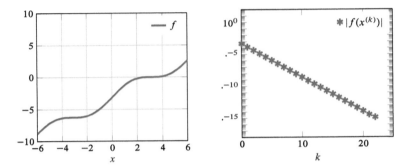

Figure 31.4. *A function with a multiple zero (left) and the rate of convergence for Newton's method (right).*

31.5 ▪ Multiple zeros

Recall that a zero ζ of a function f has multiplicity r if f and its first $r - 1$ derivatives equal zero at ζ. Theorem 31.3 considers only simple zeros and establishes a supergeometric rate of convergence in this case. At a multiple zero, Newton's method converges less rapidly.

Example 31.8. Verify that $f(x) = \sin x + x - \pi$ has a zero of multiplicity three at $x = \pi$, both symbolically and graphically. Then run Newton's method starting from $x^{(0)} = 3.0$. Classify the rate of convergence as algebraic, geometric, or supergeometric, and compare with Example 31.4.

Solution. The function and its first two derivatives are $f(x) = \sin x + x - \pi$, $f'(x) = \cos x + 1$, and $f''(x) = -\sin x$, and $f(\pi) = f'(\pi) = f''(\pi) = 0$. Hence, there is a zero of multiplicity 3 at $x = \pi$. This also seems reasonable from a graph of the function.

```
>> f = @(x) sin(x)+x-pi;
>> ax = newfig(1,2); plotfun(f,[-6 6]); grid on;
```

See the left plot of Figure 31.4 and notice that the height, slope, and curvature all appear to equal zero at $x = \pi$.

Our next task is to run Newton's method.

```
>> fprime = @(x) cos(x)+1;
>> x0 = 3.0;
>> [z,x] = newton(f,fprime,x0);
>> z
```

```
z =
    3.141580014204314
```

The computed zero is only accurate to about five significant digits. As discussed in the previous chapter, multiple zeros are ill conditioned.

The absolute residuals $|f(x^{(k)})|$ are plotted against k in the right plot of Figure 31.4.

```
>> res = arrayfun(f,x);
>> subplot(ax(2)); ylog; ylim([1e-18 1e2]);
>> plot(0:length(x)-1,abs(res),'*');
```

The convergence appears to be geometric, not supergeometric as in Example 31.4. ∎

The behavior of Newton's method near a multiple zero is illuminated by the example $f(x) = (x - \zeta)^r$. The iteration formula becomes

$$x^{(k+1)} = x^{(k)} - \frac{(x^{(k)} - \zeta)^r}{r(x^{(k)} - \zeta)^{r-1}} = x^{(k)} - \frac{1}{r}(x^{(k)} - \zeta). \tag{31.4}$$

A step of $-(x^{(k)} - \zeta)$ would take the iterate directly to the zero, but the Newton step is $\Delta x = -(x^{(k)} - \zeta)/r$. Newton's method takes a shortened step toward the zero when the multiplicity r is greater than 1.

When a desired zero is known to have multiplicity r, the Newton step can be lengthened by a factor of r to counteract the behavior just described. A method for accelerating Newton's method when the multiplicity is unknown is explored in Exercises 16–18.

Notes

Newton's method dates to the 1670's, as covered by Goldstine [34]. The method presented in this chapter is sometimes called the *Newton–Raphson method* to distinguish it from variants of Newton's method for higher-dimensional spaces.

The method of Exercises 16–18 for ameliorating multiple zeros appears in books by Traub [71] and Ralston and Rabinowitz [62].

Exercise 14 is based on an example in Moler's numerical analysis text [58, p. 121].

Exercises

Exercises 1–4: Run four steps of Newton's method by hand. Then plot the function and mark the Newton iterates on the x-axis.

1. $f(x) = 4 + 3x - 5x^2$, $x^{(0)} = -2$
2. $f(x) = 9 - x^2 + x^3 - 8x^4$, $x^{(0)} = -0.5$
3. $f(x) = -1 - x + \csc x$, $x^{(0)} = 3$
4. $f(x) = \log x - \cot x$, $x^{(0)} = 2$

Exercises 5–8: Compute the given quantity by expressing it as a zero of a function and then using newton.

5. $\sqrt[3]{10}$
6. $\log 10$
7. π

8. ϕ (the golden ratio)

9. Compute the 50 leftmost zeros of the Bessel J_0 function to full accuracy. Use newton and possibly also zerosuni or zeroscheb. The function value $J_0(x)$ can be evaluated with besselj(0,x) and, thanks to an identity, the derivative $J_0'(x)$ can be evaluated with -besselj(1,x).

10. Compute the 50 rightmost zeros of the Airy Ai function to full accuracy. Use newton and possibly also zerosuni or zeroscheb. The function value Ai(x) can be evaluated with airy(x), and the derivative Ai$'(x)$ can be evaluated with airy(1,x).

11. Let $F(x)$ be the cumulative distribution function for the standard normal distribution:
$$F(x) = \frac{1}{\sqrt{2\pi}} \int_{-\infty}^{x} e^{-\xi^2/2} \, d\xi.$$

This can also be expressed in terms of the error function erf:
$$F(x) = \frac{1}{2}\left(1 + \text{erf}\left(\frac{x}{\sqrt{2}}\right)\right).$$

The inverse cumulative distribution function is $F^{-1}(x)$. Construct a table of values $F^{-1}(x)$ with $x = 0.5, 0.6, 0.7, 0.8, 0.9, 0.95, 0.99, 0.999, 0.9999, 0.99999$ using newton and erf. Make sure that no execution of Newton's method requires more than 10 steps; explain your starting-guess strategy to achieve this.

12. The principal branch of the *Lambert W function* has domain $[-1/e, +\infty)$ and satisfies $W(x)e^{W(x)} = x$ for all x in the domain. For $x < 0$, the equation $We^W = x$ has two solutions but exactly one satisfying $W \geq -1$, and $W(x)$ is defined to be this solution. Develop a MATLAB routine to compute $W(x)$ using newton. Check that your code achieves a small absolute residual $|W(x)e^{W(x)} - x|$.

13. Find an example for which Newton's method diverges toward infinity.

14. Find a function $f(x)$ such that $f'(x)/f(x) = 1/(2x)$ for all $x \neq 0$. (There are infinitely many such functions; any one will do.) How does Newton's method behave on this function? Predict the behavior by hand and then test your prediction with newton.

15. Suppose that $f(x)$ has a zero ζ of multiplicity exactly r, that $x^{(0)}, x^{(1)}, x^{(2)}, \ldots$ converge to ζ, and that $f^{(r+1)}(x)$ exists and is continuous in an interval containing the zero and the Newton iterates. Prove that
$$\lim_{k\to\infty} \frac{x^{(k+1)} - x^{(k)}}{x^{(k)} - \zeta} = -\frac{1}{r}.$$

Explain how this generalizes (31.4). *Hint.* Replace $f(x)$ and $f'(x)$ in (31.1) by Taylor polynomials, using the Lagrange remainder of Chapter 4, Exercise 14.

16. Let f have a continuous rth derivative in a neighborhood of a number ζ, and suppose that ζ is a zero of multiplicity exactly r, i.e., $f(\zeta) = f'(\zeta) = \cdots = f^{(r-1)}(\zeta) = 0$ but $f^{(r)}(\zeta) \neq 0$. Prove
$$\lim_{x\to\zeta} \frac{f(x)}{(x-\zeta)^r} = \frac{f^{(r)}(\zeta)}{r!}.$$

17. Let (a, b) be an open interval containing a number ζ, and let f be a function defined on (a, b). Suppose (1) f has a continuous rth derivative in (a, b); (2) ζ is a zero of f of multiplicity exactly r, i.e., $f(\zeta) = f'(\zeta) = \cdots = f^{(r-1)}(\zeta) = 0$ but $f^{(r)}(\zeta) \neq 0$; and (3) f' has no zeros in (a, b) other than possibly ζ. Let

$$g(x) = \begin{cases} f(x)/f'(x) & \text{if } x \neq \zeta, \\ 0 & \text{if } x = \zeta. \end{cases}$$

Prove that ζ is a zero of g of multiplicity exactly one. *Hint.* Use Exercise 16.

18. As seen in Example 31.8, Newton's method converges comparatively slowly on $f(x) = \sin x + x - \pi$ because its zero is a multiple zero. Exercise 17 suggests that Newton's method might converge more quickly on $f(x)/f'(x)$. Try this and compare the rate of convergence with the rate seen in Example 31.8.

19. Another method for computing a single zero, specifically a sign change, of a function is *bisection*. It does not require that the function be smooth. It works as follows: Given a continuous function $f(x)$ and an interval $[a, b]$ for which $f(a)$ and $f(b)$ have different signs, it is determined whether a sign change occurs in the left half or the right half of the interval (or precisely at the midpoint). Then the process is repeated recursively on the appropriate half-interval. The process terminates once the width of the interval is below a given tolerance. Implement this method and demonstrate your implementation on $f(x) = x^5 + x^4 - 3x^3 + x^2 + x + 2$.

20. The bisection method is defined in Exercise 19. Suppose that the process terminates once the width of the interval is no greater than a given number τ. Prove that the number of subdivisions is $\lceil \log_2((b - a)/\tau) \rceil$, in which $\lceil \cdot \rceil$ is the ceiling function.

21. Another method for computing a zero of a function is the *secant method*. It works best on smooth functions, but it does not require a formula for the derivative. The user is required to supply a function $f(x)$ and a pair of distinct starting guesses $x^{(0)}$ and $x^{(1)}$ for one of the function's zeros. Each subsequent estimate $x^{(k+1)}$ is defined to be the zero, if it exists, of the linear function whose graph passes through the points $(x^{(k-1)}, f(x^{(k-1)}))$ and $(x^{(k)}, f(x^{(k)}))$. The process is terminated when the absolute residual falls below a given tolerance or when the iteration fails or the number of steps reaches a given cap. Implement this method and demonstrate your implementation on $f(x) = x^5 + x^4 - 3x^3 + x^2 + x + 2$.

Part VII

Nonlinear differential equations

Chapter 32

Collocation for nonlinear problems

Linear differential equations, seen earlier in Chapters 22–29 of this book, occupy one small but very important region of the differential equations landscape. Beyond lie the much more varied nonlinear differential equations. This chapter introduces nonlinear differential equations and a collocation method for solving them.

32.1 ▪ Nonlinear differential equations

The next few chapters consider *nonlinear differential equations*, specifically first-order equations of the form

$$u'(t) = f(t, u(t)), \tag{32.1}$$

in which $f(t, u)$ is a function of two variables, and second-order equations of the form

$$u''(t) = f(t, u(t), u'(t)), \tag{32.2}$$

in which $f(t, u, v)$ is a function of three variables.

Example 32.1. With $f(t, u) = u^2$, equation (32.1) becomes

$$u'(t) = u(t)^2.$$

Find the general solution to this nonlinear differential equation.

Solution. First, there is the simple solution $u(t) = 0$.

If $u(t) \neq 0$, then the differential equation is equivalent to

$$u^{-2} \frac{du}{dt} = 1.$$

The left-hand side reminds us of the chain rule for derivatives. By guessing and checking, we arrive at $\frac{d}{dt}[-u^{-1}] = u^{-2} \frac{du}{dt}$. Hence, the differential equation is equivalent to

$$\frac{d}{dt}[-u^{-1}] = 1.$$

Integrating both sides with respect to t gives

$$-u^{-1} = t + C,$$

in which C is an arbitrary constant. Hence, the nonzero solutions are

$$u(t) = -\frac{1}{t + C}. \qquad\blacksquare$$

The differential equation $u'(t) = u(t)^2$ is nonlinear; because of the $u(t)^2$ term, the equation cannot be written in the form $\alpha(t)u(t) + \beta(t)u'(t) = g(t)$. Thus it is outside the scope of our earlier numerical methods.

32.2 ▪ Collocation and Newton's method

The method of collocation can be extended from linear to nonlinear differential equations. As before, a continuous differential equation is replaced by a finite system of equations. Now, however, the collocation equations are nonlinear.

Definition 32.2. *Given distinct collocation nodes $s_0, \ldots, s_m$, a polynomial collocation solution for*

$$u'(t) = f(t, u(t)) \tag{32.3}$$

is a polynomial p of degree at most $m + 1$ that satisfies

$$p'(s_i) = f(s_i, p(s_i)), \quad i = 0, \ldots, m. \tag{32.4}$$

Equivalently, the residual

$$R(t) = p'(t) - f(t, p(t)) \tag{32.5}$$

equals zero at the collocation nodes.

The nonlinear collocation equations can be solved with a multivariable version of Newton's method. Suppose an estimate p is available, but it does not satisfy the collocation equations exactly. The goal is to find an adjustment Δp so that the polynomial $p + \Delta p$ more nearly satisfies the collocation equations. Ideally, the updated polynomial would satisfy

$$p'(s_i) + \Delta p'(s_i) = f(s_i, p(s_i) + \Delta p(s_i)), \quad i = 0, \ldots, m. \tag{32.6}$$

The key idea in Newton's method is to replace f by its local linear approximation. This is

$$f(t, u + \Delta u) \approx f(t, u) + f^{(0,1)}(t, u)\,\Delta u, \tag{32.7}$$

in which $f^{(0,1)}$ represents the partial derivative of f with respect to its second argument. To derive the Newton step, substitute (32.7) into (32.6) to obtain

$$p'(s_i) + \Delta p'(s_i) = f(s_i, p(s_i)) + f^{(0,1)}(s_i, p(s_i))\,\Delta p(s_i), \quad i = 0, \ldots, m,$$

or equivalently,

$$\alpha(s_i)\,\Delta p(s_i) + \Delta p'(s_i) = g(s_i), \quad i = 0, \ldots, m, \tag{32.8}$$

in which

$$\alpha(t) = -f^{(0,1)}(t, p(t)), \quad g(t) = f(t, p(t)) - p'(t). \tag{32.9}$$

The equations in (32.8) form a *linear* system of collocation equations for the unknown Δp.

If the differential equation (32.3) is paired with an initial condition $u(a) = u_a$ and the estimate p satisfies $p(a) = u_a$, then the update Δp should satisfy

$$\Delta p(a) = 0 \tag{32.10}$$

to preserve the initial condition.

Together, (32.8) and (32.10) generally determine a degree-$(m + 1)$ polynomial Δp.

The following procedure, based on the above derivations, is designed to compute a collocation solution to the first-order nonlinear IVP

$$u'(t) = f(t, u(t)), \quad u(a) = u_a \tag{32.11}$$

using Newton's method:

1. Produce a starting guess $p(t)$ that is a polynomial of degree at most $m + 1$ satisfying $p(a) = u_a$.

2. Repeat the following until the iteration converges or is aborted:

   (a) Let $\alpha(t) = -f^{(0,1)}(t, p(t))$ and $g(t) = f(t, p(t)) - p'(t)$, and compute a collocation solution $\Delta p(t)$ to the linear problem

   $$\alpha(t)\,\Delta p(t) + \Delta p'(t) = g(t), \quad \Delta p(a) = 0, \tag{32.12}$$

   using the collocation nodes $s_0, \ldots, s_m$.

   (b) Replace $p(t)$ by $p(t) + \Delta p(t)$.

Example 32.3. Consider the nonlinear IVP

$$u'(t) = tu(t)^2, \quad u(0) = 1$$

and the starting guess $p(t) = 1$. Run one step of Newton's method by hand, using collocation nodes $\mathbf{s} = (0, 0.5)$. Sample the approximate solution $p(t)$ on the grid $\mathbf{t} = (0, 0.25, 0.5)$ and its derivative $p'(t)$ on the grid $\mathbf{t}' = (0, 0.5)$. Then plot the starting guess, the updated estimate, and the exact solution $u(t) = 2/(2 - t^2)$.

Solution. The starting guess and its derivative may be represented by their samples on $\mathbf{t}$ and $\mathbf{t}'$:

$$\mathbf{p} = p|_{\mathbf{t}} = \begin{bmatrix} 1 \\ 1 \\ 1 \end{bmatrix}, \quad \mathbf{q} = p'|_{\mathbf{t}'} = \begin{bmatrix} 0 \\ 0 \end{bmatrix}.$$

The right-hand side of the differential equation and its derivative with respect to u are

$$f(t, u) = tu^2, \quad f^{(0,1)}(t, u) = 2tu.$$

Set

$$\alpha(t) = -f^{(0,1)}(t, p(t)) = -2tp(t) = -2t,$$
$$\beta(t) = 1,$$
$$g(t) = f(t, p(t)) - p'(t) = tp(t)^2 - p'(t) = t.$$

The Newton step is determined by collocation equations $\alpha(s_i)\Delta p(s_i)+\beta(s_i)\Delta p'(s_i) = g(s_i)$, $i = 0, 1$, or more specifically,

$$-2s_i\,\Delta p(s_i) + \Delta p'(s_i) = s_i, \quad i = 0, 1, \tag{32.13}$$

and the initial condition

$$\Delta p(0) = 0. \tag{32.14}$$

We expect that a unique quadratic function $\Delta p(t)$ satisfies (32.13)–(32.14), and it can be found by solving a linear system of equations.

The grids $\mathbf{s}$, $\mathbf{t}$, and $\mathbf{t}'$ are identical to those used in Example 23.3. Define $\mathbf{J_t}$, $\hat{\mathbf{K}}_{\mathbf{t},\mathbf{t}'}$, $\mathbf{E}_{\mathbf{s},\mathbf{t}}$, $\mathbf{E}_{\mathbf{s},\mathbf{t}'}$, $\mathbf{E}_{a,\mathbf{t}}$, and $\mathbf{E}_{a,\mathbf{t}'}$ as in the earlier example. In addition, noting that $\alpha(s_0) = 0$, $\alpha(s_1) = -1$, $\beta(s_0) = 1$, $\beta(s_1) = 1$, $g(s_0) = 0$, and $g(s_1) = 1/2$, let

$$\mathbf{A} = \begin{bmatrix} 0 & 0 \\ 0 & -1 \end{bmatrix} \mathbf{E}_{\mathbf{s},\mathbf{t}} = \begin{bmatrix} 0 & 0 & 0 \\ 0 & 0 & -1 \end{bmatrix}, \quad \mathbf{B} = \begin{bmatrix} 1 & 0 \\ 0 & 1 \end{bmatrix} \mathbf{E}_{\mathbf{s},\mathbf{t}'} = \begin{bmatrix} 1 & 0 \\ 0 & 1 \end{bmatrix},$$

and

$$\mathbf{g} = \begin{bmatrix} 0 \\ 1/2 \end{bmatrix}.$$

Also, set $c_0 = 1$, $c_1 = 0$, and $d = 0$. The linear system determining $\Delta\mathbf{p} = \Delta p|_{\mathbf{t}}$ and $\Delta\mathbf{q} = \Delta p'|_{\mathbf{t}'}$ is

$$\begin{bmatrix} \mathbf{J_t} & -\hat{\mathbf{K}}_{\mathbf{t},\mathbf{t}'} \\ \mathbf{A} & \mathbf{B} \\ c_0\mathbf{E}_{a,\mathbf{t}} & c_1\mathbf{E}_{a,\mathbf{t}'} \end{bmatrix} \begin{bmatrix} \Delta\mathbf{p} \\ \Delta\mathbf{q} \end{bmatrix} = \begin{bmatrix} \mathbf{0} \\ \mathbf{g} \\ d \end{bmatrix},$$

or more explicitly,

$$\left[\begin{array}{ccc|cc} -1 & 1 & 0 & -3/16 & -1/16 \\ -1 & 0 & 1 & -1/4 & -1/4 \\ \hline 0 & 0 & 0 & 1 & 0 \\ 0 & 0 & -1 & 0 & 1 \\ \hline 1 & 0 & 0 & 0 & 0 \end{array}\right] \begin{bmatrix} \Delta p(t_0) \\ \Delta p(t_1) \\ \Delta p(t_2) \\ \hline \Delta p'(t_0') \\ \Delta p'(t_1') \end{bmatrix} = \begin{bmatrix} 0 \\ 0 \\ \hline 0 \\ 1/2 \\ \hline 0 \end{bmatrix}.$$

Its solution, which can be found by Gaussian elimination, is

$$\begin{bmatrix} \Delta p(t_0) \\ \Delta p(t_1) \\ \Delta p(t_2) \\ \hline \Delta p'(t_0') \\ \Delta p'(t_1') \end{bmatrix} = \begin{bmatrix} 0 \\ 1/24 \\ 1/6 \\ \hline 0 \\ 2/3 \end{bmatrix}.$$

The Newton update is $p(t) \leftarrow p(t) + \Delta p(t)$. Equivalently, the samples $\mathbf{p}$, $\mathbf{q}$ are updated as follows, to four decimal places:

$$\mathbf{p} \leftarrow \mathbf{p} + \Delta\mathbf{p} = \begin{bmatrix} 1 \\ 1 \\ 1 \end{bmatrix} + \begin{bmatrix} 0 \\ 1/24 \\ 1/6 \end{bmatrix} = \begin{bmatrix} 1 \\ 1.0417 \\ 1.1667 \end{bmatrix},$$

$$\mathbf{q} \leftarrow \mathbf{q} + \Delta\mathbf{q} = \begin{bmatrix} 0 \\ 0 \end{bmatrix} + \begin{bmatrix} 0 \\ 2/3 \end{bmatrix} = \begin{bmatrix} 0 \\ 0.6667 \end{bmatrix}.$$

The updated approximation $p(t)$ is the quadratic polynomial whose graph intersects $(0, 1)$, $(0.25, 1.0417)$, and $(0.5, 1.1667)$, and its derivative is the linear polynomial whose graph intersects $(0, 0)$ and $(0.5, 0.6667)$.

Now we plot.

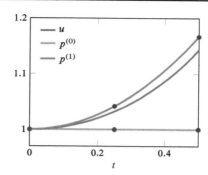

Figure 32.1. *The solution $u(t)$ to a nonlinear differential equation, a starting guess $p^{(0)}(t)$, and the approximation $p^{(1)}(t)$ after one step of Newton's method.*

```
>> newfig;
>> a = 0; b = 0.5; ts = [0; 0.25; 0.5];
>> plotfun(@(t) 2/(2-t^2),[a b]);
>> ps0 = [1; 1; 1]; p0 = interpgen(ts,ps0);
>> plotfun(p0,[a b]); plotsample(ts,ps0);
>> ps1 = [1; 1+1/24; 1+1/6]; p1 = interpgen(ts,ps1);
>> plotfun(p1,[a b]); plotsample(ts,ps1);
```
See Figure 32.1. The updated approximation appears to be an improvement. ∎

Typically, a few steps of Newton's method are required to achieve numerical convergence. Exercise 11 asks you to complete the Newton iteration for the previous example using computer code introduced in the next section.

32.3 ▪ Implementation

Nonlinear collocation is implemented by routines ivpnl1gen and ivpnl1_; see Listing 32.1. Note that the implementation leverages our earlier work on linear problems—the most computationally intensive step, the solution of a linear system of collocation equations, is accomplished by a call to ivpl1_ of Chapter 23. Also, the termination criteria parallel the criteria used in Chapter 31.

It is important to note a couple of points about the usage of ivpnl1gen. First, the user must supply the partial derivative $f^{(0,1)}(t, u)$ in the dfdu argument to ivpnl1gen. This is a necessary component of the collocation method that is not explicit in the original problem statement. Also, the user must provide a starting guess for the derivative $p'(t)$, rather than $p(t)$ itself, as argument qstart.

Remark 32.4. A natural starting guess for $p(t)$ is the linear function

$$p(t) = u_a + f(a, u_a)(t - a). \qquad (32.15)$$

This guess satisfies the initial condition $p(a) = u_a$, and it satisfies the differential equation at $t = a$: $p'(a) = f(a, u_a) = f(a, p(a))$. To use this guess in ivpnl1gen, set qstart = @(t) f(a,ua).

The aim of collocation is to make the residual $R(t) = p'(t) - f(t, p(t))$ vanish at a set of collocation nodes. For nonlinear problems, this is accomplished iteratively rather than in one fell swoop. The next two examples illustrate the procedure and end product.

Listing 32.1. Solve a first-order nonlinear IVP using collocation and Newton's method.

```
function [ps,qs] = ivpnl1gen(f,dfdu,a,ua,qstart,ss,ts,ts_,kmax,tol)

if nargin<9, kmax = 40; end
if nargin<10, tol = 1e-12; end
% form starting guess
qs = arrayfun(qstart,ts_);
ps = antiderivgen(ua,qs,ts,ts_,a);
% construct integration, resampling, and evaluation matrices
[J,K] = indefinitegen(ts,ts_);
E1 = evalmatrixgen(ss,ts);
E2 = evalmatrixgen(ss,ts_);
Ea1 = evalmatrixgen(a,ts);
Ea2 = evalmatrixgen(a,ts_);
% solve by collocation and Newton's method
[ps,qs,k] = ivpnl1_(f,dfdu,ps,qs,ss,J,K,E1,E2,Ea1,Ea2,kmax,tol);
if isempty(k)
  warning('NATE:nonlinearIvpFailure' ...
          ,'Nonlinear IVP solver failed to converge');
end
```

```
function [ps,qs,k] = ivpnl1_(f,dfdu,ps,qs,ss,J,K,E1,E2,Ea1,Ea2 ...
                                                        ,kmax,tol)

m = length(qs)-1;
% iterate
for k = 1:kmax
  ps_ = E1*ps;
  fs = arrayfun(f,ss,ps_);
  % terminate if solution found
  if all(qs==fs), return; end
  % take a Newton step
  dfdus = arrayfun(dfdu,ss,ps_);
  [dps,dqs] = ivpl1_(-dfdus,ones(m+1,1),fs-qs,[1 0],0 ...
                     ,J,K,E1,E2,Ea1,Ea2);
  ps = ps+dps;
  qs = qs+dqs;
  % signal failure on NaN or infinity
  if any(~isfinite(ps))||any(~isfinite(qs))
    k = []; return;
  end
  % terminate on numerical convergence
  de = tol*max([max(abs(ps)) max(abs(qs)) 1]);
  if max([ max(abs(dps)) max(abs(dqs)) ])<=de, return; end
end
% signal failure to converge
k = [];
```

Example 32.5. Consider the logistic model for population growth

$$u'(t) = ru(t)\left(1 - \frac{u(t)}{L}\right), \quad u(0) = u_a,$$

with time t measured in years and population u measured in thousands of individuals. The parameters and initial value specify the following:

- the *growth constant* r controls the rate at which the population can grow when

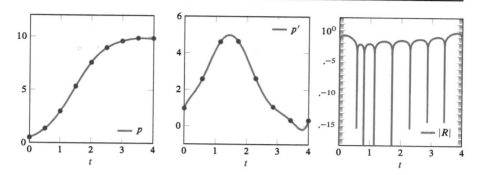

Figure 32.2. *Collocation solution for a logistic equation: population size (left), rate of growth (middle), and absolute residual (right).*

resources are plentiful;

- the *carrying capacity* L specifies the population size that the environment can stably support;

- the initial value u_a specifies the initial population size.

Set $r = 2$, $L = 10$, and $u_a = 0.5$. Solve the IVP on $0 \le t \le 4$ with ivpnl1gen, using a degree-7 uniform grid of collocation nodes and uniform interpolation grids **t** and **t'** of appropriate degrees. Plot the solution and its residual.

Solution. The problem is defined. Note that the right-hand side of the differential equation is $f(t, u) = ru(1 - u/L)$, which has the partial derivative $f^{(0,1)}(t, u) = ru(-1/L) + r(1 - u/L) = r(1 - 2u/L)$.

```
>> r = 2; L = 10;
>> f = @(t,u) r*u*(1-u/L);
>> dfdu = @(t,u) r*(1-2*u/L);
>> a = 0; b = 4;
>> ua = 0.5;
```

The grids are set.

```
>> m = 7;
>> ss = griduni([a b],m);
>> ts = griduni([a b],m+1);
>> ts_ = griduni([a b],m);
```

We'll try the starting guess suggested in Remark 32.4.

```
>> qstart = @(t) f(a,ua);
```

Newton's method is run.

```
>> [ps,qs] = ivpnl1gen(f,dfdu,a,ua,qstart,ss,ts,ts_);
>> p = interpgen(ts,ps); q = interpgen(ts_,qs);
```

The computed population size and its rate of change are plotted in the left and middle graphs of Figure 32.2.

```
>> ax = newfig(1,3); plotsample(ts,ps); plotfun(p,[a b]);
>> subplot(ax(2)); plotsample(ts_,qs); plotfun(q,[a b]);
```

The collocation solution $p(t)$ looks reasonable, given the coarseness of the grid. The population starts at the initial value $u_a = 0.5$ and grows to approximately $L = 10$ before leveling off. One infidelity is that the computed derivative is briefly negative, while the exact derivative is known to be positive at all times. We expect that this will be corrected by a finer grid in a future computation.

The defining property of collocation is that the residual equals zero on a grid. In this problem, the residual is $R(t) = p'(t) - f(t, p(t))$. See the rightmost graph in the figure.

```
>> R = @(t) q(t)-f(t,p(t));
>> subplot(ax(3)); ylog; ylim([1e-18 1e2]);
>> plotfun(@(t) abs(R(t)),[a b]);                                      ■
```

Example 32.6. Repeat Example 32.5. This time, plot the starting guess and the first eight Newton iterates. Then plot

$$\max_{i=0,\dots,m} |R(s_i)| \tag{32.16}$$

against the number of Newton steps, in which $R(t) = p'(t) - f(t, p(t))$ is the residual and $s_0, \dots, s_m$ are the collocation nodes.

Solution. The computation is set up as before.

```
>> r = 2; L = 10;
>> f = @(t,u) r*u*(1-u/L);
>> dfdu = @ (t,u) r*(1-2*u/L);
>> a = 0; b = 4;
>> ua = 0.5;
>> m = 7;
>> ss = griduni([a b],m);
>> ts = griduni([a b],m+1);
>> ts_ = griduni([a b],m);
>> qstart = @(t) f(a,ua);
```

To observe the progress of Newton iteration, ivpnl1gen is run several times from scratch, each time stopping after a prescribed number of steps.

```
>> kmax = 0:8;
>> res = nan(size(kmax));
>> ax = newfig(3,3); ax = ax';
>> for l = 1:length(kmax)
      [ps,qs] = ivpnl1gen(f,dfdu,a,ua,qstart,ss,ts,ts_,kmax(l));
      p = interpgen(ts,ps); q = interpgen(ts_,qs);
      res(l) = max(abs(arrayfun(@(t) q(t)-f(t,p(t)),ss)));
      subplot(ax(l)); plotsample(ts,ps); plotfun(p,[a b]);
   end
```

The starting guess and the first eight iterates are displayed in Figure 32.3. Observe how the function gradually relaxes into place.

The residual measure (32.16) is plotted in Figure 32.4. The rate of convergence, as a function of the number of Newton steps, appears to be supergeometric.

```
>> newfig; ylog; ylim([1e-18 1e3]);
>> plot(kmax,res,'*');                                                 ■
```

32.4 ▪ Rate of convergence

There are two questions to ask regarding the phrase *rate of convergence*. First, how quickly do the Newton iterates converge to the collocation solution? Second, how

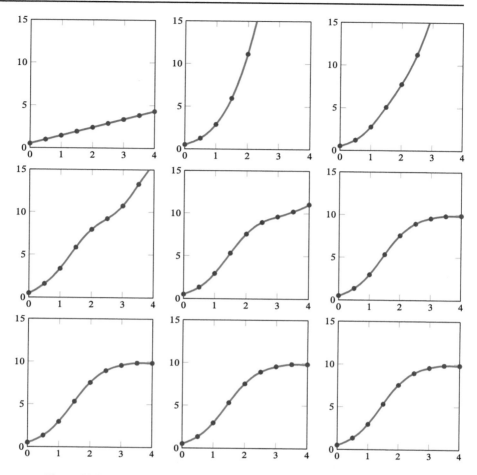

Figure 32.3. *A starting guess and eight Newton iterates for a collocation solution.*

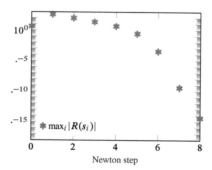

Figure 32.4. *Rate of convergence for Newton's method on a system of nonlinear collo-cation equations.*

quickly does the collocation solution converge to the exact solution as the grids are refined?

Newton's method performs basically as well on systems of collocation equations as it did on a single equation in one variable. Numerical convergence is often achieved in about a half dozen steps, and the computer code automatically detects numerical convergence. Hence, we can just run Newton's method to completion rather than performing a cost-benefit analysis of how many steps to take.

The main concern is thus the choice of grids that define the collocation solution. As before, algebraic convergence is typical for piecewise-uniform grids, and geometric convergence is possible for Chebyshev grids. In the interest of space, we will not give a detailed accuracy analysis for nonlinear problems, but we will run some experiments in the following chapters to see the power of collocation on nonlinear problems.

32.5 ▪ Failure to converge

Newton's method is not guaranteed to converge.

Example 32.7. Compute a collocation solution to

$$u'(t) = -e^t u(t)^2, \quad u(0) = 1$$

using a uniform collocation grid of degree 4 over $0 \leq t \leq 4$ and uniform interpolation grids of appropriate degrees. Use the starting guess described in Remark 32.4. Plot the starting guess and the first 11 Newton iterates.

Solution. The problem is stated, and the collocation parameters are set.

```
>> f = @(t,u) -exp(t)*u^2;
>> dfdu = @(t,u) -2*exp(t)*u;
>> a = 0; b = 4;
>> ua = 1;
>> m = 4;
>> ss = griduni([a b],m);
>> ts = griduni([a b],m+1);
>> ts_ = griduni([a b],m);
>> qstart = @(t) f(a,ua);
```

Newton iteration is run for k_{max} steps, for $k_{max} = 0, 1, 2, \ldots, 11$.

```
>> kmax = 0:11;
>> ax = newfig(4,3); ax = ax';
>> for l = 1:length(kmax)
       [ps,qs] = ivpnl1gen(f,dfdu,a,ua,qstart,ss,ts,ts_,kmax(l));
       p = interpgen(ts,ps);
       subplot(ax(l)); plotsample(ts,ps); plotfun(p,[a b]);
   end
```

The starting guess and first 11 Newton iterates are displayed in Figure 32.5. Newton iteration shows no sign of converging. ▪

In fact, the iteration in the previous example shows no sign of converging even if 100 steps are allowed.

Fortunately, there is still hope for Newton's method. In the next chapter, the domain $[a, b]$ is partitioned into short subintervals, and the IVP is solved one subinterval at a time. On a sufficiently narrow subinterval, a simple linear guess may be sufficiently accurate for Newton's method to succeed.

32.6 ▪ Conditioning

When Newton's method converges, the condition number of the linear problem (32.12) provides a good estimate for the condition number of the nonlinear problem (32.11).

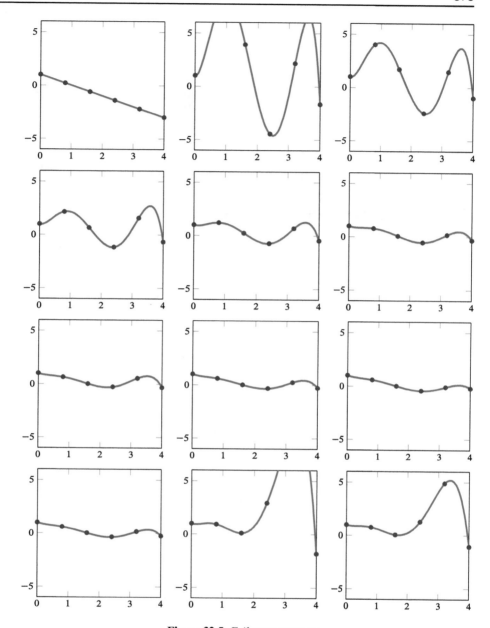

Figure 32.5. *Failure to converge.*

Intuitively, a problem with a large condition number allows significant "slack" in its solution. That is, the solution can be perturbed by a fairly large amount without significantly increasing the residual. If a Newton step is ill conditioned, then the Newton update $\Delta p(t)$ can be perturbed without significantly increasing the residual associated with (32.12). This in turn perturbs the computed $p(t)$ without much effect on the residual associated with (32.11).

Conditioning is explored further in the next chapter.

Figure 32.6. *A pendulum.*

32.7 ▪ Second-order problems

The equation

$$u''(t) = f(t, u(t), u'(t)) \tag{32.17}$$

is a second-order nonlinear differential equation.

A collocation solution $p(t)$ of degree $m + 2$ should satisfy

$$p''(s_i) = f(s_i, p(s_i), p'(s_i)), \quad i = 0, \ldots, m, \tag{32.18}$$

for a sequence of collocation nodes $s_0, \ldots, s_m$. A Newton step is derived using the local linear approximation

$$f(t, u + \Delta u, v + \Delta v) \approx f(t, u, v) + f^{(0,1,0)}(t, u, v)\, \Delta u + f^{(0,0,1)}(t, u, v)\, \Delta v.$$

Applying the local linear approximation to the collocation equations for $p(t) + \Delta p(t)$ gives

$$p''(s_i) + \Delta p''(s_i) = f(s_i, p(s_i), p'(s_i)) + f^{(0,1,0)}(s_i, p(s_i), p'(s_i))\, \Delta p(s_i)$$
$$+ f^{(0,0,1)}(s_i, p(s_i), p'(s_i))\, \Delta p'(s_i), \quad i = 0, \ldots, m,$$

or equivalently,

$$\alpha(s_i)\, \Delta p(s_i) + \beta(s_i)\, \Delta p'(s_i) + \Delta p''(s_i) = g(s_i), \quad i = 0, \ldots, m, \tag{32.19}$$

with

$$\alpha(t) = -f^{(0,1,0)}(t, p(t), p'(t)), \tag{32.20}$$
$$\beta(t) = -f^{(0,0,1)}(t, p(t), p'(t)), \tag{32.21}$$
$$g(t) = f(t, p(t), p'(t)) - p''(t). \tag{32.22}$$

A second-order equation is often accompanied by a pair of initial conditions $u(a) = u_a$, $u'(a) = v_a$. We require that the starting guess for Newton's method satisfy these initial conditions.

Routine `ivpnl2gen` computes a collocation solution to a second-order nonlinear IVP using Newton's method. The implementation is omitted here, but the routine is seen in action in the following example.

Example 32.8. Let u be a pendulum's angular displacement from vertical, as in Figure 32.6. The pendulum hangs straight down at $u = 0$, and it rotates counterclockwise as u

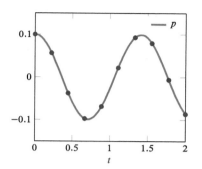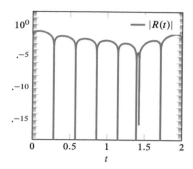

Figure 32.7. *A collocation solution (left) to the nonlinear pendulum equation and its absolute residual (right).*

increases. If the angle starts at u_a at time $t = 0$ with initial rate of change v_a, then the angle as a function of time obeys

$$u''(t) = -\frac{g}{l} \sin u(t), \quad u(0) = u_a, \quad u'(0) = v_a,$$

in which g denotes the magnitude of acceleration due to gravity near Earth's surface and l denotes the length of the pendulum arm. Set $g/l = 20$, $u_a = 0.1$, and $v_a = 0$. Solve for the angle of the pendulum over $0 \le t \le 2$. Use a uniform collocation grid of degree 7 and uniform interpolation grids of appropriate degrees. Plot the collocation solution $p(t)$ and the residual function. Discuss the graphs.

Solution. The problem is defined.

```
>> f = @(t,u,v) -20*sin(u);
>> dfdu = @(t,u,v) -20*cos(u);
>> dfdv = @(t,u,v) 0;
>> a = 0; b = 2;
>> ua = 0.1; va = 0;
```

The grids are set.

```
>> m = 7;
>> ss = griduni([a b],m);
>> ts = griduni([a b],m+2);
>> ts_ = griduni([a b],m+1);
>> ts__ = griduni([a b],m);
```

In analogy with Remark 32.4, a natural starting guess for $p(t)$ is

$$p(t) = u_a + v_a(t - a) + \frac{1}{2} f(a, u_a, v_a)(t - a)^2.$$

The second derivative $p''(t) = f(a, u_a, v_a)$ is supplied to `ivpnl2gen`.

```
>> rstart = @(t) f(a,ua,va);
```

The collocation solution is computed.

```
>> [ps,qs,rs] = ivpnl2gen(f,dfdu,dfdv,a,ua,va,rstart ...
                     ,ss,ts,ts_,ts__);
>> p = interpgen(ts,ps);
>> q = interpgen(ts_,qs);
>> r = interpgen(ts__,rs);
```

A plot of angular displacement versus time is shown in the left graph of Figure 32.7.

```
>> ax = newfig(1,2); plotsample(ts,ps); plotfun(p,[a b]);
```

The absolute value of the residual $R(t) = p''(t) - f(t, p(t), p'(t))$ is plotted in the right graph.

```
>> R = @(t) r(t)-f(t,p(t),q(t));
>> subplot(ax(2)); ylog; ylim([1e-18 1e2]);
>> plotfun(@(t) abs(R(t)),[a b]);
```
The absolute residual function falls to zero at each collocation node by design. ∎

In the previous example, the pendulum's angular displacement bears a strong resemblance to a scaled cosine function. Indeed, when a pendulum is displaced by a small angle, its motion resembles that of a mass oscillating back and forth on a spring with a simple sinusoidal displacement. However, when a pendulum is displaced farther from its lowest point, a noticeably different motion is expected. We return to the nonlinear pendulum equation in the next two chapters.

Notes

The *Newton–Kantorovich method* is a generalization of Newton's method to Banach spaces that applies to some integral equations [47, p. 524]. This is the inspiration for the method in this chapter, although the computation ultimately reduces to a fairly simple application of Newton's method in finite dimensions. Our derivation is similar to a derivation in Boyd's book [8, Appendix C].

The condition number of the linear problem (32.12) is described as an estimate for the condition number of the nonlinear problem (32.11). It is an estimate because the condition number for (32.12) is measured at $p \approx u$ rather that at u itself. Linearization by itself does not change the condition number. (See the definition of the condition number in the notes for Chapter 21.)

Exercises

Exercises 1–4: Verify that the given function is a solution to the given nonlinear IVP.

1. $u' = tu^3$, $u(0) = 1/4$; $u(t) = 1/\sqrt{16 - t^2}$
2. $u' = u(1 - u)$, $u(0) = 1/2$; $u(t) = 1/(1 + e^{-t})$
3. $u'' = (u')^2/u + u^3$, $u(0) = 1$, $u'(0) = 0$; $u(t) = \sec t$
4. $u'' = uu'$, $u(0) = 0$, $u'(0) = 1/2$; $u(t) = \tan\left(\frac{t}{2} + \pi\right)$

Exercises 5–6: An IVP on an interval $a \leq t \leq b$ is given. Run one step of Newton's method by hand using a single collocation node $s_0 = (a+b)/2$ and interpolation nodes $t_0 = a$, $t_1 = b$, and $t_0' = (a+b)/2$. Construct the starting guess for $p(t)$ suggested by Remark 32.4, state the linear system that determines the Newton step, and update $p(t)$.

5. $u' = u^2$, $u(0) = 0.6$; $0 \leq t \leq 0.5$
6. $u' = t/u$, $u(0) = 1$; $0 \leq t \leq 1$

Exercises 7–10: Solve the IVP over the given interval using ivpnl1gen or ivpnl2gen. Use a uniform grid of degree m for the collocation nodes, and sample the collocation solution and its derivative(s) on uniform grids of appropriate degrees. Verify that the residual equals zero at the collocation nodes.

7. $u' = \sec u$, $u(0) = 0$; $0 \leq t \leq 0.9$; $m = 6$
8. $u' = 1/u$, $u(0) = 1$; $0 \leq t \leq 10$; $m = 5$

9. $u'' = t - u^2, u(0) = 0, u'(0) = 0; 0 \le t \le 6; m = 8$

10. $u'' = \log u, u(0) = 1/2, u'(0) = 0; 0 \le t \le 1; m = 5$

11. Consider again the problem of Example 32.3.

   (a) Set kmax = 1 and verify that ivpnl1gen computes the same Newton update as seen earlier.

   (b) Run Newton's method to completion using ivpnl1gen. For the final approximation, verify that the residual equals zero at the collocation nodes and that the initial condition is satisfied.

12. Repeat Example 32.6, but plot $\|p'(t) - f(t, p(t))\|_\infty$ instead of (32.16). How does the resulting graph differ from Figure 32.4? Explain the reason for the difference.

13. Newton's method fails to converge in Example 32.7. Show that the method can converge on the same IVP over a shorter interval. By approximately how much must the interval be shortened?

14. An especially simple starting guess for Newton's method on $u' = f(t, u), u(a) = u_a$ is the constant function $p(t) = u_a$. Repeat Example 32.6 using this starting guess. Compare and contrast.

15. Let $f_1(\mathbf{x}), \ldots, f_n(\mathbf{x})$ be real-valued functions of the n-by-1 vector $\mathbf{x}$, and let $\mathbf{f}(\mathbf{x}) = (f_1(\mathbf{x}), \ldots, f_n(\mathbf{x}))$, a function from $\mathbb{R}^n$ to $\mathbb{R}^n$. If $\mathbf{f}$ is differentiable at $\mathbf{x}$, its local linear approximation at the point is

$$\mathbf{f}(\mathbf{x} + \Delta \mathbf{x}) \approx \mathbf{f}(\mathbf{x}) + \mathbf{J}(\mathbf{x}) \cdot \Delta \mathbf{x},$$

in which $\mathbf{J}(\mathbf{x})$ is the Jacobian matrix for $\mathbf{f}$ at $\mathbf{x}$. A Newton step for the problem $\mathbf{f}(\mathbf{x}) = \mathbf{0}$, given an estimate $\mathbf{x}^{(k)}$, is defined as follows:

$$\mathbf{J}(\mathbf{x}^{(k)}) \cdot \Delta \mathbf{x} = -\mathbf{f}(\mathbf{x}^{(k)}),$$
$$\mathbf{x}^{(k+1)} = \mathbf{x}^{(k)} + \Delta \mathbf{x}.$$

Use Newton's method to solve the simultaneous system of equations $\sin(x + y) + xy = 0$, $e^{x-y} - x - y = 0$, starting from the guess $(x, y) = (0, 0)$.

16. Duplicate and modify ivpnl2_ and ivpnl2gen to solve the nonlinear BVP

$$u''(x) = f(x, u(x), u'(x)), \quad u(a) = u_a, \quad u(b) = u_b$$

by collocation. Rather than requiring a starting guess from the user, let the initial p be the unique linear function that satisfies the boundary conditions. Demonstrate on the problem

$$u'' = 2u^3 + xu, \quad u(-4) = \sqrt{2}, \quad u(4) = 0,$$

which involves the *Painlevé II* differential equation. *Hint.* First solve Exercise 14 in Chapter 23.

Chapter 33

Nonlinear differential equations on piecewise-uniform grids

Without a good starting guess for Newton's method, the nonlinear collocation method of the previous chapter cannot be expected to converge. However, it is the basis for a reliable method of *piecewise*-polynomial collocation. With short subintervals, low-degree polynomials become reasonable starting guesses, and Newton's method can be expected to converge.

33.1 • First-order equations

First, we must define piecewise-polynomial collocation for a nonlinear problem.

Definition 33.1. *A* piecewise-polynomial collocation solution $p(t)$ *to a nonlinear differential equation*

$$u'(t) = f(t, u(t))$$

satisfies the following, in which $[a_j, b_j]$, $j = 1, \ldots, n$, is a partition of some interval $[a, b]$ and $s_{0j}, \ldots, s_{mj} \in [a_j, b_j]$ are collocation nodes on the jth subinterval:

1. *$p(t)$ is a polynomial of degree at most $m + 1$ on each subinterval $[a_j, b_j]$;*

2. *the collocation equations*

   $$p'(s_{ij}) = f(s_{ij}, p(s_{ij})), \quad i = 0, \ldots, m,$$

   are satisfied on each subinterval $[a_j, b_j]$; and

3. *$p(t)$ is continuous on $[a, b]$.*

If $s_{ij} = a_j$, then $p'(s_{ij})$ should be interpreted as the one-sided derivative from the right, and if $s_{ij} = b_j$, then $p'(s_{ij})$ should be interpreted as the one-sided derivative from the left.

When an initial condition $u(a) = u_a$ is provided, the collocation solution should be constrained to satisfy $p(a) = u_a$.

An effective way to compute a piecewise-polynomial collocation solution is by marching from left to right, solving a simple collocation problem on each subinterval before moving to the next. The first polynomial piece $p_1(t)$ is subject to the initial

379

condition $p_1(a) = u_a$, and each subsequent piece $p_j(t)$ is constrained by continuity to satisfy $p_j(a_j) = p_{j-1}(b_{j-1})$. The piecewise-polynomial function $p(t)$ defined by $p(t) = p_j(t)$ for $t \in [a_j, b_j]$ is the overall solution.

In this chapter, we use an m-by-n piecewise-uniform grid for the collocation nodes and for sampling $p'(t)$, and an $(m + 1)$-by-n piecewise-uniform grid for sampling $p(t)$. The implementation is ivpnl1uni; see Listing 33.1.

Listing 33.1. Solve a first-order nonlinear IVP on a piecewise-uniform grid.

```
function [ps,qs] = ivpnl1uni(f,dfdu,ab,ua,m,n,kmax,tol)

if nargin<7, kmax = 10; end
if nargin<8, tol = 1e-12; end
a = ab(1); b = ab(2); [as,~] = partition_([a b],n);
ss = griduni([a b],m,n);
xs = griduni([0 (b-a)/n],m+1);
[J,K,E1,E2,Ea1,Ea2] = ivp1matuni(m,(b-a)/n);
ps = nan(m+2,n); qs = nan(m+1,n);
warned = false;
% set initial condition for first subinterval
uinit = ua;
% march across subintervals
for j = 1:n
  % construct starting guess on subinterval
  vinit = f(as(j),uinit);
  qsstart = repmat(vinit,m+1,1);
  psstart = repmat(uinit,m+2,1)+vinit*xs;
  % solve on subinterval
  [ps(:,j),qs(:,j),k] = ivpnl1_(f,dfdu,psstart,qsstart,ss(:,j) ...
                          ,J,K,E1,E2,Ea1,Ea2,kmax,tol);

  if isempty(k)&&~warned
    warning('NATE:nonlinearIvpFailure' ...
          ,'Nonlinear IVP solver failed to converge');
    warned = true;
  end
  % set initial condition on next subinterval
  uinit = ps(end,j);
end
% enforce continuity
ps(1,2:end) = ps(end,1:end-1);
if m>0, qs(1,2:end) = qs(end,1:end-1); end
```

The computer code uses the linear function $p(t) = p(a_j) + f(a_j, p(a_j))(t - a_j)$ as the starting guess for Newton's method on the jth subinterval. This works well if the partition is fine enough that the solution can be reasonably well approximated by a linear function on each subinterval.

Example 33.2. Consider the logistic model for population growth

$$u'(t) = ru(t)\left(1 - \frac{u(t)}{L}\right), \quad u(0) = u_0, \tag{33.1}$$

with $r = 2$, $L = 10$, and $u_0 = 0.5$. Solve the IVP on $0 \le t \le 6$ with ivpnl1uni using $m = 1$ and $n = 8$. Plot the solution and its absolute residual.

Solution. The IVP is defined.

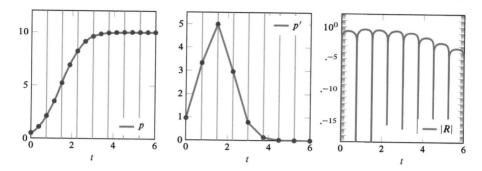

Figure 33.1. *Collocation solution for a logistic equation: population size (left), rate of growth (middle), and absolute residual (right).*

```
>> r = 2; L = 10;
>> f = @(t,u) r*u*(1-u/L);
>> dfdu = @(t,u) r*(1-2*u/L);
>> a = 0; b = 6;
>> ua = 0.5;
```

It is solved by collocation on a piecewise-uniform grid.

```
>> m = 1; n = 8;
>> [ps,qs] = ivpnl1uni(f,dfdu,[a b],ua,m,n);
>> p = interpuni(ps,[a b]);
>> q = interpuni(qs,[a b]);
```

The population size and its rate of change are plotted in Figure 33.1. Apparently, the population size approaches the carrying capacity as time proceeds.

```
>> ax = newfig(1,3); plotfun(p,[a b]);
>> plotpartition([a b],n); plotsample(griduni([a b],m+1,n),ps);
>> subplot(ax(2)); plotfun(q,[a b]);
>> plotpartition([a b],n); plotsample(griduni([a b],m,n),qs);
```

Finally, the absolute residual is plotted in the right graph of Figure 33.1.

```
>> R = @(t) q(t)-f(t,p(t));
>> subplot(ax(3)); ylog; ylim([1e-18 1e2]);
>> plotfun(@(t) abs(R(t)),[a b]);
```

The residual equals zero at every collocation node. ∎

Example 33.3. Solve the IVP of Example 33.2 again. This time, use $m = 3$ and find a value for n that gives a residual uniformly below 10^{-8}. Also measure the absolute errors on the computed population size and rate of change.

Solution. The problem is specified as before.

```
>> r = 2; L = 10;
>> f = @(t,u) r*u*(1-u/L);
>> dfdu = @(t,u) r*(1-2*u/L);
>> a = 0; b = 6;
>> ua = 0.5;
```

By trial and error, we find that the residual requirement is satisfied with $n = 270$:

```
>> m = 3; n = 270;
>> [ps,qs] = ivpnl1uni(f,dfdu,[a b],ua,m,n);
>> p = interpuni(ps,[a b]);
>> q = interpuni(qs,[a b]);
```

```
>> R = @(t) q(t)-f(t,p(t));
>> format short;
>> infnorm(R,[a b],100)
ans =
    7.9191e-09
```

The exact solution to this problem is known to be

$$u(t) = \frac{L}{1 + ((L/u_0) - 1)e^{-rt}}.$$

Therefore, the absolute errors can be measured.

```
>> u = @(t) L/(1+(L/ua-1)*exp(-r*t));
>> v = @(t) (L*r*(L/ua-1)*exp(-r*t))/(1+(L/ua-1)*exp(-r*t))^2;
>> infnorm(@(t) u(t)-p(t),[a b],100)
ans =
    1.6305e-09
>> infnorm(@(t) v(t)-q(t),[a b],100)
ans =
    8.1114e-09
```

They are comparable to the absolute residual. ∎

33.2 ▪ Second-order equations

For a second-order differential equation

$$u''(t) = f(t, u(t), u'(t)),$$

a piecewise-polynomial collocation solution $p(t)$ should (1) be a polynomial of degree at most $m + 2$ on each subinterval $[a_j, b_j]$ in a partition of the desired domain $[a, b]$, (2) satisfy the collocation equations

$$p''(s_{ij}) = f(s_{ij}, p(s_{ij}), p'(s_{ij})), \quad i = 0, \ldots, m, \quad j = 1, \ldots, n,$$

for collocation nodes $s_{0j}, \ldots, s_{mj} \in [a_j, b_j]$, and (3) have a continuous first derivative on $[a, b]$.

An IVP

$$u''(t) = f(t, u(t), u'(t)), \quad u(a) = u_a, \quad u'(a) = v_a$$

can be solved by collocation using the routine `ivpnl2uni`. This routine is similar to `ivpnl1uni`, and its implementation is omitted here. The following example demonstrates its usage.

Example 33.4. Let u be a pendulum's angular displacement from its lowest point, obeying

$$u''(t) = -\frac{g}{l} \sin u(t), \quad u(a) = u_a, \quad u'(a) = v_a,$$

as in Example 32.8. For this example, the pendulum has length satisfying $g/l = 5$, and it starts from rest at an angle of $u_0 = 0.95\pi$—almost as high as it can go. Compute the pendulum's angle over $0 \le t \le 15$ using a 3-by-500 piecewise-uniform collocation grid. Measure the absolute residual.

Solution. The model is encoded.

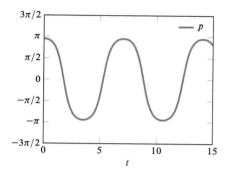

Figure 33.2. *The angle of a pendulum started near the highest possible point.*

```
>> f = @(t,u,v) -5*sin(u);
>> dfdu = @(t,u,v) -5*cos(u);
>> dfdv = @(t,u,v) 0;
>> a = 0; b = 15;
>> ua = 0.95*pi; va = 0;
```

The grid parameters are specified.

```
>> m = 3; n = 500;
```

The IVP is solved, and the solution is plotted in Figure 33.2.

```
>> [ps,qs,rs] = ivpnl2uni(f,dfdu,dfdv,[a b],ua,va,m,n);
>> p = interpuni(ps,[a b]);
>> q = interpuni(qs,[a b]);
>> r = interpuni(rs,[a b]);
>> newfig; plotfun(p,[a b]);
```

When the angle is close to $\pm\pi$, the pendulum's arm is pointing nearly straight up, and only a small component of gravity is directed tangentially to the circle that the pendulum traces. Like an amusement park ride hesitating at its peak, the pendulum nearly pauses for a moment. Thus, the graph of $u(t)$ is flatter than a sine curve at its peaks and valleys.

The motion of the pendulum is better visualized with an animation. Try running the following code on your own computer. You should see the pendulum arm swing back and forth, tracing out a portion of a circle.

```
>> newfig;
>> h = plot([0 sin(p(a))],[0 -cos(p(a))],'*-');
>> axis([-1.5 1.5 -1.5 1.5]);
>> axis square;
>> for t = a:0.05:b
     pause(0.05);
     set(h,'xdata',[0 sin(p(t))],'ydata',[0 -cos(p(t))]);
   end
```

The final task is to measure the absolute residual.

```
>> infnorm(@(t) r(t)-f(t,p(t),q(t)),[a b],100)
ans =
   1.3782e-06
```

■

33.3 ▪ Conditioning

As discussed in the previous chapter, the condition number for a nonlinear collocation problem can be estimated by the condition number of a Newton step. The Newton update $\Delta p(t)$ satisfies a linear IVP of the form

$$\alpha(t)\Delta p(t) + \Delta p'(t) = g(t), \quad \Delta p(a) = 0$$

or

$$\alpha(t)\Delta p(t) + \beta(t)\Delta p'(t) + \Delta p''(t) = g(t), \quad \Delta p(a) = 0, \quad \Delta p'(a) = 0,$$

and we already have routines for estimating the condition number of such an IVP. Routines `ivpnl1conduni` and `ivpnl2conduni`, shown in Listing 33.2, estimate the condition number of a nonlinear IVP by estimating the condition number of the appropriate Newton step above.

Listing 33.2. Estimate the condition number for a nonlinear IVP on a piecewise-uniform grid.

```
function kappa = ivpnl1conduni(dfdu,ab,m,n,p)

al = @(t) -dfdu(t,p(t));
be = @(t) 1;
kappa = ivpl1conduni(al,be,ab,[1 0],m,n);
```

```
function kappa = ivpnl2conduni(dfdu,dfdv,ab,m,n,p,q)

al = @(t) -dfdu(t,p(t),q(t));
be = @(t) -dfdv(t,p(t),q(t));
ga = @(t) 1;
kappa = ivpl2conduni(al,be,ga,ab,[ 1 0 0; 0 1 0 ],m,n);
```

The following examples demonstrate the usage and interpretation of these routines.

Example 33.5. Estimate the condition number for the logistic IVP (33.1) using the solution to Example 33.3. Comment on the implications.

Solution. The problem is solved as in Example 33.3.

```
>> r = 2; L = 10;
>> f = @(t,u) r*u*(1-u/L);
>> dfdu = @(t,u) r*(1-2*u/L);
>> a = 0; b = 6;
>> ua = 0.5;
>> m = 3; n = 270;
>> [ps,qs] = ivpnl1uni(f,dfdu,[a b],ua,m,n);
>> p = interpuni(ps,[a b]);
```

The condition number is estimated.

```
>> kappa = ivpnl1conduni(dfdu,[a b],m,n,p)
kappa =
    13.6303
```

Because the absolute condition number is not much greater than 1, the absolute error should not be much bigger than the absolute residual. This is consistent with our observations in Example 33.3. ∎

Example 33.6. Consider the nonlinear pendulum problem

$$u''(t) = -5\sin u(t), \quad u(0) = 0, \quad u'(0) = v_a.$$

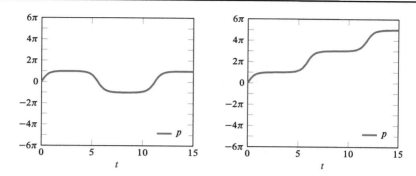

Figure 33.3. *The angle of a nonlinear pendulum with specified initial angular velocity:*
$u'(0) = 4.4720$ *(left) and* $u'(0) = 4.4722$ *(right).*

As in Example 33.4, use the domain $0 \leq t \leq 15$ and a 3-by-500 piecewise-uniform grid.

Find an initial angular velocity for which the problem's condition number is greater than 10^6. Explain physically why the condition number for this initial velocity is unusually large.

Solution. The problem is defined, with the exception of the initial angular velocity.

```
>> f = @(t,u,v) -5*sin(u);
>> dfdu = @(t,u,v) -5*cos(u);
>> dfdv = @(t,u,v) 0;
>> a = 0; b = 15;
>> ua = 0;
>> m = 3; n = 500;
```

By trial and error, the initial angular velocity $v_a = 4.472$ has been found to satisfy the condition number requirement:

```
>> va = 4.472;
>> [ps,qs,rs] = ivpnl2uni(f,dfdu,dfdv,[a b],ua,va,m,n);
>> p = interpuni(ps,[a b]);
>> q = interpuni(qs,[a b]);
>> kappa = ivpnl2conduni(dfdu,dfdv,[a b],m,n,p,q)
kappa =
    1.2980e+06
>> format long;
```

As seen in the left graph of Figure 33.3, the angle u alternates between positive and negative values, spending most of its time near $\pm\pi$. Physically, the pendulum rotates alternately counterclockwise and clockwise, lingering in a nearly vertical position at each change of direction.

```
>> ax = newfig(1,2); plotfun(p,[a b]);
```

The condition number is large because a tiny change to the initial condition can drastically change the solution. Specifically, a slight increase in the initial angular velocity is enough to send the pendulum through full counterclockwise revolutions. This produces a very different solution $u(t)$, as seen in the right plot of Figure 33.3. In this case, $u(t)$ is monotonically increasing because the pendulum always swings counterclockwise.

```
>> va = 4.4722;
>> [ps,qs,rs] = ivpnl2uni(f,dfdu,dfdv,[a b],ua,va,m,n);
>> p = interpuni(ps,[a b]);
>> subplot(ax(2)); plotfun(p,[a b]);
```

Sensitivity to initial conditions is a hallmark of ill conditioning.

Because of the size of the condition number, the absolute error could be much larger than the absolute residual. It is possible that the simulated pendulum very nearly obeys the laws of physics at all times, producing a small residual, but nevertheless follows a significantly different trajectory from the exact solution, producing a large error. ∎

Exercises

Exercises 1–4: Solve the IVP on the given interval using `ivpnl1uni`. Adjust the parameters of the grid until the absolute residual is below 10^{-6}.

1. $u' = -u^2 + 2, u(0) = 4; 0 \le t \le 5$
2. $u' = -u^2 + t, u(0) = 1; 0 \le t \le 5$
3. $u' = \sin(u - t), u(1) = 2; 1 \le t \le 10$
4. $u' = \log(t^2 + u), u(0) = 1; 0 \le t \le 10$

Exercises 5–8: Solve the IVP on the given interval using `ivpnl2uni` with $m = 3$. Increase the number of subintervals until the absolute residual is below 10^{-6}.

5. $u'' = -t + u^2, u(0) = 0, u'(0) = 0; 0 \le t \le 40$
6. $u'' = 1/(1 + u^2), u(0) = -1, u'(0) = -1; 0 \le t \le 20$
7. $u'' = \cos(u^2), u(0) = 1, u'(0) = -1; 0 \le t \le 20$
8. $u'' = -u + 5(u')^2, u(0) = 1, u'(0) = 0; 0 \le t \le 50$

9. Example 31.7 considers the orbit of a planet around a star. Derive an IVP for θ as a function of t and compute $\theta(t)$, $0 \le t \le 2\pi$, by collocation. Check that your solution agrees with the data computed in the earlier example.

10. An object is launched from the surface of a planet of radius R. Its height u in meters as a function of time t in seconds is governed by

$$u'' = -\frac{gR^2}{(u + R)^2}, \quad u(0) = 0, \quad u'(0) = v_a,$$

in which g is the magnitude of the acceleration due to gravity at the planet's surface. Take $g = 9.8 \text{ m/s}^2$, $R = 6.4 \times 10^6$ m, and $v_a = 11500$ m/s. Solve this IVP using `ivpnl2uni` with $m = 3$ and $n = 500$. Based on the numerical evidence, does it appear that the object will travel forever away from the planet, or will it eventually fall back to the surface?

11. Reformulate the model in Exercise 10 to measure distance in millions of meters and time in hours. Then solve the resulting IVP by collocation using the same values of m and n. Measure the residuals and condition numbers for both formulations. Does either computation appear to be significantly more accurate? Explain.

12. The differential equation

$$u'' = -u - 0.4u^3$$

is an instance of the *Duffing equation* for a nonlinear spring. Suppose the initial position is $u(0) = 3$ and the initial velocity is $u'(0) = 0$. Compute the period of the oscillation to at least five digits of accuracy. Provide justification.

13. The following is a model for a damped pendulum started from its lowest point:

$$u'' = -5 \sin u - 0.05 u', \quad u(0) = 0, \quad u'(0) = v_a.$$

What initial velocity is sufficient to make the pendulum complete two full revolutions? Find the minimum such velocity, accurate to two significant digits, by trial and error.

14. Try solving

$$u' = u^2, \quad u(0) = 1$$

on $0 \le t \le 5$ by collocation on a piecewise-uniform grid. What happens? Why does the computation fail?

15. A brake is applied to a mass-spring system in oscillation, slowing the mass through sliding friction. The position of the mass satisfies

$$u'' = -u - 0.1 \operatorname{sign}(u'), \quad u(0) = 1, \quad u'(0) = 0.$$

Explain why the method of this chapter has difficulty with this problem.

16. Example 33.4 contains code for animating a pendulum. Modify the code to visualize the scenarios of Example 33.6.

17. The Riccati form of Airy's differential equation is $u' = x - u^2$. The solution we desire satisfies $u(0) = -3^{1/3}\Gamma(2/3)/\Gamma(1/3)$ and $u(x) \sim -\sqrt{x}$ as $x \to +\infty$. (The asymptotic relationship is equivalent to $\lim_{x \to +\infty} u(x)/(-\sqrt{x}) = 1$.)

   (a) Solve the differential equation on $0 \le x \le 20$ with initial condition $u(0) = -3^{1/3}\Gamma(2/3)/\Gamma(1/3)$. Should you trust the solution? Explain.

   (b) Solve the differential equation from $x = 20$ to $x = 0$ with initial condition $u(20) = -\sqrt{20}$. Should you trust the solution? Explain.

Chapter 34

Nonlinear differential equations on Chebyshev grids

This chapter pursues a piecewise-Chebyshev approach to nonlinear IVPs. The main idea is to combine two strategies: (1) partition the domain into subintervals to encourage convergence in Newton's method and (2) use Chebyshev grids to enable high-degree interpolation on each subinterval.

34.1 ▪ First-order problems

Routine `ivpnl1chebpw` solves a first-order IVP

$$u'(t) = f(t, u(t)), \quad u(a) = u_a \tag{34.1}$$

using piecewise-Chebyshev collocation. The domain $[a, b]$ is partitioned into n subintervals of equal width, and a Chebyshev collocation grid of degree m is placed onto each subinterval. The IVP is then solved by collocation one subinterval at a time.

The code, in Listing 34.1, is very similar to the implementation of `ivpnl1uni` from the last chapter in the way it marches across the domain. The code makes use of routines `gridchebpw` and `interpchebpw` for working with piecewise-Chebyshev grids, which were introduced at the end of Chapter 10. Because these routines allow partitions with subintervals of different widths, the precise sequence of subintervals $[a_j, b_j]$, $j = 1, \ldots, n$, must be provided. The endpoints are stored in a matrix

$$\begin{bmatrix} a_1 & a_2 & \cdots & a_n \\ b_1 & b_2 & \cdots & b_n \end{bmatrix},$$

called `asbs` (think "a's and b's") in the computer code.

Sometimes, partitioning the domain into subintervals is not necessary, as our first example shows.

Example 34.1. Solve the IVP

$$u'(t) = ru(t)\left(1 - \frac{u(t)}{L}\right), \quad u(0) = u_0$$

on $0 \le t \le 6$ using `ivpnl1chebpw` with a single Chebyshev collocation grid of degree 20. Take $r = 2$, $L = 10$, and $u_0 = 0.5$. Plot the solution and measure the absolute residual.

Listing 34.1. Solve a first-order nonlinear IVP on a Chebyshev grid.

```
function [ps,qs,asbs] = ivpnl1chebpw(f,dfdu,ab,ua,m,n,kmax,tol)

if nargin<7, kmax = 40; end
if nargin<8, tol = 1e-12; end
a = ab(1); b = ab(2);
[as,bs] = partition_([a b],n); asbs = [as;bs];
ss = gridchebpw(asbs,m);
ts = gridchebpw(asbs,m+1);
[J,K,E1,E2,Ea1,Ea2] = ivp1matcheb(m,(b-a)/n);
ps = nan(m+2,n); qs = nan(m+1,n);
warned = false;
% set initial condition for first subinterval
uinit = ua;
% march across subintervals
for j = 1:n
  % construct starting guess on subinterval
  vinit = f(as(j),uinit);
  qsstart = repmat(vinit,m+1,1);
  psstart = repmat(uinit,m+2,1)+vinit*(ts(:,j)-as(j));
  % solve on subinterval
  [ps(:,j),qs(:,j),k] = ivpnl1_(f,dfdu,psstart,qsstart,ss(:,j) ...
                         ,J,K,E1,E2,Ea1,Ea2,kmax,tol);
  if isempty(k)&&~warned
    warning('NATE:nonlinearIvpFailure' ...
            ,'Nonlinear IVP solver failed to converge');
    warned = true;
  end
  % set initial condition on next subinterval
  uinit = ps(end,j);
end
% enforce continuity
ps(1,2:end) = ps(end,1:end-1);
qs(1,2:end) = qs(end,1:end-1);
```

Solution. The problem is defined.

```
>> r = 2; L = 10;
>> f = @(t,u) r*u*(1-u/L);
>> dfdu = @(t,u) r*(1-2*u/L);
>> a = 0; b = 6;
>> ua = 0.5;
```

It is solved by collocation on a Chebyshev grid of degree $m = 20$. Because the domain is treated as a single interval, n is set to 1.

```
>> m = 20; n = 1;
>> [ps,qs,asbs] = ivpnl1chebpw(f,dfdu,[a b],ua,m,n);
>> p = interpchebpw(ps,asbs);
>> q = interpchebpw(qs,asbs);
```

The solution is plotted in Figure 34.1. It certainly looks like Newton's method converged.

```
>> newfig; plotfun(p,[a b]);
```

Finally, the absolute residual is measured.

```
>> R = @(t) q(t)-f(t,p(t));
>> format short;
```

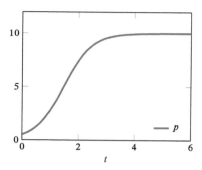

Figure 34.1. *A logistic population model solved on a Chebyshev grid.*

```
>> infnorm(R,[a b],100)
ans =
   7.0980e-04
```
The residual is fairly small and can be made much smaller by refining the grid (Exercise 9). ∎

Sometimes, though, Newton's method may fail to converge unless the domain is partitioned.

Example 34.2. Solve the following IVP on $0 \le t \le 4$ using ivpnl1chebpw with $m = 20$ and $n = 1$:

$$u'(t) = e^t \sin u(t), \quad u(0) = 6. \tag{34.2}$$

Solution. Collocation is executed.
```
>> f = @(t,u) exp(t)*sin(u);
>> dfdu = @(t,u) exp(t)*cos(u);
>> a = 0; b = 4;
>> ua = 6;
>> m = 20; n = 1;
>> [ps,qs,asbs] = ivpnl1chebpw(f,dfdu,[a b],ua,m,n);
[Warning: Nonlinear IVP solver failed to converge]
```
Newton's method fails to converge. Interpolating polynomials can still be constructed, but they are unlikely to be faithful to the true solution. Let's check the residual to see whether any accuracy has been obtained.
```
>> p = interpchebpw(ps,asbs);
>> q = interpchebpw(qs,asbs);
>> R = @(t) q(t)-f(t,p(t));
>> infnorm(R,[a b],100)
ans =
   1.7525e+21
```
Because the absolute residual is enormous, the computation is a complete failure. ∎

When Newton's method fails to converge over all of the domain at once, partitioning can help. Sometimes just one or two divisions are necessary.

Example 34.3. Repeat the previous example, but partition the domain into $n = 2$ subintervals and use collocation grids of degree $m = 20$ on each.

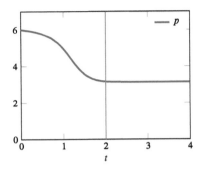

Figure 34.2. *Convergence on a piecewise-Chebyshev grid.*

Solution. Collocation is executed.

```
>> f = @(t,u) exp(t)*sin(u);
>> dfdu = @(t,u) exp(t)*cos(u);
>> a = 0; b = 4;
>> ua = 6;
>> m = 20; n = 2;
>> [ps,qs,asbs] = ivpnl1chebpw(f,dfdu,[a b],ua,m,n);
>> p = interpchebpw(ps,asbs);
```

This time, Newton's method converges. The solution is plotted in Figure 34.2.

```
>> newfig; plotfun(p,[a b]); plotpartition(asbs);
```
■

Piecewise-Chebyshev collocation is capable of very high accuracy on nonlinear IVPs. Remember that the combination of a small residual and a reasonably low condition number provide evidence of high accuracy. The condition number in this setting can be estimated by routine ivpnl1condchebpw, which is used in the next example.

Example 34.4. Solve the IVP (34.2) once more, but adjust the grid parameters to achieve an absolute residual uniformly below 10^{-12}. Also, measure the condition number and comment.

Solution. By trial and error, the pair $m = 100$, $n = 2$ is found to work.

```
>> f = @(t,u) exp(t)*sin(u);
>> dfdu = @(t,u) exp(t)*cos(u);
>> a = 0; b = 4;
>> ua = 6;
>> m = 100; n = 2;
>> [ps,qs,asbs] = ivpnl1chebpw(f,dfdu,[a b],ua,m,n);
>> p = interpchebpw(ps,asbs);
>> q = interpchebpw(qs,asbs);
>> R = @(t) q(t)-f(t,p(t));
>> infnorm(R,[a b],100)
ans =
    9.4135e-13
```

Let's check the condition number.

```
>> kappa = ivpnl1condchebpw(dfdu,asbs,m,p)
kappa =
  101.1861
```

Because the condition number is not so much bigger than 1, the error should not be much greater than the residual. ■

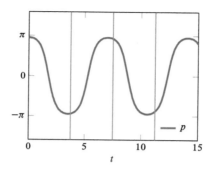

Figure 34.3. *The angle of a pendulum, computed on a sequence of Chebyshev grids.*

34.2 · Second-order problems

Routine `ivpnl2chebpw` (implementation not shown here) solves the second-order non-linear IVP

$$u''(t) = f(t, u(t), u'(t)), \quad u(a) = u_a, \quad u'(a) = v_a \qquad (34.3)$$

by piecewise-Chebyshev collocation. The parameters m and n have their same meanings from the first-order code seen earlier. The next example illustrates the usage of the second-order method and the accompanying routine `ivpnl2condchebpw` for estimating the condition number.

Example 34.5. Solve the nonlinear pendulum problem

$$u''(t) = -5 \sin u(t), \quad u(0) = 0.95\pi, \quad u'(0) = 0$$

over $0 \le t \le 15$ using `ivpnl2chebpw`. Adjust m and n to achieve an absolute residual below 10^{-12}. Comment on the accuracy, taking into account the condition number.

Solution. The problem is defined.

```
>> f = @(t,u,v) -5*sin(u);
>> dfdu = @(t,u,v) -5*cos(u);
>> dfdv = @(t,u,v) 0;
>> a = 0; b = 15;
>> ua = 0.95*pi; va = 0;
```

By trial and error, the following parameters have been found effective.

```
>> m = 100; n = 4;
>> [ps,qs,rs,asbs] = ivpnl2chebpw(f,dfdu,dfdv,[a b],ua,va,m,n);
>> p = interpchebpw(ps,asbs);
>> q = interpchebpw(qs,asbs);
>> r = interpchebpw(rs,asbs);
```

The result is plotted in Figure 34.3.

```
>> newfig; plotfun(p,[a b]); plotpartition(asbs);
```

The absolute residual is confirmed to be as small as required.

```
>> infnorm(@(t) r(t)-f(t,p(t),q(t)),[a b],100)
ans =
    1.5765e-13
```

Finally, the condition number is measured.

```
>> kappa = ivpnl2condchebpw(dfdu,dfdv,asbs,m,p,q)
kappa =
    3.2746e+04
```

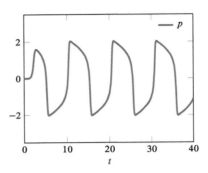

Figure 34.4. *A solution to the van der Pol equation.*

The absolute residual could be a few orders of magnitude larger than the absolute error, but it should still be very small compared to the magnitude of the solution. ∎

Solutions to the next differential equation feature regions of rapid change. This could pose numerical difficulties.

Example 34.6. The *van der Pol equation* is

$$u''(t) = -u(t) + \varepsilon(1 - u(t)^2)u'(t). \tag{34.4}$$

Set $\varepsilon = 4$ and impose the initial conditions $u(0) = 0$, $u'(0) = 0.001$. Solve the IVP by collocation on a piecewise-Chebyshev grid over $0 \le t \le 40$, achieving an absolute residual below 10^{-10}.

Solution. The problem is stated.

```
>> ep = 4;
>> f = @(t,u,v) -u+ep*(1-u^2)*v;
>> dfdu = @(t,u,v) -1+ep*(-2*u)*v;
>> dfdv = @(t,u,v) ep*(1-u^2);
>> a = 0; b = 40;
>> ua = 0; va = 0.001;
```

The IVP is solved on a piecewise-Chebyshev grid.

```
>> m = 100; n = 80;
>> [ps,qs,rs,asbs] = ivpnl2chebpw(f,dfdu,dfdv,[a b],ua,va,m,n);
>> p = interpchebpw(ps,asbs);
>> q = interpchebpw(qs,asbs);
>> r = interpchebpw(rs,asbs);
```

The solution is plotted in Figure 34.4.

```
>> newfig; plotfun(p,[a b]);
```

The parameters m and n were chosen by trial and error to achieve the residual goal. The residual is indeed below 10^{-10}:

```
>> R = @(t) r(t)-f(t,p(t),q(t));
>> infnorm(R,[a b],1000)
ans =
    7.9023e-11
>> format long;
```

∎

34.3 ▪ Adaptive subdivision

The real power of Chebyshev collocation for IVPs is achieved through adaptive subdivision, in which the width of each subinterval is automatically adjusted as it is encountered. This way, the solver can adapt to different "time scales," using narrow subintervals in regions of rapid change and wide subintervals elsewhere. The added cost of subdivision is paid only where it is needed. The implementation of adaptive subdivision is Exercises 15–16.

Notes

Our nonlinear Chebyshev method is similar to one in an article by Driscoll, Bornemann, and Trefethen [24]. Their article also considers the Painlevé II equation of Exercise 17.

Exercises

Exercises 1–4: Solve the IVP using `ivpnl1chebpw`. Adjust the parameters m and n so that Newton's method converges on each subinterval and the absolute residual is below 10^{-8}.

1. $u' = -u^2 + 2, u(0) = 4; 0 \leq t \leq 5$
2. $u' = -u^2 + t, u(0) = 1; 0 \leq t \leq 5$
3. $u' = \sin(u - t), u(1) = 2; 1 \leq t \leq 10$
4. $u' = \log(t^2 + u), u(0) = 1; 0 \leq t \leq 10$

Exercises 5–8: Solve the IVP using `ivpnl2chebpw`. Adjust the parameters m and n so that Newton's method converges on each subinterval and the absolute residual is below 10^{-8}.

5. $u'' = u(1 - u), u(0) = 1/2, u'(0) = 0; 0 \leq t \leq 15$
6. $u'' = uu', u(0) = 1, u'(0) = -0.1; 0 \leq t \leq 15$
7. $u'' = e^{-u}, u(0) = 0, u'(0) = -20; 0 \leq t \leq 2$
8. $u'' = -100u^5, u(0) = 0.1, u'(0) = 0; 0 \leq t \leq 200$

9. Solve Example 34.1 again, achieving an absolute residual below 10^{-12} without partitioning the interval.

10. The solution in Figure 34.2 appears to approach a horizontal asymptote. What is the height of this asymptote, and how could its existence be predicted before beginning the numerical computation?

11. Find an IVP not mentioned in this chapter on which `ivpnl1chebpw` fails to converge unless the domain is partitioned into subintervals ($n > 1$).

12. A penny is dropped from the top of a 100-m-high building. Its height above ground is determined by

$$u'' = -9.8 + 0.079(u')^2, \quad u(0) = 100, \quad u'(0) = 0.$$

(The drag coefficient 0.079 is determined by the known terminal velocity of 40 km/h [7].) Compute the penny's height using `ivpnl2chebpw` to an absolute residual below 10^{-10}. Determine the time T_1 when the speed reaches 10 m/s and the time T_2 when the penny strikes the ground. Plot the height and velocity over $0 \leq t \leq T_2$.

13. A skydiver jumps from a plane 1200 m above the ground. Her elevation u obeys

$$u'' = -9.8 + 0.0034(u')^2, \quad u(0) = 1200, \quad u'(0) = 0.$$

At an altitude of 800 m, she pulls the ripcord to deploy her parachute, at which point the physics of the problem change, and her elevation satisfies

$$u'' = -9.8 + \frac{0.39}{1 + ((0.39/0.0034) - 1)e^{-3.3(t-T_1)}}(u')^2,$$

in which T_1 is the time at which the ripcord is pulled. Compute the elevation of the skydiver using `ivpnl2chebpw`, obtaining an absolute residual below 10^{-8}. Graph the position, velocity, and acceleration from time $t = 0$ to the moment that she touches the ground. *Suggestion.* Break the time domain into three subintervals: rapid fall, parachute deployment, and slow descent. (For more information on parachute problems, see [52, 61, 79].)

14. Compute the global maximum value of the solution to the van der Pol equation in Example 34.6 without using `infnorm`.

15. Develop an adaptive routine for solving the first-order nonlinear IVP (34.1) on a given interval. The routine should first attempt to solve the IVP over the entire interval using a single Chebyshev collocation grid of degree 32. If Newton's method fails to converge or if the absolute residual is estimated to be above a given tolerance, then the interval should be divided in half and the same process should be applied recursively on each half. The recursion should be limited by a maximum depth to ensure that the computation terminates. Test the routine on the problem $u' = \sin(t^2) - u^2$, $u(0) = 1$, $0 \leq t \leq 15$. Plot the collocation solution p and delineate the polynomial pieces using `plotpartition`.

16. Develop an adaptive routine for solving the second-order nonlinear IVP (34.3) on a given interval, analogous to the solution to the previous exercise. Test the routine on the van der Pol equation (34.4) with $\varepsilon = 10$ and initial conditions $u(0) = 0$, $u'(0) = 0.001$. Plot the collocation solution $p(t)$ for $0 \leq t \leq 40$ and delineate the polynomial pieces using `plotpartition`.

17. Exercise 16 in Chapter 32 considers a method for solving a second-order nonlinear BVP on a general collocation grid. Specialize this routine to work with Chebyshev grids. (Treat the problem domain as a single interval, rather than attempting a piecewise method.) Demonstrate on the BVP

$$u'' = 2u^3 + xu, \quad u(-18) = 3, \quad u(18) = 0,$$

seeking an absolute residual below 10^{-10}.

Appendix A

Interpolation with repeated nodes

The theory of Lagrange interpolating polynomials in Chapter 6 generalizes to a setting in which nodes may be repeated, constraining not just the value of an interpolating polynomial but also one or more of its derivatives.

A.1 ▪ Multiplicities

For a polynomial $p(x)$, a *root of multiplicity k* is a number ζ for which $(x - \zeta)^k$ divides $p(x)$. That is, there must exist another polynomial $q(x)$ such that $p(x) = (x - \zeta)^k q(x)$.

We want to extend this notion to functions other than polynomials. We define a *zero of multiplicity k* for a function $f(x)$ to be any number ζ for which $f(\zeta) = f'(\zeta) = \cdots = f^{(k-1)}(\zeta) = 0$. The motivation is the following theorem.

Theorem A.1. *Let $p(x)$ be a polynomial. Then $(x - \zeta)^k$ divides $p(x)$ if and only if $p(\zeta) = p'(\zeta) = \cdots = p^{(k-1)}(\zeta) = 0$.*

Proof. Let $p(x)$ be a polynomial of degree m. Expand $p(x)$ in a Taylor series about $x = \zeta$:

$$p(x) = \sum_{i=0}^{m} \frac{p^{(i)}(\zeta)}{i!} (x - \zeta)^i.$$

This can be broken into two convenient terms:

$$p(x) = \sum_{i=0}^{k-1} \frac{p^{(i)}(\zeta)}{i!} (x - \zeta)^i + (x - \zeta)^k \sum_{i=k}^{m} \frac{p^{(i)}(\zeta)}{i!} (x - \zeta)^{i-k}.$$

The second term is divisible by $(x - \zeta)^k$. Hence, the entire polynomial $p(x)$ is divisible by $(x - \zeta)^k$ if and only if the first term, a polynomial of degree at most $k - 1$, is identically zero. This is true if and only if $p^{(i)}(\zeta) = 0$ for $i = 0, \ldots, k - 1$. $\quad\square$

When the zeros of a function are *counted with multiplicity*, a zero of multiplicity k contributes k to the count. For example, $p(x) = (x - 5)(x - 7)^2$ has three zeros counting multiplicities: 5, 7, and 7.

The notion of a multiple zero enables a broader notion of interpolation.

Definition A.2. *Let $f(x)$ be a function and let $p(x)$ be a polynomial of degree at most m. If $f(x) - p(x)$ has distinct zeros $x_0, \ldots, x_l$ with multiplicities $k_0, \ldots, k_l$, respectively, and $k_0 + \cdots + k_l = m + 1$, then $p(x)$ is an interpolating polynomial for $f(x)$ at the nodes, counting multiplicities.*

Example A.3. Show that a Taylor polynomial

$$p(x) = \sum_{i=0}^{m} \frac{f^{(i)}(x_0)}{i!}(x - x_0)^i$$

is an interpolating polynomial for $f(x)$ with a single node of multiplicity $m + 1$.

Solution. The Taylor polynomial $p(x)$ is a polynomial of degree m, and it is constructed so that $p^{(i)}(x_0) = f^{(i)}(x_0)$, $j = 0, \ldots, m$. Thus, $f(x) - p(x)$ has a zero of multiplicity $m + 1$ at $x = x_0$. ∎

Theorem A.4. *Let $x_0, \ldots, x_l$ be distinct real numbers and $k_0, \ldots, k_l$ be positive integers. If a function $f(x)$ is $(k_i - 1)$-times differentiable at $x = x_i$, $i = 0, \ldots, l$, then there exists a unique interpolating polynomial $p(x)$ for $f(x)$ at the nodes, counting multiplicities.*

Proof. The proof of existence is skipped because it is not needed in the rest of the book.

For uniqueness, suppose that $p(x)$ and $q(x)$ are two interpolating polynomials for the given data. On one hand, $p(x) - q(x)$ is a polynomial of degree at most $m = k_0 + \cdots + k_l - 1$ because it is the difference of polynomials of degree at most m. On the other hand, $p(x) - q(x) = (f(x) - q(x)) - (f(x) - p(x))$ has zeros $x_0, \ldots, x_l$ with multiplicities $k_0, \ldots, k_l$, respectively, and therefore it is divisible by

$$\prod_{i=0}^{l} (x - x_i)^{k_i},$$

which is a polynomial of degree $m + 1$. The only possibility is that $p(x) - q(x)$ is identically zero, and therefore $p(x)$ and $q(x)$ are actually the same polynomial. □

A.2 ▪ Interpolation error with repeated nodes

Many results on simple zeros or nodes extend naturally to repeated zeros or nodes.

Lemma A.5. *Suppose $f \in C^1[a, b]$ has $m + 1$ zeros counting multiplicities. Then f' has at least m zeros counting multiplicities. Furthermore,*

1. *strictly between any two adjacent but distinct zeros of f lies a zero of f', and*

2. *any zero of f of multiplicity $k \geq 2$ is also a zero of f' of multiplicity $k - 1$.*

Proof. Suppose f has distinct zeros $x_1, \ldots, x_l$ of multiplicities $k_1, \ldots, k_l$, respectively.

By definition, every zero of f of multiplicity two or more is a zero of f' of multiplicity one less. In addition, Rolle's theorem guarantees a zero of f' strictly between every pair of adjacent distinct zeros of f. Interlacing is proved.

The zeros of f' that coincide with the zeros of f have multiplicities that sum to $\sum_{i=1}^{l}(k_i - 1) = m + 1 - l$, and there are $l - 1$ zeros of f' guaranteed by Rolle's theorem. Thus, the number of zeros of f' counting multiplicities is at least $(m + 1 - l) + (l - 1) = m$. □

Theorem A.6. *Suppose* $f \in C^r[a, b]$ *has* $m + 1$ *zeros counting multiplicities. Then* $f^{(r)}$ *has at least* $m + 1 - r$ *zeros counting multiplicities. In particular, if* $f \in C^r[a, b]$ *has* $r + 1$ *zeros counting multiplicities, then* $f^{(r)}$ *has at least one zero.*

Proof. The proof is by induction on r.

The base case $r = 1$ is proved by Lemma A.5.

For $r > 1$, Lemma A.5 shows that f' has at least m zeros counting multiplicities. Then induction shows that the $(r - 1)$th derivative of f', i.e., $f^{(r)}$, has at least $m - (r - 1) = m + 1 - r$ zeros counting multiplicities. □

Lemma A.7. *Lemma 6.9 remains true if repetition is allowed in the list of zeros* $\zeta_0, \ldots, \zeta_m$.

Proof. Change the proof of Lemma 6.9 to appeal to Theorem A.6 instead of Theorem 6.8. □

Theorem A.8. *Theorem 6.6 remains true if repetition is allowed in the list of nodes* $x_0, \ldots, x_m$.

Proof. Change the proof of Theorem 6.6 to appeal to Lemma A.7 instead of Lemma 6.9. □

Example A.9. Let $p(x)$ be the degree-m Taylor polynomial for a function $f(x)$ about $x = x_0$, and suppose that f has a continuous $(m + 1)$th derivative between x_0 and x. Use Theorem A.8 to find an expression for the error $f(x) - p(x)$.

Solution. The Taylor polynomial is an interpolating polynomial with a single node x_0 of multiplicity $m + 1$. (See Example A.3.) Hence,

$$f(x) - p(x) = \frac{f^{(m+1)}(\eta_x)}{(m + 1)!}(x - x_0)^{m+1}$$

for some η_x between x_0 and x. ■

This provides a second proof of the Lagrange remainder for a Taylor polynomial expansion, first introduced in Exercises 13–14 of Chapter 4.

Notes

The book by Davis has a wealth of information on various polynomial interpolation schemes [21].

Appendix B

Complex functions

The performance of a numerical method can be affected by the shape of a function in the complex plane, even if the problem statement only concerns the real line. The primary consideration is the location of singularities—places where the function is undefined or not complex differentiable. In this appendix, we briefly review complex functions and locate singularities in a couple of examples.

B.1 ▪ Complex numbers

The *imaginary unit* i satisfies $i^2 = -1$. A *complex number* z is one that can be written in the form

$$z = x + yi$$

for real numbers x and y. The *real part* of z is re $z = x$, the *imaginary part* is im $z = y$, and the *absolute value* is $|z| = \sqrt{x^2 + y^2}$.

Addition of complex numbers is straightforward:

$$(x_1 + y_1 i) + (x_2 + y_2 i) = (x_1 + x_2) + (y_1 + y_2)i.$$

Multiplication follows from the identity $i^2 = -1$:

$$(x_1 + y_1 i)(x_2 + y_2 i) = x_1 x_2 + x_1 y_2 i + x_2 y_1 i + y_1 y_2 i^2$$
$$= (x_1 x_2 - y_1 y_2) + (x_1 y_2 + x_2 y_1)i.$$

Every complex number can be represented by a point in the *complex plane*, in which the horizontal axis is the real axis and the vertical axis is the imaginary axis. For example, the complex number $-2 + 4i$ is represented by the point $(-2, 4)$, as in Figure B.1.

In the MATLAB language, an imaginary number can be entered by appending the letter i. For example, the complex number $-2 + 4i$ can be entered as follows:

```
>> -2+4i
ans =
 -2.000000000000000 + 4.000000000000000i
```

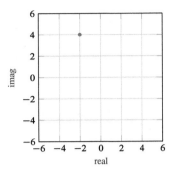

Figure B.1. *The complex number* $-2 + 4\mathrm{i}$ *represented as a point in the complex plane.*

B.2 ▪ Complex functions

Let f be a function mapping complex numbers to complex numbers, i.e., a *complex function*. For example,

$$f(z) = z^2$$

is a complex function. This definition can also be written as follows by expressing z and $f(z)$ in terms of their real and imaginary parts:

$$\begin{aligned}
f(x + y\mathrm{i}) &= (x + y\mathrm{i})^2 \\
&= x^2 + 2xy\mathrm{i} + y^2\mathrm{i}^2 \\
&= (x^2 - y^2) + 2xy\mathrm{i}.
\end{aligned}$$

Because the input z to a complex function lies in a plane of two real dimensions and the output $f(z)$ also lies in a plane of two real dimensions, a complete graph of f requires *four* real dimensions. Hence, a complex function is not easy to visualize. One way to get around this problem is to plot the real and imaginary parts of $f(z)$ separately. This can be done using provided MATLAB routines `plotrealpart` and `plotimagpart`, which the next example illustrates.

Example B.1. Plot the real and imaginary parts of $f(z) = z^2$ over the domain $z = x + y\mathrm{i}$ with $-2 \le x \le 2$ and $-2 \le y \le 2$.

Solution. As shown above, the function $f(z) = z^2$ can also be expressed as

$$f(x + y\mathrm{i}) = (x^2 - y^2) + 2xy\mathrm{i}.$$

The real and imaginary parts are

$$\operatorname{re} f(x + y\mathrm{i}) = x^2 - y^2, \quad \operatorname{im} f(x + y\mathrm{i}) = 2xy.$$

Both of these quadratic functions are plotted in Figure B.2.

```
>> f = @(z) z^2;
>> ax = newfig(1,2);
>> plotrealpart(f,[-2 2 -2 2 -8 8]);
>> xlabel('x'); ylabel('y');
>> subplot(ax(2));
>> plotimagpart(f,[-2 2 -2 2 -8 8]);
>> xlabel('x'); ylabel('y');
```

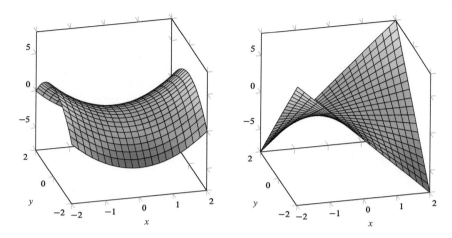

Figure B.2. *Real (left) and imaginary (right) parts of the function* $f(z) = z^2$.

The final argument to `plotrealpart` or `plotimagpart` specifies the ranges of the axes in the order [xmin xmax ymin ymax vmin vmax], with vmin and vmax referring to the minimum and maximum values on the vertical axis.

On a real argument x, the function is $f(x) = x^2$, with real part x^2 and imaginary part 0. This more familiar behavior appears along the slices $y = 0$ in the plots. ∎

B.3 · The complex derivative

The derivative of a complex function $f(z)$ is

$$f'(z) = \lim_{h \to 0} \frac{f(z+h) - f(z)}{h}.$$

This looks just like the derivative of a real function, but now h is complex. For the limit to exist, the limiting value must be the same regardless of the path that h takes toward zero. If the limit does not exist, then $f(z)$ is not complex differentiable at z.

Many rules for the real derivative extend naturally to the complex derivative, including the following:

$$\frac{d}{dz}[c] = 0 \quad (c \text{ a constant}),$$

$$\frac{d}{dz}[z^k] = kz^{k-1} \quad (k \text{ a nonzero integer}),$$

$$\frac{d}{dz}[e^z] = e^z,$$

$$\frac{d}{dz}[\log z] = \frac{1}{z},$$

$$\frac{d}{dz}[cf(z)] = cf'(z),$$

$$\frac{d}{dz}[f(z) + g(z)] = f'(z) + g'(z),$$

$$\frac{d}{dz}[f(z)g(z)] = f'(z)g(z) + f(z)g'(z),$$

$$\frac{d}{dz}[f(g(z))] = f'(g(z))g'(z).$$

A *neighborhood* of a point z is an open disk centered at z, i.e., the set of complex numbers w satisfying $|w - z| < \delta$ for some $\delta > 0$. A complex function is *analytic at a point z* if it is complex differentiable in some neighborhood of z. A *complex domain* is a connected open subset of the complex plane. A function is *analytic in a domain D* if it is analytic at every point in D. When a function is analytic in a domain, its derivative is necessarily analytic in that domain as well. It follows that an analytic function is infinitely differentiable and therefore very smooth.

B.4 ▪ Singularity of a complex function

If a complex function f fails to be complex differentiable at a point z but any neighborhood of z contains at least one point where f is complex differentiable, then f is said to have a *singularity* at z. Singularities are often easy to spot, appearing as punctures or tears or creases in graphs.

Example B.2. Locate all singularities of $f(z) = 1/z$. Illustrate by graphing the real and imaginary parts of the function.

Solution. The function is undefined at $z = 0$, but it exists and is complex differentiable, with derivative $f'(z) = -1/z^2$, at every other point in the complex plane. Hence, there is exactly one singularity, and it is located at $z = 0$.

In any neighborhood of $z = 0$, the function is unbounded. To see this, note that

$$f(x + 0\mathrm{i}) = \frac{1}{x}$$

and

$$f(0 + y\mathrm{i}) = \frac{1}{y\mathrm{i}} = \frac{\mathrm{i}}{y\mathrm{i}^2} = -\frac{1}{y}\mathrm{i}.$$

Hence, re $f(z)$ diverges to infinity as z approaches 0 along the real axis, and im $f(z)$ diverges to infinity as z approaches 0 along the imaginary axis. Graphs of the real and imaginary parts, shown in Figure B.3, confirm this.

```
>> f = @(z) 1/z;
>> ax = newfig(1,2);
>> plotrealpart(f,[-6 6 -6 6 -2 2]);
>> xlabel('x'); ylabel('y');
>> subplot(ax(2));
>> plotimagpart(f,[-6 6 -6 6 -2 2]);
>> xlabel('x'); ylabel('y');
```

The function $f(z) = 1/z$ is said to have a *pole* at $z = 0$. ∎

Example B.3. Locate all singularities of the natural logarithm function $f(z) = \log z$. Illustrate by graphing the real and imaginary parts of the function.

Solution. Every complex number $z = x + y\mathrm{i}$ can also be written in polar form $z = re^{\theta\mathrm{i}}$, in which the absolute value r is a nonnegative real number and the *phase* θ is in the interval $(-\pi, \pi]$. By Euler's identity $e^{\theta\mathrm{i}} = (\cos\theta) + (\sin\theta)\mathrm{i}$, the relationship between Cartesian and polar coordinates is

$$x = r\cos\theta, \quad y = r\sin\theta.$$

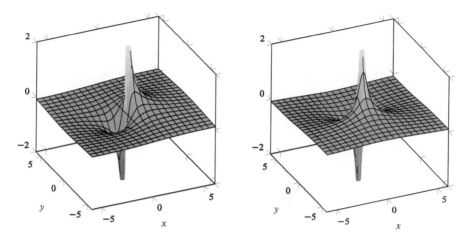

Figure B.3. *Real (left) and imaginary (right) parts of the function $f(z) = 1/z$.*

The complex logarithm function satisfies

$$\log(re^{\theta i}) = \log r + \log(e^{\theta i}) = \log r + \theta i. \qquad (B.1)$$

Note that the imaginary part of $\log z$ equals the phase of z.

The complex logarithm is discontinuous where the phase function is discontinuous. Imagine a path through the complex plane that crosses the negative real axis. As the imaginary part y of $z = x + yi$ crosses from positive to negative, the phase θ in $z = re^{\theta i}$ jumps from π to $-\pi$. Considering (B.1), the complex logarithm function inherits this discontinuity. Thus, $\log z$ is singular at every $z = x + yi$ for which $x \leq 0$ and $y = 0$.

The behavior described above is immediately evident in a graph. See especially the right plot in Figure B.4.

```
>> f = @(z) log(z);
>> ax = newfig(1,2);
>> plotrealpart(f,[-4 4 -4 4 -4 4]);
>> xlabel('x'); ylabel('y');
>> subplot(ax(2));
>> plotimagpart(f,[-4 4 -4 4 -4 4]);
>> xlabel('x'); ylabel('y');
```

We say that there is a *branch cut* along the negative real axis.   ∎

A particular kind of singularity called a *removable singularity* is benign. An example is provided by $f(z) = (e^z - 1)/z$. This function is undefined at $z = 0$, but it has a limit as z approaches 0, as shown below. The hole in the function can be "patched," and the resulting function is not only continuous but also analytic. Because the singularity can be removed in this fashion, it is known as a removable singularity.

Theorem B.4. *Let $f(z)$ be a complex function, let D be an open set of the complex plane, and let $z_0 \in D$. Suppose that $f(z)$ is analytic on $D \setminus \{z_0\}$ and that $L = \lim_{z \to z_0} f(z)$ exists. Then*

$$g(z) = \begin{cases} f(z) & \text{if } z \neq z_0, \\ L & \text{if } z = z_0 \end{cases}$$

is analytic on all of D.

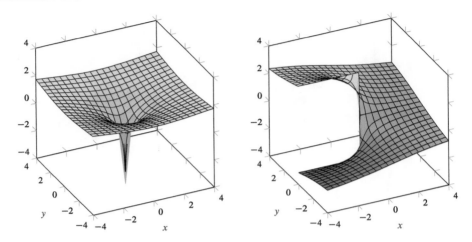

Figure B.4. *Real (left) and imaginary (right) parts of the natural logarithm function.*

Example B.5. Show that the function $f(z) = (e^z - 1)/z$ has a removable singularity at $z = 0$. Remove it and plot the real and imaginary parts of the resulting function.

Solution. The function $f(z)$ is defined and complex differentiable everywhere except at $z = 0$. Furthermore, by l'Hôpital's rule,

$$\lim_{z \to 0} \frac{e^z - 1}{z} = \lim_{z \to 0} \frac{e^z}{1} = e^0 = 1.$$

The function

$$g(z) = \begin{cases} (e^z - 1)/z & \text{if } z \neq 0, \\ 1 & \text{if } z = 0 \end{cases}$$

is analytic everywhere.

The function is plotted in Figure B.5 with its patched hole indicated by a black dot. The real and imaginary parts of $g(z)$ appear to be smooth, as expected.

```
>> g = @(z) iif(z==0,@()1,@()(exp(z)-1)/z);
>> ax = newfig(1,2);
>> plotrealpart(g,[-2 2 -2 2 -4 4]);
>> plot3(0,0,1,'k.');
>> xlabel('x'); ylabel('y');
>> subplot(ax(2));
>> plotimagpart(g,[-2 2 -2 2 -4 4]);
>> plot3(0,0,0,'k.');
>> xlabel('x'); ylabel('y');
```                                               ∎

It is often convenient to assume that all removable singularities have been removed. For example, given the function $f(z) = (e^z - 1)/z$ of the previous example, we may simply write $f(0) = 1$ instead of explicitly constructing and naming the function g.

Notes

Theorem B.4 is proved, for example, in Flanigan's textbook [31, p. 239].

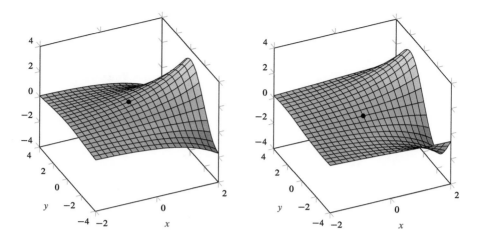

Figure B.5. *Real (left) and imaginary (right) parts of $(e^z - 1)/z$, with the singularity at $z = 0$ patched.*

Appendix C

Interpolation in the complex plane

The purpose of this appendix is to bound the rate of convergence of Chebyshev interpolation on analytic functions. We make the simplifying assumption that the interpolation interval is $[-1, 1]$ rather than a more general $[a, b]$. The ultimate conclusion here is that if the function to be interpolated is analytic and bounded within a Bernstein ellipse $E_\rho(-1, 1)$, then the interpolation error is bounded by $C\rho^{-m}$ for some constant C. The specific constant $C = 4M/(\rho - 1)$ in Theorem 11.3 is proved, for example, in [76].

The development is organized into two sections. The first section considers general regions of the complex plane and arbitrary grids of distinct nodes. The second section considers Bernstein ellipses and Chebyshev grids specifically.

Familiarity with contour integrals and the residue theorem is assumed.

C.1 ▪ The Lagrange interpolation problem

Consider a complex function $f(z)$ and a grid $z_0, \ldots, z_m$ of distinct nodes, possibly complex. Let γ be a simple closed contour enclosing the nodes, and suppose that $f(z)$ is analytic within the contour and continuous within and on the contour. Contour integrals along γ are taken in the counterclockwise direction.

Let p be the interpolating polynomial for the points $(z_i, f(z_i))$, $i = 0, \ldots, m$. The Lagrange basis polynomials

$$l_i(z) = \prod_{\substack{k=0,\ldots,m \\ k \neq i}} (z - z_k)$$

and the associated monic polynomial

$$l(z) = \prod_{k=0}^{m} (z - z_k)$$

are defined as in Chapters 6 and 10 but now are considered over the complex plane rather than just the real line. As before, the interpolating polynomial can be expressed as

$$p(z) = \sum_{i=0}^{m} \frac{f(z_i)}{l_i(z_i)} l_i(z),$$

but it can also be expressed in terms of a contour integral, as the following theorem shows.

Theorem C.1. *The interpolating polynomial satisfies*

$$p(z) = \frac{1}{2\pi i} \int_\gamma \frac{l(\zeta) - l(z)}{l(\zeta)(\zeta - z)} f(\zeta) \, d\zeta \tag{C.1}$$

for z within γ.

Proof. Let $F(z)$ equal the expression on the right-hand side of (C.1).
First, consider $z = z_i$. Because $l(z_i) = 0$,

$$F(z_i) = \frac{1}{2\pi i} \int_\gamma \frac{l(\zeta) - l(z_i)}{l(\zeta)(\zeta - z_i)} f(\zeta) \, d\zeta = \frac{1}{2\pi i} \int_\gamma \frac{f(\zeta)}{\zeta - z_i} \, d\zeta.$$

The residue theorem implies $F(z_i) = f(z_i) = p(z_i)$.

Next, consider z not equal to any of the nodes. Then the integrand in (C.1) has poles at $\zeta = z, z_0, z_1, \ldots, z_m$. The residue theorem implies

$$F(z) = \frac{l(z) - l(z)}{l(z)} f(z) + \sum_{i=0}^m \frac{l(z_i) - l(z)}{l_i(z_i)(z_i - z)} f(z_i).$$

The first term equals zero, and the second is simplified by noting that $l(z_i) = 0$ and $l(z)/(z - z_i) = l_i(z)$:

$$F(z) = \sum_{i=0}^m \frac{-l(z)}{l_i(z_i)(z_i - z)} f(z_i) = \sum_{i=0}^m \frac{l_i(z)}{l_i(z_i)} f(z_i).$$

This is the Lagrange form of the interpolating polynomial, so $F(z) = p(z)$.
In summary, $p(z) = F(z)$ for all z within the contour.   □

Corollary C.2. *The interpolation error satisfies*

$$f(z) - p(z) = \frac{l(z)}{2\pi i} \int_\gamma \frac{f(\zeta)}{l(\zeta)(\zeta - z)} \, d\zeta \tag{C.2}$$

for z within γ.

Proof. By the residue theorem,

$$f(z) = \frac{1}{2\pi i} \int_\gamma \frac{f(\zeta)}{\zeta - z} \, d\zeta.$$

Subtract from this the contour integral for $p(z)$ given by the previous theorem:

$$f(z) - p(z) = \frac{1}{2\pi i} \int_\gamma \left(\frac{1}{\zeta - z} - \frac{l(\zeta) - l(z)}{l(\zeta)(\zeta - z)} \right) f(\zeta) \, d\zeta$$

$$= \frac{1}{2\pi i} \int_\gamma \frac{l(z)}{l(\zeta)(\zeta - z)} f(\zeta) \, d\zeta.   □$$

C.2 ▪ Chebyshev interpolation in the complex plane

For the remainder of the appendix, consider the more specific setting in which the interpolation nodes are the Chebyshev nodes $x_0, \ldots, x_m$ on $[-1, 1]$ and the contour γ is a Bernstein ellipse $E_\rho(-1, 1)$, with $\rho > 1$.

To find a bound on $|f(z) - p(z)|$, we would like to consult (C.2) and apply an upper bound on $|l(z)|$ and a lower bound on $|l(\zeta)|$.

The relationship between $l(z)$ and Chebyshev polynomials is useful, particularly (10.6) and (10.10). However, we must think carefully about how to interpret the identity

$$l(x) = 2^{-m+1}(x^2 - 1)U_{m-1}(x)$$

in the complex plane. How can the definition

$$U_m(x) = \frac{\sin((m+1)\cos^{-1} x)}{\sqrt{1-x^2}} \tag{C.3}$$

be extended from the real variable x to a complex variable z? The equation

$$z = \frac{1}{2}(w + w^{-1}), \tag{C.4}$$

because it is equivalent to the quadratic equation $w^2 - 2zw + 1 = 0$, associates to each complex number z a pair of complex numbers

$$w_\pm = z \pm \sqrt{z^2 - 1}$$

(with the exceptions of $z = 1 \iff w = 1$ and $z = -1 \iff w = -1$). Note that the solutions w_+ and w_- are reciprocal:

$$w_+ w_- = (z + \sqrt{z^2 - 1})(z - \sqrt{z^2 - 1}) = z^2 - (z^2 - 1) = 1.$$

The definition of the Chebyshev polynomial in the complex plane is

$$U_m(z) = \frac{w^{m+1} - w^{-(m+1)}}{w - w^{-1}}. \tag{C.5}$$

There are two objections that one might raise. First, (C.5) does not specify whether to use w_+ or w_-. The reason is that it does not matter. Because w_+ and w_- are reciprocal,

$$\frac{w_-^{m+1} - w_-^{-(m+1)}}{w_- - w_-^{-1}} = \frac{w_+^{-(m+1)} - w_+^{m+1}}{w_+^{-1} - w_+} = \frac{w_+^{m+1} - w_+^{-(m+1)}}{w_+ - w_+^{-1}}.$$

Second, $U_m(x)$ was already defined in (C.3). We would not want to give the same name to two different functions. The following proposition shows that this is not a problem.

Proposition C.3. *For $z \in [-1, 1]$, equations (C.3) and (C.5) are equivalent, i.e.,*

$$\frac{\sin((m+1)\cos^{-1} z)}{\sqrt{1-z^2}} = \frac{w^{m+1} - w^{-(m+1)}}{w - w^{-1}}.$$

Proof. Suppose $z \in [-1, 1]$, and let $\theta = \cos^{-1} z \in [0, \pi]$. The original definition for $U_m(z)$ gives

$$\frac{\sin((m+1)\cos^{-1} z)}{\sqrt{1-z^2}} = \frac{\sin((m+1)\theta)}{\sqrt{1-\cos^2 \theta}} = \frac{\sin((m+1)\theta)}{\sin \theta}.$$

Now define $w = e^{i\theta}$. Note that z and w are associated by (C.4):

$$\frac{1}{2}(w + w^{-1}) = \frac{e^{i\theta} + e^{-i\theta}}{2} = \cos\theta = z.$$

The new definition for $U_m(z)$ gives

$$\frac{w^{m+1} - w^{-(m+1)}}{w - w^{-1}} = \frac{e^{i(m+1)\theta} - e^{-i(m+1)\theta}}{e^{i\theta} - e^{-i\theta}} = \frac{2i\sin((m+1)\theta)}{2i\sin(\theta)}$$

$$= \frac{\sin((m+1)\theta)}{\sin\theta}. \qquad \square$$

Proposition C.4. *The function $U_m(z)$ defined by (C.5) is a polynomial in the complex plane of degree m with leading coefficient 2^m.*

Proof. The proof is by induction.

The base cases are $m = 0$ and $m = 1$. We find

$$U_0(z) = \frac{w - w^{-1}}{w - w^{-1}} = 1$$

and

$$U_1(z) = \frac{w^2 - w^{-2}}{w - w^{-1}} = \frac{(w - w^{-1})(w + w^{-1})}{w - w^{-1}} = w + w^{-1} = 2z.$$

For the induction step, we are inspired by the proof of Proposition 10.7 to check the recurrence relation $U_m(z) = 2zU_{m-1}(z) - U_{m-2}(z)$. Indeed,

$$2zU_{m-1}(z) - U_{m-2}(z) = (w + w^{-1})\frac{w^m - w^{-m}}{w - w^{-1}} - \frac{w^{m-1} - w^{-(m-1)}}{w - w^{-1}}$$

$$= \frac{w^{m+1} - w^{-(m+1)}}{w - w^{-1}} = U_m(z).$$

The function $U_m(z)$ is a polynomial, its degree is one greater than the degree of $U_{m-1}(z)$, and its leading coefficient is twice as large. $\square$

Proposition C.5. *If $x_0, \ldots, x_m$ are the nodes of the Chebyshev grid of degree m on $[-1, 1]$, then*

$$l(z) = 2^{-m+1}(z^2 - 1)U_{m-1}(z) \tag{C.6}$$

on the complex plane.

Proof. On each side of (C.6) is a monic polynomial of degree $m + 1$ with zeros $x_0, \ldots, x_m$. $\square$

Lemma C.6. *We have the following bounds:*

1. *If $x \in [-1, 1]$, then $|l(x)| \leq 2^{-m+1}$.*

2. *If z is on the Bernstein ellipse $E_\rho(-1, 1)$, then $|U_{m-1}(z)| \geq \frac{\rho^m - \rho^{-m}}{\rho + \rho^{-1}}$.*

Proof. First, for $x \in [-1, 1]$,

$$|l(x)| = 2^{-m+1}\sqrt{1-x^2}\,|\sin(m\cos^{-1}x)| \le 2^{-m+1}.$$

Second, let $w = \rho e^{i\theta}$, with $\rho > 1$. This is associated in a one-to-one fashion with

$$z = \frac{1}{2}(w + w^{-1}) = \frac{1}{2}(\rho e^{i\theta} + \rho^{-1}e^{-i\theta}) = \left(\frac{\rho + \rho^{-1}}{2}\right)\cos\theta + i \times \left(\frac{\rho - \rho^{-1}}{2}\right)\sin\theta,$$

which lies on the Bernstein ellipse $E_\rho(-1, 1)$ (Exercises 17–18 in Chapter 11). Thus,

$$U_{m-1}(z) = \frac{w^m - w^{-m}}{w - w^{-1}} = \frac{\rho^m e^{im\theta} - \rho^{-m}e^{-im\theta}}{\rho e^{i\theta} - \rho^{-1}e^{-i\theta}}$$

and

$$|U_{m-1}(z)| \ge \frac{\rho^m - \rho^{-m}}{\rho + \rho^{-1}}. \qquad \square$$

Lemma C.7. *Let $\zeta(t) : [0, 1] \to \gamma$ be a parameterization of γ and suppose $|f(z)| \le M$ for z on $E_\rho(-1, 1)$. For $x \in [-1, 1]$,*

$$|f(x) - p(x)| \le \frac{M}{2\pi}\left(\int_0^1 \frac{|\zeta'(t)|}{|\zeta(t)^2 - 1|\,|\zeta(t) - x|}\,dt\right)\frac{\rho + \rho^{-1}}{\rho^m - \rho^{-m}}.$$

Proof. By Corollary C.2,

$$f(x) - p(x) = \frac{l(x)}{2\pi i}\int_0^1 \frac{f(\zeta(t))}{l(\zeta(t))(\zeta(t) - x)}\zeta'(t)\,dt.$$

By Proposition C.5 and Lemma C.6,

$$\frac{|l(x)|}{|l(\zeta(t))|} \le \frac{2^{-m+1}}{2^{-m+1}|\zeta(t)^2 - 1|(\rho^m - \rho^{-m})/(\rho + \rho^{-1})} = \frac{1}{|\zeta(t)^2 - 1|}\frac{\rho + \rho^{-1}}{\rho^m - \rho^{-m}}.$$

Applying also $|f(\zeta(t))| \le M$, we find

$$|f(x) - p(x)| \le \frac{|l(x)|}{2\pi}\int_0^1 \frac{|f(\zeta(t))|\,|\zeta'(t)|}{|l(\zeta(t))|\,|\zeta(t) - x|}\,dt$$

$$\le \frac{1}{2\pi}\int_0^1 \frac{M\,|\zeta'(t)|}{|\zeta(t)^2 - 1|\,|\zeta(t) - x|}\frac{\rho + \rho^{-1}}{\rho^m - \rho^{-m}}\,dt. \qquad \square$$

Theorem C.8. *Suppose that f is analytic within the ellipse $E_\rho(-1, 1)$ and continuous within and on the ellipse. For $x \in [-1, 1]$, the interpolation error satisfies*

$$|f(x) - p(x)| \le C\rho^{-m} \tag{C.7}$$

for a number C that is independent of m.

Proof. Take

$$\zeta(t) = \left(\frac{\rho + \rho^{-1}}{2}\right) \cos(2\pi t) + i \times \left(\frac{\rho - \rho^{-1}}{2}\right) \sin(2\pi t).$$

Then $|\zeta'(t)|$ is bounded, and we leave it as an exercise to prove that

$$\int_0^1 \frac{|\zeta'(t)|}{|\zeta(t)^2 - 1| \, |\zeta(t) - x|} \, dt$$

is bounded above by a finite constant C_1. Also,

$$\frac{\rho + \rho^{-1}}{\rho^m - \rho^{-m}} = \frac{\rho + \rho^{-1}}{1 - \rho^{-2m}} \rho^{-m} \le \frac{\rho + \rho^{-1}}{1 - \rho^{-2}} \rho^{-m}.$$

Because f is continuous within and on $E_\rho(-1, 1)$, which is a compact space, the absolute value of the function is bounded by some number M. By Lemma C.7,

$$|f(x) - p(x)| \le \left(\frac{M}{2\pi} C_1 \frac{\rho + \rho^{-1}}{1 - \rho^{-2}}\right) \rho^{-m}. \qquad \square$$

Notes

Davis provides a brief and direct introduction to interpolation in the complex plane [21]. Walsh provides a more succinct proof of Theorem C.1 [80, pp. 50–51].

The Chebyshev polynomials are treated in depth by Mason and Handscomb [51].

Appendix D

Additional proofs for Newton–Cotes quadrature

In Chapter 16, it is demonstrated that Newton–Cotes quadrature performs especially well on grids of even degree. The reason, elaborated below, is that Newton–Cotes quadrature on a grid of even degree m produces an identical value to a related quadrature scheme on a grid of degree $m + 1$. The accuracy from the more-refined grid is obtained at the cost of the lower-degree grid.

D.1 ▪ A single interval

Let $m = 2k$ be a positive even number, and let

$$x_0, \ldots, x_{2k} \tag{D.1}$$

be a uniform grid on $[a, b]$. We will compare this grid with the grid formed by doubling the multiplicity of the middle node:

$$x_0, \ldots, x_{k-1}, x_k, x_k, x_{k+1}, \ldots, x_{2k}. \tag{D.2}$$

(See Appendix A for discussion of repeated nodes.)

Lemma D.1. *Let $q(x)$ be the polynomial interpolant for a function $g(x)$ on the grid (D.1), and let $r(x)$ be the polynomial interpolant for $g(x)$ on the grid (D.2). Then $\int_a^b q(x)\,dx = \int_a^b r(x)\,dx$.*

Proof. The two polynomial interpolants are related in a simple way. Let

$$l(x) = \prod_{i=0}^{2k}(x - x_i)$$

and $\alpha = (g'(x_k) - q'(x_k))/l'(x_k)$. We claim that $r(x) = q(x) + \alpha l(x)$. To see this, note that $q(x) + \alpha l(x)$ is a polynomial of degree at most $2k + 1$, that

$$q(x_i) + \alpha l(x_i) = g(x_i) + 0 = g(x_i), \quad i = 0, \ldots, 2k,$$

and that

$$q'(x_k) + \alpha l'(x_k) = q'(x_k) + \frac{g'(x_k) - q'(x_k)}{l'(x_k)} l'(x_k) = g'(x_k).$$

Both $r(x)$ and $q(x) + \alpha l(x)$ are interpolating polynomials for $g(x)$ on the grid (D.2). By Theorem A.4, they must be equal.

The equality of the integrals follows from symmetry. Let $c = (a+b)/2$, $t = x - c$, and $t_i = x_i - c$. We argue that

$$l(c + t) = \prod_{i=0}^{2k} ((c + t) - (c + t_i)) = \prod_{i=0}^{2k} (t - t_i)$$

is an odd function of t. Note that $t_i = -t_{2k-i}$. Thus,

$$l(c - t) = \prod_{i=0}^{2k} (-t - t_i) = \prod_{i=0}^{2k} (-t + t_{2k-i}) = (-1)^{2k+1} \prod_{i=0}^{2k} (t - t_{2k-i}) = -l(c + t).$$

For the integral, we find

$$\int_a^b l(x)\, dx = \int_{-(b-a)/2}^{(b-a)/2} l(c + t)\, dt = 0$$

and therefore

$$\int_a^b r(x)\, dx = \int_a^b (q(x) + \alpha l(x))\, dx = \int_a^b q(x)\, dx. \qquad \square$$

D.2 ▪ A partitioned interval

In Newton–Cotes quadrature, an interval $[a, b]$ is partitioned into subintervals $[a_j, b_j]$, $j = 1, \ldots, n$, of equal width and a uniform grid is placed on each subinterval. When $m = 2k$ is even, the grid on the jth subinterval is

$$x_{0j}, \ldots, x_{2k,j}.$$

Then an interpolant $q \in P_{mn}[a, b]$ is constructed to match $g(x)$ on the grid, and $q(x)$ is integrated in place of $g(x)$.

To analyze the accuracy of this procedure, the middle node in each subinterval is doubled, as in Lemma D.1. On the jth subinterval, the odd-degree grid is

$$x_{0j}, \ldots, x_{k-1,j}, x_{kj}, x_{kj}, x_{k+1,j}, \ldots, x_{2k,j}. \tag{D.3}$$

Proof of Theorem 16.1. Suppose $q(x) \in P_{mn}[a, b]$ interpolates $g(x)$ on an m-by-n piecewise-uniform grid over $[a, b]$. Let $N = (m \vee 1)n$. That

$$\left| \int_a^b g(x)\, dx - \int_a^b q(x)\, dx \right| \le CN^{-(m+1)}$$

for a constant C was proved as (16.1). The remaining statement is that, when m is even, the bound can be improved to

$$\left| \int_a^b g(x)\, dx - \int_a^b q(x)\, dx \right| \le CN^{-(m+2)}.$$

Construct $r \in P_{m+1,n}[a, b]$ to interpolate $g(x)$ on the piecewise-defined grid (D.3). By Lemma D.1, we have $\int_{a_j}^{b_j} q(x) \, dx = \int_{a_j}^{b_j} r(x) \, dx$ for $j = 1, \ldots, n$, and therefore

$$\int_a^b g(x) \, dx - \int_a^b q(x) \, dx = \int_a^b (g(x) - r(x)) \, dx.$$

By Theorem A.8, on each subinterval $[a_j, b_j]$, we have

$$g(x) - r(x) = \frac{g^{(m+2)}(\eta_x)}{(m+2)!} (x - x_{kj}) \prod_{i=0}^{2k} (x - x_{ij})$$

for some η_x. Let $h = (b - a)/N$. By (8.2) and (8.3),

$$|g(x) - r(x)| \le \frac{\|g^{(m+2)}\|_\infty}{(m+2)!} \left(\frac{b-a}{2n} \right) (m! h^{m+1}) = \frac{\|g^{(m+2)}\|_\infty (b-a) h^{m+1}}{2(m+2)(m+1)n}$$

on each subinterval and therefore

$$\|g - r\|_\infty \le \frac{\|g^{(m+2)}\|_\infty (b-a)^{m+2}}{2(m+2)(m+1)n N^{m+1}} \le \frac{\|g^{(m+2)}\|_\infty (b-a)^{m+2}}{2(m+2)} N^{-(m+2)}$$

over the entire interval $[a, b]$. Altogether,

$$\left| \int_a^b g(x) \, dx - \int_a^b q(x) \, dx \right| \le \int_a^b |g(x) - r(x)| \, dx$$

$$\le (b-a) \|g - r\|_\infty \le \frac{\|g^{(m+2)}\|_\infty (b-a)^{m+3}}{2(m+2)} N^{-(m+2)}. \qquad \square$$

Bibliography

[1] M. ABRAMOWITZ AND I. A. STEGUN, *Handbook of mathematical functions with formulas, graphs, and mathematical tables*, vol. 55 of National Bureau of Standards Applied Mathematics Series, for sale by the Superintendent of Documents, U.S. Government Printing Office, Washington, D.C., 1964. (Cited on p. 86)

[2] K. E. ATKINSON, *The numerical solution of integral equations of the second kind*, Cambridge University Press, Cambridge, UK, 1997. (Cited on pp. 268, 331)

[3] Z. BATTLES AND L. N. TREFETHEN, *An extension of MATLAB to continuous functions and operators*, SIAM J. Sci. Comput., 25 (2004), pp. 1743–1770. (Cited on p. x)

[4] S. BERNSTEIN, *Sur la limitation des valeurs d'un polynôme $P_n(x)$ de degré n sur tout un segment par ses valeurs en $(n + 1)$ points du segment*, Izv. Akad. Nauk SSSR, 8 (1931), pp. 1025–1050. (Cited on p. 132)

[5] S. N. BERNSTEIN, *Sur l'ordre de la meilleure approximation des fonctions continues par des polynômes de degré donné*, Mém. Cl. Sci. Acad. Roy. Belg., 4 (1912), pp. 1–103. (Cited on p. 132)

[6] J.-P. BERRUT AND L. N. TREFETHEN, *Barycentric Lagrange interpolation*, SIAM Rev., 46 (2004), pp. 501–517. (Cited on p. 75)

[7] L. BLOOMFIELD, *How dangerous is a penny falling from the Empire State Building?* http://howeverythingworks.org/wordpress/2012/03/13/question-1595/ (12 September 2018). (Cited on p. 395)

[8] J. P. BOYD, *Chebyshev and Fourier spectral methods*, Dover Publications, Inc., Mineola, NY, second ed., 2001. (Cited on pp. 115, 376)

[9] J. P. BOYD, *Finding the zeros of a univariate equation: Proxy rootfinders, Chebyshev interpolation, and the companion matrix*, SIAM Rev., 55 (2013), pp. 375–396. (Cited on p. 344)

[10] H. BRUNNER, *Collocation methods for Volterra integral and related functional differential equations*, vol. 15 of Cambridge Monographs on Applied and Computational Mathematics, Cambridge University Press, Cambridge, UK, 2004. (Cited on p. 268)

[11] F. BUCK, *Behold! The midpoint rule is better than the trapezoidal rule for concave functions*, College Math. J., 16 (1985), p. 56. (Cited on p. 191)

[12] A.-L. CAUCHY, *Sur les fonctions interpolaires*, in Œuvres complètes, vol. 5, Cambridge University Press, Cambridge, UK, 2009, pp. 409–424. (Cited on p. 75)

[13] M. M. CHAWLA, *Error estimates for the Clenshaw-Curtis quadrature*, Math. Comp., 22 (1968), pp. 651–656. (Cited on p. 208)

[14] P. L. CHEBYSHEV, *Théorie des mécanismes connus sous le nom de parallélogrammes*, Mém. Acad. Sci. St.-Pétersb., 7 (1854), pp. 539–568. (Cited on p. 115)

[15] P. L. CHEBYSHEV, *Sur les questions de minima qui se rattachent à la représentation approximative des fonctions*, Mém. Acad. Imp. Sci. St.-Pétersb., 7 (1859), pp. 199–291. (Cited on p. 115)

[16] C. W. CLENSHAW AND A. R. CURTIS, *A method for numerical integration on an automatic computer*, Numer. Math., 2 (1960), pp. 197–205. (Cited on p. 199)

[17] E. A. CODDINGTON, *An introduction to ordinary differential equations*, Prentice–Hall Mathematics Series, Prentice–Hall, Inc., Englewood Cliffs, NJ, 1961. (Cited on p. 258)

[18] E. A. CODDINGTON AND N. LEVINSON, *Theory of ordinary differential equations*, McGraw–Hill Book Company, Inc., New York-Toronto-London, 1955. (Cited on p. 321)

[19] R. M. CORLESS, *Generalized companion matrices in the Lagrange basis*, in Proceedings of Encuentros de Álgebra Computacional y Aplicaciones (EACA) 2004, L. Gonzalez-Vega and T. Recio, eds., 2004, pp. 317–322. (Cited on p. 344)

[20] R. M. CORLESS, *On a generalized companion matrix pencil for matrix polynomials expressed in the Lagrange basis*, in Symbolic-Numeric Computation, Trends Math., Birkhäuser, Basel, 2007, pp. 1–15. (Cited on p. 344)

[21] P. J. DAVIS, *Interpolation and approximation*, Dover Publications, Inc., Mineola, NY, 1975. (Cited on pp. 75, 399, 414)

[22] *Digital Library of Mathematical Functions*. http://dlmf.nist.gov/, release 1.0.19 of 2018-06-22, F. W. J. Olver, A. B. Olde Daalhuis, D. W. Lozier, B. I. Schneider, R. F. Boisvert, C. W. Clark, B. R. Miller and B. V. Saunders, eds. (Cited on p. 345)

[23] T. A. DRISCOLL, *Automatic spectral collocation for integral, integro-differential, and integrally reformulated differential equations*, J. Comput. Phys., 229 (2010), pp. 5980–5998. (Cited on p. 311)

[24] T. A. DRISCOLL, F. BORNEMANN, AND L. N. TREFETHEN, *The chebop system for automatic solution of differential equations*, BIT, 48 (2008), pp. 701–723. (Cited on pp. 311, 395)

[25] T. A. DRISCOLL AND N. HALE, *Rectangular spectral collocation*, IMA J. Numer. Anal., 36 (2016), pp. 108–132. (Cited on pp. 142, 269, 311)

[26] A. EDELMAN AND H. MURAKAMI, *Polynomial roots from companion matrix eigenvalues*, Math. Comp., 64 (1995), pp. 763–776. (Cited on p. 344)

[27] H. EHLICH AND K. ZELLER, *Auswertung der Normen von Interpolationsoperatoren*, Math. Ann., 164 (1966), pp. 105–112. (Cited on pp. 115, 132)

[28] D. ELLIOTT, *Truncation errors in two Chebyshev series approximations*, Math. Comp., 19 (1965), pp. 234–248. (Cited on p. 132)

[29] J. F. EPPERSON, *On the Runge example*, Amer. Math. Monthly, 94 (1987), pp. 329–341. (Cited on p. 103)

[30] L. FEJÉR, *Mechanische Quadraturen mit positiven Cotesschen Zahlen*, Math. Z., 37 (1933), pp. 287–309. (Cited on p. 199)

[31] F. J. FLANIGAN, *Complex variables: harmonic and analytic functions*, Dover Publications, Inc., Mineola, NY, 1983. (Cited on p. 406)

[32] B. FORNBERG, *A practical guide to pseudospectral methods*, vol. 1 of Cambridge Monographs on Applied and Computational Mathematics, Cambridge University Press, Cambridge, UK, 1996. (Cited on p. 103)

[33] G. FROBENIUS, *Theorie der linearen Formen mit ganzen Coefficienten*, J. Reine Angew. Math., 86 (1879), pp. 146–208. (Cited on p. 344)

[34] H. H. GOLDSTINE, *A history of numerical analysis from the 16th through the 19th century*, Springer-Verlag, New York–Heidelberg, 1977. Studies in the History of Mathematics and Physical Sciences, Vol. 2. (Cited on pp. 183, 357)

[35] G. H. GOLUB AND C. F. VAN LOAN, *Matrix computations*, Johns Hopkins Studies in the Mathematical Sciences, Johns Hopkins University Press, Baltimore, MD, fourth ed., 2013. (Cited on p. 244)

[36] I. J. GOOD, *The colleague matrix, a Chebyshev analogue of the companion matrix.*, Quart. J. Math. Oxford Ser. (2), 12 (1961), pp. 61–68. (Cited on p. 344)

[37] L. GREENGARD, *Spectral integration and two-point boundary value problems*, SIAM J. Numer. Anal., 28 (1991), pp. 1071–1080. (Cited on pp. 269, 321)

[38] W. W. HAGER, *Condition estimates*, SIAM J. Sci. Statist. Comput., 5 (1984), pp. 311–316. (Cited on p. 331)

[39] E. HAIRER, S. P. NØRSETT, AND G. WANNER, *Solving ordinary differential equations. I*, vol. 8 of Springer Series in Computational Mathematics, Springer-Verlag, Berlin, second ed., 1993. Nonstiff problems. (Cited on p. 302)

[40] E. HAIRER AND G. WANNER, *Solving ordinary differential equations. II*, vol. 14 of Springer Series in Computational Mathematics, Springer-Verlag, Berlin, 2010. Stiff and differential-algebraic problems, second revised edition, paperback. (Cited on p. 302)

[41] J. HAVIL, *The irrationals*, Princeton University Press, Princeton, NJ, 2012. (Cited on p. 37)

[42] T. HAWKINS, *The mathematics of Frobenius in context: A journey through 18th to 20th century mathematics*, Sources and Studies in the History of Mathematics and Physical Sciences, Springer, New York, 2013. (Cited on p. 344)

[43] N. J. HIGHAM, *Accuracy and stability of numerical algorithms*, SIAM, Philadelphia, second ed., 2002. (Cited on p. 244)

[44] N. J. HIGHAM, *The numerical stability of barycentric Lagrange interpolation*, IMA J. Numer. Anal., 24 (2004), pp. 547–556. (Cited on p. 75)

[45] *IEEE standard for binary floating-point arithmetic*, ANSI/IEEE Std 754-1985, 1985. (Cited on p. 59)

[46] *IEEE standard for floating-point arithmetic*, IEEE Std 754-2008, 2008. (Cited on p. 59)

[47] L. V. KANTOROVICH AND G. P. AKILOV, *Functional analysis*, Pergamon Press, Oxford–Elmsford, NY, second ed., 1982. Translated from the Russian by Howard L. Silcock. (Cited on pp. 268, 376)

[48] J. L. LAGRANGE, *Leçons élémentaires sur les mathématiques données à l'École normale en 1795*, in Œuvres de Lagrange, J.-A. Serret, ed., vol. 7, Gauthier-Villars, Paris, 1877, pp. 183–287. (Cited on p. 75)

[49] P. W. LAWRENCE AND R. M. CORLESS, *Stability of rootfinding for barycentric Lagrange interpolants*, Numer. Algorithms, 65 (2014), pp. 447–464. (Cited on p. 344)

[50] A. LOEWY, *Die Begleitmatrix eines linearen homogenen Differentialstatusdruckes*, Nachr. Ges. Wiss. Göttingen, Math.-Phys. Kl., 1917 (1917), pp. 255–263. (Cited on p. 344)

[51] J. C. MASON AND D. C. HANDSCOMB, *Chebyshev polynomials*, Chapman & Hall/CRC, Boca Raton, FL, 2003. (Cited on pp. 115, 414)

[52] D. B. MEADE, *ODE models for the parachute problem*, SIAM Rev., 40 (1998), pp. 327–332. (Cited on p. 396)

[53] E. MEIJERING, *A chronology of interpolation: From ancient astronomy to modern signal and image processing*, Proc. IEEE, 90 (2002), pp. 319–342. (Cited on pp. 75, 86)

[54] C. MÉRAY, *Observations sur la légitimité de l'interpolation*, Ann. Sci. École Norm. Sup. (3), 1 (1884), pp. 165–176. (Cited on p. 102)

[55] C. MÉRAY, *Nouveaux exemples d'interpolations illusoires*, Bull. Sci. Math., 20 (1896), pp. 266–270. (Cited on p. 103)

[56] C. MOLER, *The origins of MATLAB*.
http://www.mathworks.com/company/newsletters/articles/the-origins-of-matlab.html
(28 July 2015). (Cited on p. 27)

[57] C. MOLER, *Cleve's corner: ROOTS—of polynomials, that is*, The MathWorks Newsletter, 5 (1991), pp. 8–9. (Cited on p. 344)

[58] C. B. MOLER, *Numerical computing with MATLAB: revised reprint*, SIAM, Philadelphia, 2004. (Cited on p. 357)

[59] S. OLVER AND A. TOWNSEND, *A fast and well-conditioned spectral method*, SIAM Rev., 55 (2013), pp. 462–489. (Cited on p. 269)

[60] G. PETERS AND J. H. WILKINSON, *On the stability of Gauss-Jordan elimination with pivoting*, Comm. ACM, 18 (1975), pp. 20–24. Collection of articles honoring Alston S. Householder. (Cited on p. 224)

[61] R. PHOEBUS AND C. REILLY, *Differential equations and the parachute problem*. https://mse.redwoods.edu/darnold/math55/DEproj/sp04/coleron/paper1.pdf (2 July 2018). (Cited on p. 396)

[62] A. RALSTON AND P. RABINOWITZ, *A first course in numerical analysis*, Dover Publications, Inc., Mineola, NY, second ed., 2001. (Cited on p. 357)

[63] R. D. RIESS AND L. W. JOHNSON, *Error estimates for Clenshaw-Curtis quadrature*, Numer. Math., 18 (1971/72), pp. 345–353. (Cited on p. 208)

[64] C. RUNGE, *Über empirische Funktionen und die Interpolation zwischen äquidistanten Ordinaten*, Z. Math. Phys., 46 (1901), pp. 224–243. (Cited on p. 102)

[65] H. E. SALZER, *Lagrangian interpolation at the Chebyshev points $x_{n,v} \equiv \cos(v\pi/n)$, $v = 0(1)n$; some unnoted advantages*, Comput. J., 15 (1972), pp. 156–159. (Cited on p. 115)

[66] W. SPECHT, *Die Lage der Nullstellen eines Polynoms. III*, Math. Nachr., 16 (1957), pp. 369–389. (Cited on p. 344)

[67] W. SPECHT, *Die Lage der Nullstellen eines Polynoms. IV*, Math. Nachr., 21 (1960), pp. 201–222. (Cited on p. 344)

[68] J. STOER AND R. BULIRSCH, *Introduction to numerical analysis*, Springer-Verlag, New York-Heidelberg, 1980. Translated from the German by R. Bartels, W. Gautschi, and C. Witzgall. (Cited on p. 321)

[69] E. TADMOR, *The exponential accuracy of Fourier and Chebyshev differencing methods*, SIAM J. Numer. Anal., 23 (1986), pp. 1–10. (Cited on p. 132)

[70] W. J. TAYLOR, *Method of Lagrangian curvilinear interpolation*, J. Research Nat. Bur. Standards, 35 (1945), pp. 151–155. (Cited on p. 75)

[71] J. F. TRAUB, *Iterative methods for the solution of equations*, Prentice–Hall, Inc., Englewood Cliffs, NJ, 1964. (Cited on p. 357)

[72] L. N. TREFETHEN, *The definition of numerical analysis*, SIAM News, 25 (1992). (Cited on p. 8)

[73] L. N. TREFETHEN, *Spectral methods in MATLAB*, vol. 10 of Software, Environments and Tools, SIAM, Philadelphia, 2000. (Cited on p. 311)

[74] L. N. TREFETHEN, *Is Gauss quadrature better than Clenshaw–Curtis?*, SIAM Rev., 50 (2008), pp. 67–87. (Cited on p. 208)

[75] L. N. TREFETHEN, *Six myths of polynomial interpolation and quadrature*, Math. Today (Southend-on-Sea), 47 (2011), pp. 184–188. (Cited on p. 115)

[76] L. N. TREFETHEN, *Approximation theory and approximation practice*, SIAM, Philadelphia, 2013. (Cited on pp. 115, 132, 409)

[77] L. N. TREFETHEN AND D. BAU, III, *Numerical linear algebra*, SIAM, Philadelphia, 1997. (Cited on p. 224)

[78] L. N. TREFETHEN AND J. A. C. WEIDEMAN, *Two results on polynomial interpolation in equally spaced points*, J. Approx. Theory, 65 (1991), pp. 247–260. (Cited on p. 103)

[79] USAF ACADEMY, *Student Handbook for Airmanship 490: Basic Free Fall Parachuting*, Colorado Springs, CO, May 1990. (Cited on p. 396)

[80] J. L. WALSH, *Interpolation and approximation by rational functions in the complex domain*, fifth edition. American Mathematical Society Colloquium Publications, Vol. XX, American Mathematical Society, Providence, RI, 1969. (Cited on p. 414)

[81] E. WARING, *Problems concerning interpolations*, Philos. Trans. Roy. Soc. London, 69 (1779), pp. 59–67. (Cited on p. 75)

[82] J. A. C. WEIDEMAN, *Numerical integration of periodic functions: A few examples*, Amer. Math. Monthly, 109 (2002), pp. 21–36. (Cited on p. 191)

[83] J. A. C. WEIDEMAN AND S. C. REDDY, *A MATLAB differentiation matrix suite*, ACM Trans. Math. Software, 26 (2000), pp. 465–519. (Cited on pp. 154, 155)

[84] B. D. WELFERT, *Generation of pseudospectral differentiation matrices I*, SIAM J. Numer. Anal., 34 (1997), pp. 1640–1657. (Cited on p. 154)

[85] E. T. WHITTAKER AND G. N. WATSON, *A course of modern analysis*, Cambridge Mathematical Library, Cambridge University Press, Cambridge, UK, 1996. An introduction to the general theory of infinite processes and of analytic functions; with an account of the principal transcendental functions, reprint of the fourth (1927) edition. (Cited on p. 258)

[86] J. H. WILKINSON, *The algebraic eigenvalue problem*, Clarendon Press, Oxford, UK, 1965. (Cited on p. 223)

[87] J. H. WILKINSON, *Rounding errors in algebraic processes*, Dover Publications, Inc., Mineola, NY, 1994. Reprint of the 1963 original [Prentice–Hall, Englewood Cliffs, NJ]. (Cited on p. 244)

[88] S. XIANG AND F. BORNEMANN, *On the convergence rates of Gauss and Clenshaw–Curtis quadrature for functions of limited regularity*, SIAM J. Numer. Anal., 50 (2012), pp. 2581–2587. (Cited on p. 208)

Index